Lecture Notes in Computer Science 16528

The series Lecture Notes in Computer Science (LNCS), including its subseries Lecture Notes in Artificial Intelligence (LNAI) and Lecture Notes in Bioinformatics (LNBI), has established itself as a medium for the publication of new developments in computer science and information technology research, teaching, and education.

LNCS enjoys close cooperation with the computer science R & D community, the series counts many renowned academics among its volume editors and paper authors, and collaborates with prestigious societies. Its mission is to serve this international community by providing an invaluable service, mainly focused on the publication of conference and workshop proceedings and postproceedings. LNCS commenced publication in 1973.

Deisy Chaves · Manuel Forero Vargas ·
Oswaldo Rojas Camacho
Editors

Progress in Pattern Recognition, Image Analysis, Computer Vision, and Applications

28th Iberoamerican Congress, CIARP 2025
Bogotá, Colombia, November 25–28, 2025
Proceedings, Part I

Editors
Deisy Chaves
Universidad del Valle
Cali, Colombia

Manuel Forero Vargas
Universidad de Ibague
Ibagué, Colombia

Oswaldo Rojas Camacho
Universidad Nacional de Colombia
Bogotá, Colombia

ISSN 0302-9743 ISSN 1611-3349 (electronic)
Lecture Notes in Computer Science
ISBN 978-3-032-23160-4 ISBN 978-3-032-23161-1 (eBook)
https://doi.org/10.1007/978-3-032-23161-1

This Springer imprint is published by the registered company Springer Nature Switzerland AG
The registered company address is: Gewerbestrasse 11, 6330 Cham, Switzerland

Preface

The 28th Ibero-American Congress on Pattern Recognition (CIARP 2025) successfully convened as the annual international conference dedicated to advancing research in pattern recognition, computer vision, artificial intelligence, data mining, and all related disciplines. This prestigious event, renowned for fostering a rich integration of theoretical developments and practical applications, stands as a cornerstone for international collaboration and knowledge exchange within the Ibero-American and global scientific community. The conference once again enjoyed the official endorsement of the International Association for Pattern Recognition (IAPR) and was organized by the Colombian Association for Pattern Recognition (ACORP.IA), a chapter of the Colombian Computer Society. It also received invaluable support from the Universidad Nacional de Colombia and its Faculty of Engineering in Bogotá.

True to its mission, CIARP 2025 brought together leading researchers, academics, and professionals from across the globe to present, discuss, and disseminate cutting-edge advances in mathematical methodologies and computational techniques for Artificial Intelligence and Pattern Recognition. The comprehensive scientific program spanned a wide array of specialized topics, including but not limited to Artificial Intelligence, Biometrics, Cognitive and Humanoid Vision, Computer Vision, Deep Learning and Transfer Learning, Data Mining and Big Data, Natural Language Processing, Document Analysis, Image Processing and Analysis, Neural Networks, Remote Sensing Applications, and Statistical Pattern Recognition, alongside their transformative applications in industry, health, robotics, and telecommunications. The congress served as a dynamic platform for sharing research experiences, disseminating novel insights, and forging collaborative partnerships among specialized research groups worldwide.

Consistent with its profoundly international character, CIARP 2025 attracted 110 full-paper submissions from 11 countries, of which 52 were accepted for inclusion in these proceedings, resulting in a competitive acceptance rate of 47%. The strong Ibero-American participation was evident, with contributions from Colombia, Brazil, Mexico, Cuba, Chile, Peru, Uruguay, Ecuador, and Costa Rica. The conference also welcomed research from other regions, including Europe (Greece) and beyond (Saudi Arabia), thereby enriching the cross-cultural and academic dialogue. Every submission underwent a rigorous double-blind peer-review process by an average of three qualified reviewers, being and papers were accepted that achieved scientific quality scores exceeding the overall mean rating. The paper evaluation was done by a geographically diverse Program Committee of distinguished experts, to ensure the highest standards of quality, originality, and impartiality.

CIARP 2025 was hosted on the campus of the Universidad Nacional de Colombia in the vibrant city of Bogotá, from November 25 to 28, 2025. The event featured engaging single-track paper presentation sessions, and distinguished keynote lectures. Confirmed keynote speakers included prominent international figures such as Ioannis

Pitas (University of Thessaloniki - Greece), Cesar Armando Beltrán (Pontificia Universidad Católica del Perú - Peru), Oscar Corcho (Universidad Politécnica de Madrid - Spain), and Emanuele Trucco (University of Dundee - UK), alongside leading national researchers from Colombia, including Elizabeth Leon, Jonatan Gomez Perdomo, and Fabio González from the host institution.

The conference continued its tradition of recognizing scientific excellence by conferring special awards, including prizes for the IAPR Best Paper and the IAPR Best Student Paper. The organization of CIARP 2025 was a joint effort led by the Organizing Committee, including General Chairs Oswaldo Rojas Camacho and Manuel Guillermo Forero, and supported by the entire Local Committee from the Universidad Nacional de Colombia. We extend our sincere gratitude to the Faculty of Engineering, the Extension and Research Institute (IEI), and the Continuing Education Unit (UECP) for their indispensable contributions to the conference's success. We would like to give special thanks to Deisy Chaves, a member of the Program Committee Chair team and one Editor of the Proceedings, for her dedication and leadership in coordinating the scientific program and overseeing the editorial process, which were key to ensuring the high quality and standards of CIARP 2025. Our thanks also go to the Springer LNCS team for their professional support in producing the proceedings. Finally, we are deeply grateful to all the authors who submitted their valuable work to CIARP 2025 and to the attendees who made this edition a fruitful and memorable scientific gathering.

November 2025

Deisy Chaves
Manuel Forero Vargas
Oswaldo Rojas Camacho

Organization

General Chairs

Manuel Forero Vargas	Universidad de Ibagué, Colombia
Oswaldo Rojas Camacho	Universidad Nacional de Colombia, Colombia

Program Committee Chairs

Deisy Chaves	Universidad del Valle, Colombia
María Trujillo	Universidad del Valle, Colombia
Sergio A. Velastin	Queen Mary University of London, UK/Universidad Carlos III de Madrid, Spain

Steering Committee

Iván Mauricio Cabezas Trollano	Universidad San Buenaventura, Colombia
Oscar J. Suarez	Universidad de Pamplona, Colombia

Local Committee

Elkin Aponte	Universidad Nacional de Colombia, Colombia
Libia Denise Cangrejo Aljure	Universidad Nacional de Colombia, Colombia
Karen Julieth Castillo	Universidad Nacional de Colombia, Colombia
Javier Santiago Clavijo Del Valle	Universidad Nacional de Colombia, Colombia
Iván Darío Gil Chaves	Universidad Nacional de Colombia, Colombia
Valentina Jaimes	Universidad Nacional de Colombia, Colombia
Sonia Esperanza Monroy Varelax	Universidad Nacional de Colombia, Colombia
Nicolás Pinilla	Universidad Nacional de Colombia, Colombia
Andrea Rojas	Universidad Nacional de Colombia, Colombia
Diego Alexander Tibaduiza Burgos	Universidad Nacional de Colombia, Colombia

Program Committee

João Almeida	Universidade Federal do Maranhão, Brazil
Pedro Alejandro Amador Rodríguez	Universidad del Valle, Colombia
Jesús Alexander Aranda Bueno	Universidad del Valle, Colombia
Fray Luis Becerra Suarez	Universidad Privada Norbert Wiener, Peru
Oscar Fernando Bedoya Leyva	Universidad del Valle, Colombia
Cesar Beltran	Pontificia Universidad Católica del Perú, Peru
Jose Bernal	Friedrich-Alexander Universität Erlangen-Nürnberg, Germany
Alessandro Bof de Oliveira	Universidade Federal do Pampa, Brazil
Rodolfo Bojorque Chasi	Universidad Politécnica Salesiana, Ecuador
Isabela Borlido Barcelos	Pontifical Catholic University of Minas Gerais, Brazil
John Willian Branch Bedoya	Universidad Nacional de Colombia, Colombia
Pedro H. Bugatti Federal	University of São Carlos, Brazil
Lázaro Bustio-Martínez	Universidad Iberoamericana, Mexico
Cristina Bustos	Universitat Oberta de Catalunya, Spain
Iván Mauricio Cabezas Trollano	Universidad San Buenaventura, Colombia
Fábio Cappabianco	Universidade Federal de São Paulo, Brazil
Leidy Johanna Carvajal Ortiz	Universidad del Valle, Colombia
Ángela Casanova Granada	Universidad del Valle, Colombia
Pedro Segundo Castañeda Vargas	Universidad Nacional Toribio Rodríguez de Mendoza, Peru
Alexander Ceron	Universidad Militar Nueva Granada, Colombia
Pedro Couto	Universidade de Trás-os-Montes e Alto Douro, Portugal
Deisy Chaves	Universidad del Valle, Colombia
René Cruz Guerrero	Instituto Tecnológico Superior del Oriente del Estado de Hidalgo, Mexico
Jorge Ernesto Espinosa Oviedo	Politécnico Colombiano Jaime Isaza Cadavid, Colombia
Manuel Forero Vargas	Universidad de Ibagué, Colombia
Yaima Filiberto Cabrera	AMV Soluciones, Spain
Giorgio Fumera	University of Cagliari, Italy
Carlos Eric Galvan-Tejada	Universidad Autónoma de Zacatecas, Mexico
Sonia Gouveia	Universidade de Aveiro, Portugal
Fabio González	Universidad Nacional de Colombia, Colombia
Jhon Mauro Gomez Benitez	Universidad del Valle, Colombia
Valia Guerra	Universidad de La Laguna, Spain
Raul Ernesto Gutierrez de Pineres Reyes	Universidad del Valle, Colombia

David Alberto Guzmán Ardila	Universidad del Valle, Colombia
Pablo A. Henríquez	Universidad Diego Portales, Chile
Ruber Hernández García	Universidad Católica del Maule, Chile
Alexandra La Cruz Puente	Universidad de Ibague, Colombia
Manuel S. Lazo Cortés	Instituto Tecnológico de Tlalnepantla, Mexico
Mauricio Alberto Ledesma Villegas	Universidad del Valle, Colombia
Marcos Lévano	Universidad Católica de Temuco, Chile
José Libreros	Technische Universität Ilmenau, Germany
Josep López Camuñas	Universitat Oberta de Catalunya, Spain
Fabricio Martins Lopes	Universidade Tecnológica Federal do Paraná, Brazil
Cristian Martinez	Universidad Nacional de Salta, Argentina
José Francisco Martínez-Trinidad	Instituto Nacional de Astrofísica, Óptica y Electrónica, Mexico
Heber Ivan Mejia Cabrera	Universidad Señor de Sipán, Peru
Carlos Mera	Universidad de Antioquia, Colombia
Domingo Mery	Pontificia Universidad Católica de Chile, Chile
Miguel Moctezuma-Flores	Universidad Nacional Autonóma de México, Mexico
José Manuel Molina López	Universidad Carlos III de Madrid, Spain
Annette Morales-González	Centro de Aplicaciones de Tecnologías de Avanzada, Cuba
Pedro Moreno	Universidad del Valle, Colombia
Agustín Moreno Cañadas	Universidad Nacional de Colombia, Colombia
Carlos Vicente Niño Rondón	Universidad Francisco de Paula Santander, Colombia
Stiven Díaz Noguera	Universidad de la Costa, Colombia
Felipe Ocampo Osorio	Fundación Valle del Lili, Colombia
Mauricio Orozco Alzate	Universidad Nacional de Colombia, Colombia
Kálmán Palágyi	University of Szeged, Hungary
João Paulo Papa	São Paulo State University, Brazil
Javier Pérez Escamilla	Instituto Tecnológico Superior del Occidente del Estado de Hidalgo, Mexico
Aldo-Eleazar Pérez-Ramos	Instituto Tecnológico de Oaxaca, Mexico
Adrián Pérez-Suay	Universitat de València, Spain
Lloy Pinedo Tuanama	Universidad Privada Norbert Wiener, Peru
Bahbibi Rahmatullah	Universiti Pendidikan Sultan Idris, Malasia
Heilym Camila Ramirez Rico	Universidad Técnica Federico Santa María, Chile
Pedro Real Jurado	Universidad de Sevilla, Spain
Edgar Roman-Rangel	Instituto Tecnológico Autónomo de México, Mexico
Oswaldo Rojas Camacho	Universidad Nacional de Colombia, Colombia

Lihki Jose Rubio Ortega	Universidad del Norte, Colombia
Priscila Saito	Universidade Federal de Sao Carlos, Brazil
Guillermo Sánchez Díaz	Universidad Autónoma de San Luis Potosí, Mexico
Sebastian Scotti	Knowlix, Germany
Carlo Sansone	University of Naples Federico II, Italy
Eduard Sierra Ballen	Universidad Militar Nueva Granada, Colombia
Catarina Silva	University of Coimbra, Portugal
Oscar J. Suarez	Universidad de Pamplona, Colombia
Nicolás Torres	Universidad Técnica Federico Santa María, Chile
Emmanuele Trucco	University of Dundee, UK
Maria Trujillo	Universidad del Valle, Colombia
Edwin Valarezo	Escuela Superior Politécnica del Litoral, Ecuador
Alejandro Valdés Villada	Universidad del Valle, Colombia
Fabián Stiven Valencia Córdoba	Universidad del Valle, Colombia
Sergio A. Velastin	Queen Mary University of London, UK/Universidad Carlos III de Madrid, Spain

Contents

Do Superpixel Segmentation Methods Influence Deforestation Image Classification?

Hugo Resende[1(✉)], Fabio A. Faria[1,2], Eduardo B. Neto[1], Isabela Borlido[3], Victor Sundermann[3], Silvio Jamil F. Guimarães[3], and Álvaro L. Fazenda[1]

[1] Institute of Science and Technology, Universidade Federal de São Paulo, Avenida Cesare Mansueto Giulio Lattes, n° 1201 - Eugênio de Mello, São José dos Campos, SP 12247-014, Brazil
{hresende,alvaro.fazenda}@unifesp.br
[2] AIPS, INESC-ID, Instituto Superior Tecnico, Universidade de Lisboa, Lisbon, Portugal
fabio.faria@tecnico.ulisboa.pt
[3] Computer Science Department, Pontifícia Universidade Católica de Minas Gerais, Belo Horizonte, MG, Brazil
{isabela.borlido,vgmsundermann}@sga.pucminas.br, sjamil@pucminas.br

Abstract. Image segmentation is a crucial step in various visual applications, including environmental monitoring through remote sensing. In the context of the ForestEyes project, which combines citizen science and machine learning to detect deforestation in tropical forests, image segments are used for labeling by volunteers and subsequent model training. Traditionally, the Simple Linear Iterative Clustering (SLIC) algorithm is adopted as the segmentation method. However, recent studies have indicated that other superpixel-based methods outperform SLIC in remote sensing image segmentation, and might suggest that they are more suitable for the task of detecting deforested areas. In this sense, this study investigated the impact of the four best segmentation methods, together with SLIC, on the training of classifiers for the target application. Initially, the results showed little variation in performance among segmentation methods, even when selecting the top five classifiers using the PyCaret AutoML library. However, by applying a classifier fusion approach (ensemble of classifiers), noticeable improvements in balanced accuracy were observed, highlighting the importance of both the choice of segmentation method and the combination of machine learning-based models for deforestation detection tasks.

Keywords: Superpixel Methods · Remote Sensing · Deforestation Detection · Ensemble Learning · ForestEyes Project

D. Chaves et al. (Eds.): CIARP 2025, LNCS 16528, pp. 1–15, 2026.
https://doi.org/10.1007/978-3-032-23161-1_1

1 Introduction

Image segmentation plays a relevant role in a wide range of applications, including the analysis and interpretation of visual data, such as medical diagnosis, environmental monitoring, precision agriculture, and the generation of data for training machine learning models [2,7,12,22,25]. Good segmentation accurately identifies homogeneous regions within an image by grouping pixels with similar characteristics. Well-defined segments enable the extraction of more reliable information, reduce noise and inconsistencies in analysis, and support decision-making in automated processes. On the other hand, poorly executed segmentation can lead to the loss of important details, confusion between objects of interest and background or other classes, and significantly compromise the performance of classification or detection models [17,29].

Several segmentation strategies have been developed, including deep learning approaches such as semantic, instance, and panoptic segmentation [13,14,16]. Although these methods achieve high performance in various domains, they require large amounts of annotated data and significant computational resources [19]. As an alternative, superpixel-based methods stand out for their simplicity, efficiency, and independence from supervised training. They group similar pixels while preserving important boundaries and structures in the image, making them useful as a preprocessing step. However, they still face challenges such as sensitivity to texture and lighting, as well as difficulty in determining the optimal number of segments [3].

The segments generated by superpixel-based methods can be leveraged in citizen science projects focused on environmental monitoring, such as the ForestEyes project [9,10]. This initiative combines citizen science and machine learning efforts to monitor tropical forests, particularly in detecting areas showing signs of deforestation. In the project, campaigns are organized on the Zooniverse platform [26,27], where non-expert volunteers analyze and label segments previously extracted from satellite images. These labeled segments are then used as training data for machine learning models aiming to automate the detection of deforested areas [24]. In this context, the choice of the segmentation algorithm is crucial, as it must produce well-defined regions, clearly separate the classes of interest, and generate regular and visually coherent polygons, which helps human cognition and increases the reliability of the labels provided by volunteers.

Designing, creating, and conducting a citizen science campaign are not trivial tasks, as they require careful planning of activities and continuous engagement of volunteers. Each task, represented by a segment to be evaluated, demands time and attention from participants, which can make the process lengthy and costly, especially in campaigns involving numerous segments. In light of this challenge, alternative labeling sources become valuable, such as the official deforestation maps provided by the Amazon Deforestation Monitoring Project (PRODES) [15]. Although these data do not fully replace human contributions, they can help accelerate the evaluation process and provide a reliable reference for training and validating machine learning models.

In the ForestEyes project campaigns, the Simple Linear Iterative Clustering (SLIC) [1] algorithm has been used as the standard segmentation method. However, recent studies by [4] and [23] have shown, through quantitative analyses using both traditional superpixel evaluation metrics and citizen science-based metrics, that among the 22 main superpixel segmentation algorithms found in the literature, there are alternatives that outperform SLIC for the specific goals of the ForestEyes project, suggesting more deeply investigations regarding citizen science and automatic classification. In this context, the present study aimed to investigate the influence of the four best-performing methods identified in those previous studies, along with SLIC, on the training of machine learning models for deforestation detection tasks.

2 Background

Fig. 1. The pipeline of the ForestEyes project. Highlighted is the target module of this work.

In this section, the concepts and main characteristics of the ForestEyes project and the superpixel-based segmentation algorithms, which are the focus of this study, will be presented.

2.1 ForestEyes Project

The ForestEyes project is structured around five core modules, as illustrated in Fig. 1. The **Pre-processing Module** is responsible for tasks related to the acquisition, preparation, and segmentation of remote sensing data. Satellite imagery is sourced from platforms such as EarthExplorer (for Landsat-8) and the Copernicus Browser (for Sentinel-2). Additionally, deforestation maps from the PRODES project are incorporated, which span from August of one year to July of the next. Data selection requires specifying parameters such as the region of interest and the acquisition period. Once collected, the data undergo preparation using tools such a regular GIS. At this stage, non-relevant regions, identified through PRODES, are excluded, enabling the segmentation process to concentrate solely on forested and recently deforested areas. The regions of interest are also cropped, and spatial alignment between satellite images and PRODES data is performed to ensure pixel-level correspondence. For segmentation, the project employs a superpixel-based approach. The current method in use is MaskSLIC, though alternative techniques are under evaluation, as discussed in [4] and [23].

In the **Citizen Science Module**, classification campaigns are created and deployed on the Zooniverse platform. Each task displays the same image segment using different false-color compositions to help volunteers distinguish between forested and deforested areas more easily. These compositions are selected based on proven methods from the literature and validated by image-processing experts. Metadata is generated to organize the sequential display of tasks and compositions. During campaign setup, the logic of volunteer participation is defined, including parameters such as the minimum number of responses per task, visual layout, instructional content, and usability features. Once launched, campaigns can be monitored in real-time, enabling ongoing support and ensuring the smooth progress of volunteer activity.

In the **Organization and Selection Module**, responses collected during the citizen science campaigns are analyzed to determine the final classification for each task. The dataset includes metadata such as volunteer identifiers, the tasks completed, and the time taken for each classification. Based on this information, majority voting is applied to derive consensus labels, while response variability is quantified using entropy measures. Response time is also evaluated as a proxy for task complexity. These metrics are instrumental in identifying inconsistencies, assessing annotation reliability, and informing refinements for future campaigns. Upon completion of this analysis, the module defines the training and testing sets that will be used in the subsequent Machine Learning Module.

The **Machine Learning Module** consists of a series of steps ranging from the extraction of Haralick texture features from the labeled segments to the application of one or more classification models. With the training and testing sets defined in the previous module, the process begins with feature extraction, which serves as input for the learning phase. Next, a classifier, or a set of classifiers previously used in the ForestEyes project, is trained using these features. Finally, the trained model(s) are evaluated and applied, completing the supervised learning cycle within the context of the project.

Finally, the **Post-processing Module** is responsible for decision-making activities based on the deforestation detection results obtained in the machine learning module. This module interprets the outcomes and transforms them into actionable information or recommendations. Additionally, future developments are expected to include mechanisms that enable reliable automatic labeling of samples, among other features aimed at enhancing the system's autonomy and scalability.

2.2 Superpixel-Based Segmentation Algorithms

The algorithm **Compactness-Controlled Contour-Relaxed Superpixels (CRS)** is an approach for image pre-segmentation that aims to partition an image into a predefined number of regions (superpixels). The main feature of this method is its ability to simultaneously maximize the internal texture homogeneity within each superpixel and the adherence of contours to the visual boundaries of the image, supported by a Gibbs-Markov random field model. Unlike traditional methods, which limit homogeneity to smooth regions or similar intensity values, CRS remains effective even in highly textured scenes. The algorithm optimizes an energy function, which depends on only a few parameters defined by the chosen statistical model. The deformation of superpixels in response to image content is controlled by a single parameter that regulates the compactness of the segments. Extensive experimental evaluations have shown that CRS outperforms well-established superpixel algorithms in the literature in terms of boundary recall and undersegmentation error while offering competitive or superior runtime performance [8].

Eikonal-based Region Growing Clustering (ERGC) was developed to facilitate the segmentation of anatomical structures in medical images, especially in scenarios that demand speed, such as radiation therapy planning. The technique clusters pixels based on the solution of an Eikonal equation, where each pixel receives a value representing its travel time or geodesic distance from initial seeds. The function that regulates this propagation is computed from the dissimilarity between the pixel features and those of the expanding region, allowing segmentation that is sensitive to the image content. The Fast Marching method is used to accelerate the computation, ensuring efficiency even for large volumes. ERGC also enables control over the compactness of the generated regions, which contributes to the creation of regular and homogeneous segments [5].

Efficient Topology Preserving Segmentation (ETPS) algorithm was developed to provide fast, regular, and topologically coherent segmentation, suitable for applications requiring real-time processing, such as autonomous driving scenarios. It combines a coarse-to-fine energy update strategy, inspired by the SEEDS algorithm, with an efficient optimization approach that significantly reduces the number of boundary updates throughout the process. Unlike previous methods that require multiple iterations to achieve good results, ETPS can reach high-quality energy minima with just a single iteration, while maintaining connectivity between segmented regions. The algorithm also incorporates shape regularization by penalizing boundary length and can integrate additional information, such as stereo data, to enhance segmentation in stereo image pairs. Evaluations on datasets such as BSD and KITTI show that ETPS outperforms popular approaches like SLIC and SEEDS in both segmentation quality and execution speed [28].

Rooted Spanning Segmentation (RSS) is a graph-based image segmentation algorithm in which regularly distributed seed pixels serve as roots. Segmentation is performed by extending paths originating from these roots, based on cost functions that consider both the maximum difference and the range of values between pixels. In this way, it balances color similarity and spatial proximity. The number and placement of segmented regions are controlled through seed selection, while path-based propagation ensures coherence, contour adherence, and adaptability to image structures. Uniformity and compactness of the resulting regions are modulated by a scaling factor, and connectivity is inherently preserved through the region-growing process. Among its main features is scalability, as it can handle different types of data in a unified manner. RSS also has promising applications, such as extracting regions based on seeds and cost thresholds, which are useful for tasks like network extraction and global object proposals. Finally, studies also indicate its potential in convolutional neural network-based semantic segmentation, due to its ability to operate on high-dimensional representations extracted by such networks [6].

Simple Linear Iterative Clustering (SLIC) is one of the most popular segmentation algorithms due to its simplicity, efficiency, and the quality of its results. It works by grouping pixels based on color similarity and spatial proximity, using a variation of the k-means algorithm. To achieve this, it operates in a combined five-dimensional space, three for color (in the CIELAB space) and two for pixel position, which allows the generation of compact and regular superpixels. One of SLIC's main advantages is the direct control over the number of segmented regions, as well as its ability to produce results that align well with image boundaries. Although the algorithm does not natively guarantee connectivity among pixels within each superpixel, this issue can be addressed with a simple post-processing step. Due to its high computational efficiency, SLIC is widely used as a preprocessing step in various computer vision tasks, such as semantic labeling, tracking, and object proposal generation [1].

3 Experiments

This section presents the details of the experiments conducted in this study. In particular, it describes the construction of the dataset and the results obtained by applying the PyCaret AutoML library and building the classifier ensembles proposed in [11].

3.1 Dataset Construction

Like depicted in [4,25] and [23] nine remote sensing images from the Landsat-8 satellite (also referred to as study areas) were collected, representing a geographic region of approximately 8,514 hectares, located near the Xingu River Basin in the state of Pará, Brazil. This region is known for its recurrent deforestation hotspots. For each of these images, the corresponding PRODES deforestation map was obtained, matching the acquisition period (year 2022). After collection, the images underwent a preparation process in a GIS, which included cropping the regions of interest, performing spatial alignment between the Landsat-8 and PRODES data at the pixel level, and creating masks for segmentation. Next, the principal component (PCA) of each image was extracted to serve as the basis for the segmentation step. Based on these preparations, the five superpixel-based segmentation algorithms presented in Subsect. 2.2 were applied, using a fixed parameter of 6,000 superpixels for each method.

Once the segments were generated, a set of useful segments was created for each segmentation method. These segments, as defined in the ForestEyes project, are those containing more than 70 pixels (given the 30-m spatial resolution of Landsat-8) and having over 70% homogeneity in terms of forest or deforestation pixels. From this set, 45 segments with perfect homogeneity (100%) were selected for each target class (forest and deforestation), per segmentation method. Additionally, another 45 segments per class were selected with homogeneity greater than 70%, ordered increasingly, to ensure a diverse training set containing both ideal and more heterogeneous examples. As a result, 180 training segments were selected per segmentation method, totaling 900 training segments across the five methods. It is important to highlight that, since SLIC is the baseline method used in the ForestEyes project, segments generated by this algorithm were used as reference, and corresponding geographic regions were identified in the outputs of the other segmentation methods (largest intersection of pixels within segments). An example of such equivalent segmented regions is shown in Fig. 2. Finally, the remaining useful segments for each method, those not used in training, were assigned to the respective test sets. Since the dataset was built from nine different study areas and each algorithm produces a different segmentation of the same region, the number of test samples varied slightly among the segmentation methods (CRS = 19,039, ERGC = 19,009, ETPS = 19,034, RSS = 18,999, and SLIC = 19,044 segments).

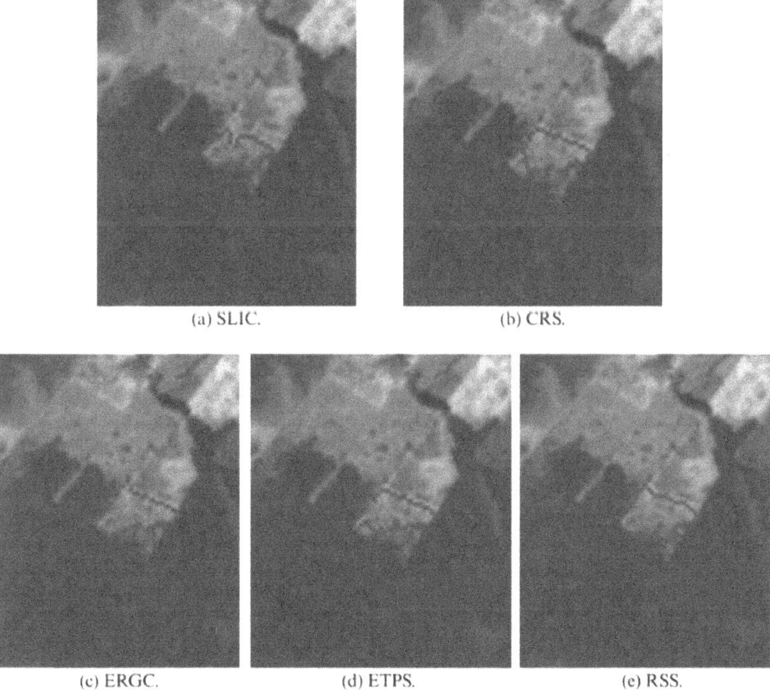

Fig. 2. Examples of deforestation segment, from the same geographic area, for each one of the segmentation methods. To visualize the segments, the false-color image composed of the $4-6-7$ bands was used to highlight forest and deforestation regions.

After defining the segments for the training and testing sets, 13 Haralick features were extracted for each segment, considering four directions (0°, 45°, 90°, and 135°) in the gray-level co-occurrence matrix for bands 6-4-3-1, following the same experimental protocol adopted in [20]. For each of the 13 features, the arithmetic mean of the values obtained across the four directions was computed. As a result, each segment was represented by a feature vector containing 13 attributes, which were then used as input for training and evaluating the machine learning models.

3.2 Results and Discussion

The **first stage**, serving as a baseline, of the experiments focused on using the PyCaret AutoML library to identify the best classifiers for each training dataset generated by the different segmentation methods (CRS, ERGC, ETPS, RSS, and SLIC). Initially, the PyCaret automatically evaluates 14 classification algorithms, including some of the most well-known in the literature, such as Ridge Classifier (RC), Linear Discriminant Analysis (LDA), Logistic Regression (LR), Gradient Boosting Classifier (GBC), Decision Tree Classifier (DT), Stochastic Gradient Descent Classifier (SGD), Extra Trees Classifier (ET), Light Gradient Boosting Machine (LGBM), Support Vector Machine (SVM), Quadratic Discriminant Analysis (QDA), Random Forest (RF) and k-Nearest Neighbors Classifier (KNN). Each model is assessed using 5-fold cross-validation, and the results are summarized in a performance table that includes several key metrics such as accuracy, area under the ROC curve (AUC), recall, precision, F1-score, Kappa, MCC, and training time (TT). Based on the aforementioned metrics, PyCaret generates an overall performance ranking, prioritizing classifiers that demonstrate a good balance between accuracy and statistical robustness.

For each superpixel segmentation method, this ranking is used to select the top five machine learning models. These selected classifiers are then submitted to an additional hyperparameter optimization step using PyCaret's automatic fine-tuning feature, which aims to further enhance model performance. At the end of this process, the best-performing classifiers, with their optimized hyperparameters, are finalized for each dataset associated with each segmentation method. As an example, for the dataset generated by the SLIC segmentation method, the best performing classifier after fine-tuning was Logistic Regression, with the following hyperparameters: C = 8.426, class_weight = 'balanced', penalty = 'l2', solver = 'lbfgs', and max_iter = 1000, among other parameters that remained at their default values. It is noteworthy that the values obtained align with those recommended in the official scikit-learn documentation [21], indicating that the optimization process performed by PyCaret adheres to established best practices and guidelines from the literature.

Once the hyperparameters of the top five classifiers for each segmentation method were tuned, the next step was to test them on the respective test sets associated with each method. Due to the imbalanced classes of the test sets, to evaluate the performance of the machine learning models at this stage, the balanced accuracy metric was adopted. As shown in Table 1, the family of linear classifiers, such as Ridge Classifier (RC) and Logistic Regression (LR), stood out as effective classification solutions, regardless of the superpixel method studied in this paper. However, it is important to note that there were no significant differences among the best-performing classifiers, indicating that the evaluated segmentation methods are equally suitable for processing this type of image and for training machine learning models.

Table 1. Balanced accuracy results (%) of the top five classifiers for each superpixel method.

Spx Methods	RC	LDA	LR	GBC	Other
CRS	**86.00**	84.91	85.78	79.87	DT (73.98)
ERGC	84.43	84.05	**85.50**	78.07	SGD (85.13)
ETPS	**86.10**	83.89	85.99	79.15	ET (77.89)
RSS	**85.10**	84.27	84.62	–	LGBM (76.96), SGD (84.82)
SLIC	83.92	78.88	**84.60**	–	DT (77.16), LGBM (78.84)

Given this scenario, the **second stage** involves selecting the best classifier (top 1) obtained for each superpixel method (CRS, ERGC, ETPS, RSS, and SLIC) and testing it on the test sets of the other segmentation methods. This strategy is referred to as the *cross-method* scenario, and its results can be observed in Table 2. In this experiment, the classifier names are indicated alongside their corresponding segmentation methods (e.g., RC/CRS indicates that the Ridge Classifier was the best machine learning model for the CRS segmentation method).

Table 2. Balanced accuracy (%) for individual classifiers (top one for each superpixel method) trained on a given superpixel method and tested on others, and performance of ensemble strategies.

Spx	Individual Classifier/Spx Method					Classifier Ensemble Strategies				
	RC/	LR/	RC/	RC/	LR/	UMDA		MV		Rel. Gain
Methods	CRS	ERGC	ETPS	RSS	SLIC	Acc.	#C	Acc.	#C	(UMDA × Top1)
CRS	86.00	86.60	**86.90**	78.60	85.00	**88.84**	12	88.50	25	2.2
ERGC	76.40	**85.50**	79.30	84.50	84.50	**87.38**	10	86.06	25	2.2
ETPS	83.60	**86.60**	86.10	79.40	86.30	**87.43**	10	86.62	25	1.0
RSS	77.00	82.60	80.80	**85.10**	84.20	**85.12**	12	83.94	25	0.0
SLIC	79.10	**84.90**	81.70	77.80	84.60	**86.17**	14	85.07	25	1.5
Avg.	80.42	**85.24**	82.96	81.08	84.92	**86.99**	12	86.04	25	–
Std. Dev.	4.21	**1.65**	3.36	3.45	0.82	1.41	2	1.71	0	–

Overall, the results show that the classifiers maintain good performance even when applied to data from different segmentation methods. In particular, the Logistic Regression classifier trained on ERGC segments achieved the highest average *cross-method* accuracy (85.24%) and the lowest standard deviation (1.65), according to Table 2, indicating greater stability across tests. It is important to highlight that even after the *cross-method* scenario, considering the accuracy values obtained, a slight improvement can be seen. However, given the average values and standard deviations, it can be seen that the best classifiers tend to present very similar values.

It is important to emphasize that, although the top-1 classifiers of each superpixel method did not show statistically significant differences in balanced accuracy, the computational cost associated with each segmentation technique varies considerably. In one of the analyzed areas, for instance, CRS required 3.99 s to complete the processing, whereas ERGC took 1.7 s, ETPS only 0.33 s, RSS 0.48 s, and SLIC 0.80 s. Therefore, ETPS demonstrated the best performance in terms of execution time, being approximately 12 times faster than CRS, which proved to be the most computationally demanding method.

Furthermore, it was observed that the CRS superpixel method produced segments that were easier to classify, regardless of the training set used, even when it was generated from any of the other methods evaluated in this study. This behavior indicates greater consistency of CRS in supporting the classification stage. In addition, the pair LR/ERGC achieved the highest balanced accuracy average among all the combinations analyzed, showing a slight advantage of this pair compared to the others.

Since this fact was observed, the next stage of the experiments consisted of verifying the existence of possible complementarity of information between the classifiers composed of different segmentation methods. Therefore, in this work, the diversity measure known as the *correlation coefficient* ρ ($COR(c_i, c_j)$) is computed. Given $COR(c_i, c_j) = \frac{ad-bc}{\sqrt{(a+b)(c+d)(a+c)(b+d)}}$, the value a is the percentage of examples that both classifiers c_i and c_j classified correctly in a validation set. Values b and c are the percentage of examples that c_j hit and c_i missed and vice-versa. The value d is the percentage of examples that both classifiers missed. A lower score between a pair of classifiers c_i and c_j indicates higher diversity [18]. As can be seen in Fig. 3, most $COR(c_i, c_j)$ scores are below 0.4 (shades of blue), suggesting that the individual classifiers provide complementary information. Therefore, the use of classifier fusion strategies (ensemble of classifiers) may be a viable way to improve classification results on the target task.

Finally, for the **third and last stage** of the experiments, a framework for building ensembles of classifiers proposed in the literature has been employed [11]. This framework brings together many bioinspired optimization algorithms (e.g., Particle Swarm Optimization, Ant Colony Optimization, Genetic Algorithms, and Artificial Bee Colony algorithms) for classifier selection and fusion. Among the algorithms available in this framework is the Univariate Marginal Distribution Algorithm (UMDA), which achieved the best performance

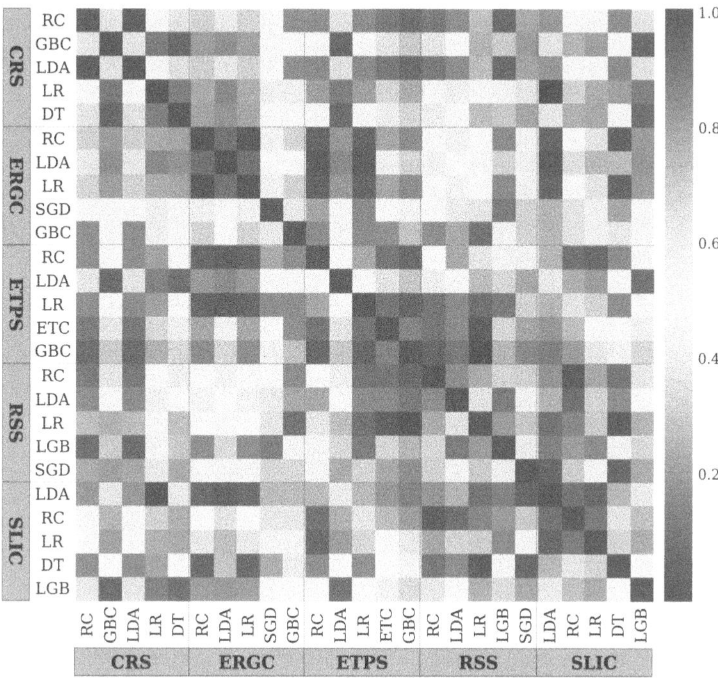

Fig. 3. Relationship matrix computed among all 25 classifiers using $COR(c_i, c_j)$ measure.

across all comparisons and was selected as the most effective classifier ensemble strategy. UMDA is a type of Estimation of Distribution Algorithm (EDA) that generates new individuals by sampling from univariate normal distributions, which are updated in each iteration based on the top-performing individuals from the previous generation.

Table 2 also presents the balanced accuracy results obtained by the UMDA strategy, alongside the well-known majority voting (MV) fusion method from the literature. Additionally, the number of classifiers (#C) selected by UMDA is reported. Overall, the UMDA strategy outperformed individual classifiers, achieving a relative gain of up to 2.2%. Moreover, it outperformed the MV strategy while using significantly fewer classifiers, less than 50% of the total used by MV.

A final analysis of the impact of the superpixel methods that compose the ensemble of classifiers based on UMDA strategy can be observed in Table 3. Among the main findings, it should be noted that the LC/RSS classifier showed a probability of 100% (CI) being selected for the ensemble of classifiers, even when evaluating samples from different superpixel methods. Furthermore, when analyzing the contribution of classifiers per superpixel method (CS), it is observed that, in the case of SLIC, its classifiers reached a probability of 60% in the ensemble of classifiers. Finally, although not explicitly reported in Table 3, it was verified that all classifiers were selected at least once using the UMDA strategy to compose the final classification solution of some superpixel methods, which reinforces the idea of complementarity among them.

Table 3. Top-3 classifiers/superpixel method with the highest probabilities (Pr., in %) of being part of the ensemble of classifiers based on UMDA strategy. The probabilities refer to each classifier isolated by superpixel method (*CI*) and to all classifiers of a given superpixel method (*CS*).

Pr.	CRS			ERGC			ETPS			RSS			SLIC		
–	GBC	LR	DT	LC	LDA	GBC	LC	LR	ETC	LC	LGB	SGD	LDA	RC	DT
CI	60	60	40	40	40	80	60	40	80	100	40	40	60	80	80
CS	40			40			48			44			60		

4 Conclusion

This study investigated the influence of five superpixel-based segmentation methods on training machine learning models for classifying remote sensing images focused on deforestation detection. Based on the top-performing methods identified in previous studies [4] and [23], the analysis was conducted in three stages: (i) ranking and fine-tuning the top five classifiers for each segmentation method using the PyCaret library; (ii) evaluating generalization through cross-method testing, where classifiers trained on one method were tested on data segmented by others; and (iii) applying ensemble strategies based on bioinspired algorithms to improve accuracy through model combination.

The results showed that, despite known differences among the segmentation methods in terms of perceptual behavior [23], all achieved similar performance in supporting model training, demonstrating their suitability for remote sensing tasks. This apparent equivalence raised an important question regarding the need to continue investigating segmentation methods even when they produce similar accuracies.

To address this question, we analyzed the diversity among classifiers using the correlation $COR(c_i, c_j)$ metric. The results (Fig. 3) revealed that classifiers based on different segmentation methods capture complementary patterns, which would not be explored if only a single method were used. To demonstrate the benefits of this diversity, we conducted experiments with classifier ensembles, which confirmed that combining segmentation methods improves detection performance in tropical deforestation areas. It is important to note that the ensemble is not a core contribution of this work but serves as evidence that maintaining segmentation diversity is essential to fully exploit the potential of the data.

As future works, we plan a new citizen science campaign using data segmented by the same five methods investigated in this study. The goal is to compare the effectiveness of volunteer annotations across different segmentation methods using PRODES reference labels as ground truth. This will allow a deeper analysis of the impact of segmentation quality on the accuracy of human annotations. Furthermore, it is planned to combine segmentation masks from different segmenters in order to generate segments that better capture complementary spatial patterns and boundaries, thus providing richer representations for subsequent classification tasks. Finally, we highlight that future investigations may include additional classifier performance metrics, such as Precision, Recall, F1-score, AUC-ROC, AUC-PR, MCC, and Cohen's Kappa, to complement the analysis presented in this study.

Acknowledgments. The authors would like to thank the UNIFESP and the IFSULDEMINAS for their support; the Coordination for the Improvement of Higher Education Personnel (CAPES) for financial assistance; and the National Laboratory for Scientific Computing (LNCC) for providing the HPC computational resources of the Santos Dumont Supercomputer (SDumont). This work was partially financially supported by the Conselho Nacional de Desenvolvimento Científico e Tecnológico – CNPq – (Grants 306573/2022-9, 442950/2023-3 and 407242/2021-0), the Fundação de Amparo a Pesquisa do Estado de Minas Gerais – FAPEMIG – (APQ-01079-23 and APQ-05058-23). This research is part of INCT Future Internet for Smart Cities, funded by CNPq (#465446/2014-0), and FAPESP (#2014/50937-1, #2015/24485-9, #2017/25908-6, #2018/23908-1, #2019/26702-8, #2023/00811-0, #2023/00782-0, #2024/01115-0).

Disclosure of Interests. The authors have no competing interests to declare that are relevant to the content of this article.

References

1. Achanta, R., Shaji, A., Smith, K., Lucchi, A., Fua, P., Süsstrunk, S.: Slic superpixels compared to state-of-the-art superpixel methods. IEEE Trans. Pattern Anal. Mach. Intell. **34**(11), 2274–2282 (2012). https://doi.org/10.1109/TPAMI.2012.120
2. Azad, R.: Medical image segmentation review: the success of u-net. IEEE Trans. Pattern Anal. Mach. Intell. **46**(12), 10076–10095 (2024). https://doi.org/10.1109/TPAMI.2024.3435571
3. Barcelos, I.B., Belém, F.D.C., João, L.D.M., Patrocínio Jr, Z.K.D., Falcão, A.X., Guimarães, S.J.F.: A comprehensive review and new taxonomy on superpixel segmentation. ACM Comput. Surv. **56**(8), 1–39 (2024)
4. Borlido, I., et a;.: How to identify good superpixels for deforestation detection on tropical rainforests. IEEE Geosci. Remote Sens. Lett. (2024). https://doi.org/10.1109/LGRS.2024.3454973
5. Buyssens, P., Gardin, I., Ruan, S.: Eikonal based region growing for superpixels generation: application to semi-supervised real time organ segmentation in CT images. IRBM **35**(1), 20–26 (2014). https://doi.org/10.1016/j.irbm.2013.12.007. https://www.sciencedirect.com/science/article/pii/S1959031813001437
6. Chai, D.: Rooted spanning superpixels. Int. J. Comput. Vision **128**(12), 2962–2978 (2020)

7. Charisis, C., Argyropoulos, D.: Deep learning-based instance segmentation architectures in agriculture: a review of the scopes and challenges. Smart Agric. Technol. **8**, 100448 (2024) https://doi.org/10.1016/j.atech.2024.100448. https://www.sciencedirect.com/science/article/pii/S2772375524000534
8. Conrad, C., Mertz, M., Mester, R.: Contour-relaxed superpixels. In: Heyden, A., Kahl, F., Olsson, C., Oskarsson, M., Tai, X.-C. (eds.) EMMCVPR 2013. LNCS, vol. 8081, pp. 280–293. Springer, Heidelberg (2013). https://doi.org/10.1007/978-3-642-40395-8_21
9. Dallaqua, F.B.J.R., Fazenda, Á.L., Faria, F.A.: Foresteyes project: can citizen scientists help rainforests? In: 2019 15th International Conference on eScience (eScience), pp. 18–27. IEEE (2019)
10. Fazenda, A.L., Faria, F.A.: Foresteyes: citizen scientists and machine learning-assisting rainforest conservation. Commun. ACM **67**(8), 95–96 (2024)
11. Ferreira, A.R., de Rosa, G.H., Papa, J.P., Carneiro, G., Faria, F.A.: Creating classifier ensembles through meta-heuristic algorithms for aerial scene classification. In: 2020 25th International Conference on Pattern Recognition (ICPR), pp. 415–422 (2021). https://doi.org/10.1109/ICPR48806.2021.9412938
12. van der Graaf, J.W., et al.: Lumbar spine segmentation in MR images: a dataset and a public benchmark. Sci. Data **11**(1), 264 (2024). https://doi.org/10.1038/s41597-024-03090-w
13. Guo, Y., Liu, Y., Georgiou, T., Lew, M.S.: A review of semantic segmentation using deep neural networks. Int. J. Multimedia Inf. Retr. **7**, 87–93 (2018)
14. Hafiz, A.M., Bhat, G.M.: A survey on instance segmentation: state of the art. Int. J. Multimedia Inf. Retr. **9**(3), 171–189 (2020)
15. INPE: PRODES - Project for Monitoring Deforestation in the Legal Amazon by Satellite (2024). https://www.obt.inpe.br/prodes/index.php. acessado em setembro/2024
16. Kirillov, A., He, K., Girshick, R., Rother, C., Dollár, P.: Panoptic segmentation. In: Proceedings of the IEEE/CVF conference on computer vision and pattern recognition. pp. 9404–9413 (2019)
17. Küchler, J., Kröll, D., Schoenen, S., Witte, A.: Uncertainty estimates for semantic segmentation: providing enhanced reliability for automated motor claims handling. Mach. Vis. Appl. **35**(4), 66 (2024)
18. Kuncheva, L.I., Whitaker, C.J.: Measures of diversity in classifier ensembles and their relationship with the ensemble accuracy. Mach. Learn. **51**, 181–207 (2003)
19. Minaee, S., Boykov, Y., Porikli, F., Plaza, A., Kehtarnavaz, N., Terzopoulos, D.: Image segmentation using deep learning: a survey. IEEE Trans. Pattern Anal. Mach. Intell. **44**(7), 3523–3542 (2021)
20. Neto, E.B., et al.: A satellite band selection framework for amazon forest deforestation detection task. In: Proceedings of the Genetic and Evolutionary Computation Conference, GECCO '24, pp. 1300–1308. Association for Computing Machinery, New York (2024)
21. Pedregosa, F., et al.: Scikit-learn: machine learning in python. J. Mach. Learn. Res. **12**, 2825–2830 (2011)
22. Peng, M., et al.: Advanced image segmentation for precision agriculture using CNN-gat fusion and fuzzy c-means clustering. Comput. Electron. Agric. **226**, 109431 (2024). https://doi.org/10.1016/j.compag.2024.109431
23. Resende, H., et al.: Exploring superpixel segmentation methods in the context of citizen science and deforestation detection. In: Proceedings of the International Geoscience and Remote Sensing Symposium (IGARSS) (2025)

24. Resende, H., Álvaro L. Fazenda, Cappabianco, F.A., Faria, F.A.: Increasing the reliability of citizen science campaign data for deforestation detection in tropical forests. Future Gener. Comput. Syst. 108081 (2025). https://doi.org/10.1016/j.future.2025.108081. https://www.sciencedirect.com/science/article/pii/S0167739X25003759
25. Resende, H., Neto, E.B., Cappabianco, F.A.M., Fazenda, A.L., Faria, F.A.: Sampling strategies based on wisdom of crowds for amazon deforestation detection. In: 2024 37th SIBGRAPI Conference on Graphics, Patterns and Images (SIBGRAPI), pp. 1–6 (2024). https://doi.org/10.1109/SIBGRAPI62404.2024.10716332
26. Simpson, R., Page, K.R., De Roure, D.: Zooniverse: observing the world's largest citizen science platform. In: Proceedings of the 23rd International Conference on World Wide Web, pp. 1049–1054 (2014)
27. Smith, A., Lynn, S., Lintott, C.: An introduction to the zooniverse. In: Proceedings of the AAAI Conference on Human Computation and Crowdsourcing, vol. 1, pp. 103–103 (2013)
28. Yao, J., Boben, M., Fidler, S., Urtasun, R.: Real-time coarse-to-fine topologically preserving segmentation. In: Proceedings of the IEEE Conference on Computer Vision and Pattern Recognition (CVPR) (2015)
29. Zhang, C., Pan, T.Y., Chen, T., Zhong, J., Fu, W., Chao, W.L.: Learning with free object segments for long-tailed instance segmentation. In: European Conference on Computer Vision, pp. 655–672. Springer, Heidelberg (2022). https://doi.org/10.1007/978-3-031-20080-9_38

Comparative Study of Image Descriptors and Dimensionality Reduction Methods for the Prediction of Sugarcane Crop Yield

Luiz Antonio Falaguasta Barbosa[1,2](✉), Ivan Rizzo Guilherme[1], Daniel Carlos Guimarães Pedronette[1], and Bruno Tisseyre[3]

[1] São Paulo State University – UNESP, Rio Claro 13506-900, Brazil
luiz.falaguasta@unesp.br, ivan.guilherme@unesp.br, daniel.pedronette@unesp.br

[2] Brazilian Agricultural Research Corporation – EMBRAPA, Digital Agriculture – CNPTIA, Campinas 13083-886, Brazil

[3] ITAP, University Montpellier, Institut Agro, INRAE, 34060 Montpellier, France
bruno.tisseyre@supagro.fr

Abstract. Accurate crop yield prediction is crucial for effective farm management, efficient harvest scheduling, and informed resource allocation. While traditional models often rely on Vegetative Indices (VIs) and weather data, other factors such as cultivar, management practices, and soil conditions also influence yield. This study examines the integration of Visual Features (VFs), extracted using pre-trained Convolutional Neural Networks (CNNs) and Vision Transformers (ViTs), with VIs and Light Detection and Ranging (LiDAR) data to enhance yield prediction accuracy. To optimize the feature space, dimensionality reduction techniques like Principal Component Analysis (PCA) and Uniform Manifold Approximation and Projection (UMAP) were applied. Predictive models were trained using multispectral and LiDAR data collected by Unmanned Aerial Vehicles (UAVs) over two sugarcane experimental fields in Australia. Results show that combining VFs with Normalized Difference Vegetation Index (NDVI) variants outperforms models using other VIs and LiDAR, highlighting the potential of deep learning-based VFs. Larger datasets may further enhance the performance of VF-based models.

Keywords: Crop yield prediction · Sugarcane · Visual features extraction · Dimensionality reduction

1 Introduction

Crop yield prediction is typically carried out using dependable variables based on VIs or by combining them with weather data collected on various dates. The objective of predicting crop yield is to inform decisions in farm management, such as determining the optimal time to harvest.

Weather, cultivar type, management practices, and soil composition all contribute to a scenario that defines the yield. The results of this scenario can be

D. Chaves et al. (Eds.): CIARP 2025, LNCS 16528, pp. 16–29, 2026.
https://doi.org/10.1007/978-3-032-23161-1_2

captured by the vigor of plants by their canopy architecture and color, and VFs, VIs, and LiDAR can, therefore, identify these characteristics.

This work utilizes pre-trained CNN and Vision Transformer (ViT) models to explore alternative crop yield prediction techniques for extracting VFs. Therefore, it analyzes different feature extractors, different dimensionality reductions, and the combination of VFs, VIs, and LiDAR as source data for prediction. It employs a dataset from an experiment conducted by researchers at the Commonwealth Scientific and Industrial Research Organisation (CSIRO) utilizing a UAV with LiDAR and multispectral imaging sensors [1]. This research, conducted by CSIRO, aimed to monitor two sugarcane field trials subjected to varying Nitrogen (N) fertilization regimes in the Wet Tropics region of Australia. Six surveys were performed at 42-day intervals to evaluate crop growth dynamics, including canopy height, plant density, and VIs. During each survey, predictive models were developed to estimate biomass at harvest at a high spatial resolution ($2\,\text{m} \times 2\,\text{m}$ collection squares randomly positioned within each plot). The study assessed the predictive performance of models incorporating visual features, multispectral features (comprising different VIs), LiDAR-derived attributes, and a fusion of both data modalities, comparing their effectiveness.

Despite the widespread use of traditional VIs for crop yield prediction, VFs have been gaining increasing interest in the research community. Recent studies [2] evaluated pre-trained deep learning models for remotely sensed imagery from satellite and aerial datasets. The authors assessed Self-Supervised Learning (SSL) pre-trained models on seven satellite and aerial imagery datasets with varying resolutions. Their research demonstrated that following the preprocessing steps used during pre-training, including image sizing and normalization, significantly improved performance on downstream tasks. A Residual Neural Network with 50 layers (ResNet-50) pre-training remains a strong baseline for transfer learning in satellite imagery, achieving notable accuracy improvements on datasets such as So2Sat and EuroSAT [3].

Numerous approaches exist for extracting VFs and performing dimensionality reduction; however, determining the most appropriate method for a given scenario remains a complex challenge. While applying VFs is relatively common in classification tasks [4], their use in regression—particularly for crop yield prediction—has received limited attention. In contrast, VIs are widely employed in yield modeling due to their established correlation with plant health and productivity. Nevertheless, leveraging VFs may provide complementary information that enhances the predictive performance and deepens understanding of crop dynamics. Similarly, selecting an appropriate dimensionality reduction strategy must balance the trade-offs between dataset size, computational cost, and the retention of informative features. To the best of our knowledge, the literature has not yet systematically explored and evaluated the joint application of VFs and dimensionality reduction techniques for predicting crop yields.

The most closely related study identified in the literature that utilizes a pre-trained model on remote sensing imagery for VF extraction is presented by [2]. The authors employed various classification datasets comprising multispectral aerial imagery to extract feature representations using ResNet-50 with different initialization weights and alternative methods. However, their work does not assess the integration of VF extraction with dimensionality reduction techniques. In the present study, ResNet-101 [5] and Self-Distillation with No Labels, Version 2 (DINOv2) [6] models were employed as pre-trained CNN and ViT to extract VFs from the dataset established in the experiment conducted by [1]. Feature extraction was performed on full plot-level images using both techniques and on 80 × 80-pixel patches using ResNet-101. Dimensionality reduction was applied using PCA and UMAP. The resulting VFs, both standalone and combined with VIs and LiDAR-derived features, were then used as features for regressor training and crop yield estimation.

This work evaluates the use of isolated VFs and their use combined with VIs and LiDAR data. The work also considers combining different VFs (pre-trained CNN and ViT) with varying dimensionality reductions, comparing its outcomes with the results obtained using only VIs and LiDAR in [7,8]. In this sense, the main contributions of this work are:

- the use of VFs evaluation with dimensionality reduction (PCA and UMAP);
- the combination of VFs, VIs, and LiDAR;
- the evaluation of different VFs extractions (ResNet-101 and DINOv2).

The remainder of this paper is organized as follows: Sect. 2 discusses the proposed methodology. Sections 3 and 4 present the VFs and dimensionality reduction techniques, respectively. The experimental evaluation is discussed in Sect. 5, while Sects. 6 and 7 present the results and conclusions, respectively.

2 Proposed Methodology

This work focuses on exploiting VFs extracted from aerial imagery for crop yield prediction as a way of improving outcomes instead of only using standard features like VIs, meteorological data, or management data. It also considers combining VFs with VIs and LiDAR data, utilizing dimensionality reduction techniques to explore other complementary tools and achieve better results.

Inspired by a recent comparative study that analyzed feature representation extraction from remotely sensed imagery datasets used for classification tasks under different image preprocessing steps, including image sizing variation and image normalization, among others [2], pre-trained models based on CNN and ViT were utilized for VF extraction. Regression models suggested by recent work [7] were employed for crop yield prediction on a publicly available sugarcane dataset [1] to evaluate if the use of VFs as feature predictors would be more feasible to estimate sugarcane crop yield.

Using ResNet-101, extracting features from each full plot-level area produced 2,048 features, while using patches 80 × 80 pixels produced 143,360 features. When DINOv2 was used to extract features, it was applied to each plot as a single image, producing 768 features. Since it did not perform well for patches of 80 × 80 pixels, it was explored to combine VIs and LiDAR only to plot as a single image.

2.1 Experimental Area and Dataset

Experimental Area. This study utilizes data from an experiment conducted by CSIRO researchers [1] using UAV-mounted LiDAR and multispectral sensors to monitor two 1-hectare sugarcane fields in the Wet Tropics of Australia, each under five N fertilization rates (0, 70, 110, 150, and 190 kg N/ha) in a randomized complete block design with four replicates (Fig. 1). Six surveys (Fig. 2) were conducted at 42-day intervals to assess crop growth through canopy height, plant density, and VIs. For each survey, models were developed to predict at-harvest biomass at a 2×2 m resolution within randomly selected plot areas. Model performance was evaluated using multispectral features (70 features based on statistical measures like maximum, minimum, average, standard deviation, and percentiles calculated over the ten VIs: NDVI, Normalized Difference Red Edge Index (NDRE), Green NDVI (GNDVI), Enhanced Vegetation Index (EVI), Modified Anthocyanin Content Index (MACI), Optimized Soil Adjusted Vegetation Index (OSAVI), Simplified Canopy Chlorophyll Content Index (SCCCI), Transformed Chlorophyll Absorption and Reflectance Index (TCARI), Triangular Greenness Index (TGI), and Visible Atmospherically Resistant Index (VARI)), LiDAR-derived features, and their fusion and compared against a benchmark model based on NDVI. The study was conducted on two fields with distinct soil drainage characteristics—well-drained (Site #1) and poorly drained (Site #2)—that receive over 4,000 mm of rainfall annually, primarily between December and March. Fertilizer treatments were applied 52 and 55 days after the previous crop's harvest at Sites #1 and #2, respectively.

In [7], various regressors were tested and tuned using grid search [9] to optimize predictions on the same dataset from [1]. The dataset was split into training and testing sets. A grid search was applied to the training set to determine the optimal hyperparameters for the models, including SVR (`kernel = 'rbf', C = 10, coef0 = 0.01, degree = 3, gamma = 'scale'`), Random Forest (`n_estimators = 200, max_features = 'sqrt', max_depth = 3, random_state = 18`), Gradient Boosting (`learning_rate = 0.01, max_depth = 4, n_estimators = 100, subsample = 0.5`), and AdaBoost (`learning_rate = 0.1, loss = 'exponential', n_estimators = 50`) as described in [7]. AdaBoost yielded the best results and was utilized to make the final predictions. After that, VF extraction was performed on these features to combine them with VIs and LiDAR in the prediction.

Dataset. The dataset was composed of features from VIs, such as maximum NDVI, average NDVI, four principal components of statistical measures from multispectral indices (Multispec_PCs), three principal components from LiDAR data (LiDAR_PCs), and the fusion of multispectral and LiDAR principal components (Multispec_PCs + LiDAR_PCs) collected on six surveys, as detailed in Table 1. From VIs, at each survey, 70 features based on statistical measures, such as maximum, minimum, average, standard deviation, and percentiles, were calculated from the ten VIs, including the NDVI. In contrast, LiDAR data was based

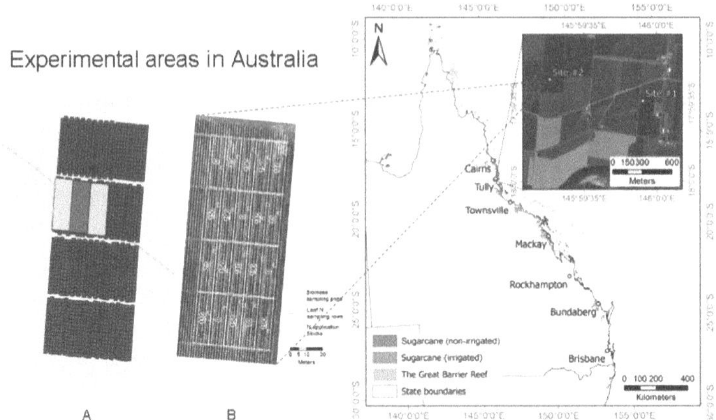

Fig. 1. Experimental fields are shown as Site #1 and Site #2 in the right-high corner of the northeast Australia map, item A shows the orthomosaic of 4 queues containing plots sampled by yellow and grey rectangles (plot borders), and item B has the zoomed Site #2 showing details of plots, including the 2 m × 2 m collection squares, already mentioned in Sect. 1 in yellow. (Color figure online)

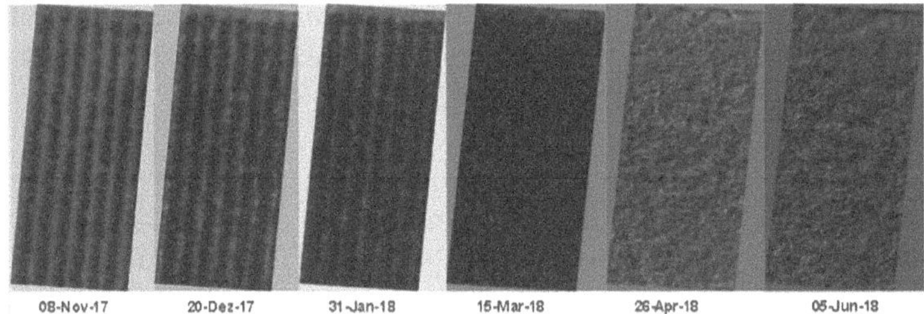

Fig. 2. Sample of aerial imagery of one plot in an experimental field showing the development of the crop throughout the six available surveys.

on 48 other statistical measures [1]. Beyond these features, the other VFs produced by the extraction using ResNet-101 from the full plot-level image and 80 × 80 pixel patches, as well as DINOv2 from the full plot-level image, were also conducted by configuring the VFs of the dataset. As the dataset comprises 20 plots in each of the two fields for six surveys, it contains 240 observations.

2.2 Data Workflow

Figure 3 shows the workflow to manipulate images and produce the VFs employed as part of the features.

Before extracting features, the image files and shapefiles that delimit the borders of plots were made available by the author of [1]: a single shapefile for each experimental field and images for each experimental field on six dates. The files passed through a process that comprehends (1) the separation of each

Table 1. Description of each feature used in the prediction.

Features names	Feature meaning
feat_1, ..., feat_2048	VFs extracted using ResNet-101 from plot size image
feat_1, ..., feat_143360	VFs extracted using ResNet-101 from 80×80 pixel patches image
feat_1, ..., feat_768	VFs extracted using DINOv2 from plot size image
max_NDVI	Maximum NDVI
avg_NDVI	Average NDVI
Multispec_PCs	Principal components of multispectral features
LiDAR_PCs	Principal components of LiDAR features

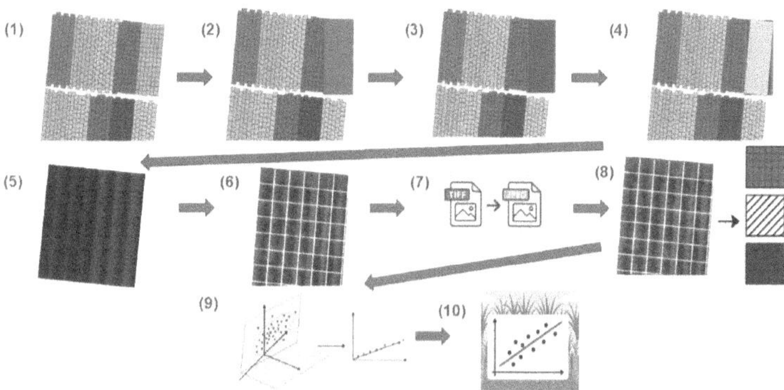

Fig. 3. Workflow to evaluate outcomes obtained with different dimensionality reductions and feature extractions based on steps: (1) dividing field shapefile in shapefiles for each plot, (2) definition of a rectangle to define the pixels of each plot, (3) adjustment of the rectangle width, (4) and adjustment of the rectangle angle, (5) removing NIR and RedEdge bands, (6) cutting field in separated plot images and 80×80 pixels, (7) transforming TIFF to JPEG, (8) extracting features from JPEG images using ResNet-101 and DINOv2, (9) applying dimensionality reduction using PCA and UMAP, (10) making Adaboost regression using VFs.

field shapefile in plot shapefiles, (2) followed by the definition of a rectangle based on the centroid position of each shapefile per plot used to divide the images, (3) their width adjustment, (4) and their angle adjustment, as described in first four steps of Fig. 3. The original experimental field image consists of an orthomosaic created from a series of images captured by a UAV.

The data had to be treated for the feature extraction, feeding the workflow with full plot-level images. Hence, it was divided into patches of 80×80 pixels, and consequently, each was used as input to the ResNet-101 pre-trained model in the same way that the full plot images were. The exact process was performed using DINOv2, which utilized only full-plot-level images.

2.3 Analysis of Pre-trained Models

In [2], the impact of image pre-processing techniques on the performance of SSL models applied to remote sensing imagery is investigated. The authors highlight that standard SSL pre-training typically utilizes larger patch sizes (e.g., 224 × 224 pixels), whereas evaluations on remote sensing datasets often involve smaller patch sizes (e.g., 32 × 32 pixels). This discrepancy and variations in image normalization methods can lead to suboptimal model performance.

The study demonstrates significant improvements in downstream task performance by systematically applying the exact image sizing and normalization procedures used during pre-training across seven diverse datasets of satellite and aerial imagery. Notably, adhering to these pre-processing steps resulted in a 32.28% increase in overall accuracy on the So2Sat random split dataset and an 11.16% improvement on the EuroSAT dataset. These findings underscore the importance of consistent pre-processing and reaffirm the competitiveness of ResNet pre-training as a baseline for transfer learning tasks in satellite imagery.

Furthermore, the paper [2] provides comprehensive benchmark results using various simple baseline methods across seven datasets, establishing an initial benchmark suite for remote sensing imagery. The authors emphasize that meticulous attention to pre-processing details, particularly image resizing and normalization, is crucial when evaluating SSL models on remote sensing data.

This work compared different scenarios to evaluate the impact of image descriptors and dimensionality reduction on crop yield prediction performance. PCA and UMAP were employed for dimensionality reduction in these tests, and ResNet-101 and DINOv2 were used as pre-trained CNN and ViT to extract VFs from the images. After that, the prediction based solely on VFs was verified, followed by its combination with VIs and LiDAR data features.

3 Visual Feature Extraction

Visual feature extraction identifies and represents key information from an image for classification, recognition, and object detection [10]. Instead of analyzing raw pixel values, which are high-dimensional and computationally expensive, VFs provide a compact and meaningful representation of an image. They can be categorized into three main types: low-level, mid-level, and high-level features [11].

This work utilizes CNN and ViT, resulting from the use of ResNet-101 and DINOv2 employed. The first was chosen because it is a classical model, but extracting the features took a considerable amount of time. ViT was also employed in this extraction because it is a state-of-the-art attention mechanism that requires less time.

3.1 CNN and ViT Features

In [2], it is demonstrated that despite being a traditional network, ResNet achieves excellent results. This is evidenced by the overall accuracy (multiclass) or mean average precision (multilabel). Otherwise, ViTs [12] are a class of deep learning models designed for image processing and computer vision tasks, and unlike traditional CNNs, which rely on spatially constrained convolutional filters, ViTs leverage self-attention mechanisms from the Transformer architecture to process images as sequences of patches.

ViTs have achieved impressive results on various computer vision tasks over the past several years, and recently, DINOv2 [6] has been one of the most promising models, as it addresses a self-supervised issue. DINOv2 is a self-supervised ViT model that learns high-quality visual representations without requiring labeled data.

4 Dimensionality Reduction

Dimensionality reduction is a critical technique in data science, serving as a fundamental tool for data visualization and as a preprocessing step for machine learning applications [13]. The increasing complexity and volume of modern datasets necessitate dimensionality reduction methods that are both computationally scalable and adaptable to diverse data types. These techniques can generally be categorized into two main approaches: those that aim to preserve the global pairwise distance structure among all data samples and those that prioritize retaining local distance relationships over global structures.

Despite recent advances in modern dimensionality reduction techniques—such as autoencoders (including variants like Denoising and Variational Autoencoders) and topological approaches like PHATE [14] and Diffusion Maps [15]—these methods have not proven as suitable for the scenario addressed in this study. Methods like ISOMAP, TriMAP, and PaCMAP [16–18], while offering various balances between local and global structure preservation, also exhibit instability under data variation and have less consistent performance evidence when applied to datasets with a limited number of samples and features already extracted from deep neural networks. Similarly, autoencoders and techniques like BYOL [19] require large amounts of data and substantial computational resources to learn effective latent representations—conditions that contrast with the nature of the dataset used in this study, which consists of features already extracted using CNNs and DINOv2. Therefore, PCA and UMAP were chosen, as they have demonstrated robust performance and reduced processing time in widely adopted benchmarks [20], including cases where PCA outperforms UMAP [21]. In [20], a set of classical datasets like Iris, Breast cancer, Wine, Digits, and others had their dimension of features reduced by different techniques. PCA and UMAP were the best for a larger number of datasets and with a lower time to process the reduction, which justifies the choice of utilizing both these methods in this work. And, in some contexts, PCA can outperform UMAP [21].

PCA and UMAP were used to reduce the dataset's dimensionality by selecting the best VFs to predict crop yield. PCA was used in two ways: for the full plot-level image and for the divided plot image, which was segmented into patches of 80 × 80 pixels. UMAP was used only with full plot-level images.

Extracting 80 × 80-pixel patches resulted in 143,360 features per plot. This high dimensionality arises from the number of image patches generated, which was standardized by resizing each plot to a uniform size based on the average dimensions of all plots across both experimental fields. Specifically, 70 image patches were created for each plot, and from each patch, 2,048 features were extracted and subsequently concatenated, yielding a final feature vector of 70 × 2,048 = 143,360 features per plot.

To define the number of components to be used by PCA, the Eigenvalues [22] were calculated to discover how many features would be used to make the predictions, resulting in a reduction from 2,048 to 142 features for full plot-level size and from 143,360 to 239 features for patches of 80 × 80 pixels. When UMAP was used, a grid search was performed to determine the optimal hyperparameters, which were `n_neighbors=15, min_dist=0.1`, resulting in a reduction from 768 to 60 features.

Additionally, two alternative strategies were evaluated for selecting VFs in combination with VIs and/or LiDAR data: (1) selecting the same number of VFs as the number of features derived from VIs and/or LiDAR, and (2) using empirically defined quantities of VFs (e.g., 40, 90, 140, 190, 240). However, the most accurate results were achieved when the number of features was determined through the Eigenvalue-based criterion for PCA and grid search optimization for UMAP, as previously discussed.

4.1 Principal Components Analysis

Principal Component Analysis is a widely used dimensionality reduction technique that transforms high-dimensional data into a lower-dimensional space while preserving as much variance as possible [23]. It achieves this by identifying new axes, called principal components, which are orthogonal to each other and capture the directions of maximum variance in the dataset. The first principal component accounts for the highest variance, while each subsequent component captures the next highest variance under the constraint of being orthogonal to the previous ones.

4.2 Unified Manifold Approximation and Projection

Unified Manifold Approximation and Projection is a state-of-the-art nonlinear dimensionality reduction technique designed to preserve the underlying manifold structure of high-dimensional data while maintaining computational efficiency. Based on principles from Riemannian geometry and algebraic topology [24], UMAP models data as a weighted graph where points are connected based on local neighborhood similarities.

5 Discussion and Limitations

One aspect of this comparative study is how the dimensionality reductions are performed. When using PCA and UMAP in a small dataset, as is the case with the dataset used in this work, predictions tend to yield better results with PCA. This occurs because UMAP tends to overfit when applied to small datasets. In contrast, PCA performs better with data that has thousands of variables and a limited number of examples, resulting in improved variable selection.

Another aspect is how VFs can improve the results. When VFs are extracted, they are produced by the textured, color, and canopy architecture of sugarcane, delivering information contained in the images captured from the field. These elements are responsible for the capacity to estimate crop yield through the VFs.

Additionally, various types of source features were utilized to make the predictions. The comparison to [2] shows this work has better results when VFs are combined with NDVI variations than when VFs are used alone; what does not happen for the max R^2 when the combination of VFs is with LiDAR_PCs, and for both min R^2 and max R^2 for Multispec_PCs and Multispec_PCs + LiDAR_PCs. In situations where only NDVI is available, it is a good option to combine it with VFs. The worst outcomes of combining VFs with Multispec_PCs or LiDAR_PCs also indicate that this combination of techniques involving dimensionality reduction and VFs should be explored in depth in conjunction with larger datasets. The results are described in Table 2 and Table 3.

Despite the limited number of observations available for analysis and the vast quantity of features, resulting in a somewhat unexpressive outcome, significant improvements were achieved in most scenarios when image plot sizes were used in conjunction with PCA dimensionality reduction.

To mitigate the lack of observations that hinders UMAP from yielding more effective results, methods such as data augmentation or the use of a larger dataset should be employed. This way, the quantity of features produced by the extraction of VFs would be supported by a substantial amount of data, providing sufficient variability to enable the models to converge.

Even with the insufficient quantity of data to explore these techniques, finding the best arrangement to decrease the amount of feature predictors, through the use of dimensionality reduction and the combination of VIs and LiDAR with VFs, the feature extractors combined with PCA can bring more robust outcomes, and possibly even better results to UMAP because it fits better when applied to a larger dataset. It will be done as future work.

6 Results

Considering the results achieved using only VFs extracted from ResNet-101 and reduced by PCA and UMAP, the min R^2 and max R^2 decreased drastically when UMAP was employed. This is why the cells with corresponding values to UMAP in the min R^2 and max R^2 columns are underlined (Table 2). Because of this lack of performance, UMAP was not tested by combining VFs with VIs and/or LiDAR data.

Table 2. Performance results in yield prediction from VFs extracted by **ResNet-101** under dimensionality reduction using PCA and UMAP, comparing results obtained in this work with those in [7]. The other blocks compare each feature model from [7] with its correspondence in conjunction with VFs. The red values spot the worst values obtained using UMAP, and the arrows by the side of R^2 values from the combination of VFs + VIs + LiDAR data features indicate if they are better or worse than values obtained without VFs features.

Patch size	Features	Dimensionality reduction	min R^2	max R^2
Plot size	2,048	PCA	0.40	0.65
80 × 80	143,360	PCA	0.45	0.65
Plot size	2,048	UMAP	0.07	0.28
80 × 80	143,360	UMAP	0.02	0.14
Plot size	max_NDVI [7]	PCA	0.38	0.60
Plot size	2,048 + max_NDVI	PCA	**0.55** ↑	**0.67** ↑
80 × 80	143,360 + max_NDVI	PCA	0.42 ↑	0.67 ↑
Plot size	avg_NDVI [7]	PCA	0.40	0.60
Plot size	2,048 + avg_NDVI	PCA	**0.45** ↑	**0.68** ↑
80 × 80	143,360 + avg_NDVI	PCA	0.44 ↑	0.68 ↑
Plot size	Multispec_PCs [7]	PCA	**0.62**	**0.81**
Plot size	2,048 + Multispec_PCs	PCA	0.57 ↓	0.72 ↓
80 × 80	143,360 + Multispec_PCs	PCA	0.57 ↓	0.71 ↓
Plot size	LiDAR_PCs [7]	PCA	0.50	**0.80**
Plot size	2,048 + LiDAR_PCs	PCA	**0.52** ↑	0.71 ↓
80 × 80	143,360 + LiDAR_PCs	PCA	0.46 ↓	0.71 ↓
Plot size	Multispec_PCs + LiDAR_PCs [7]	PCA	**0.62**	**0.88**
Plot size	2,048 + Multispec_PCs + LiDAR_PCs	PCA	0.61 ↓	0.76 ↓
80 × 80	143,360 + Multispec_PCs + LiDAR_PCs	PCA	0.61 ↓	0.76 ↓

After evaluating only VFs as features, experiments were conducted to check if using VFs combined with VIs and/or LiDAR data would improve R^2. In this sense, a comparison was made, and when compared to the use of only VFs, the min R^2 and max R^2 increased compared to NDVI variations; however, this was not confirmed when combined with other VIs and/or LiDAR with VFs.

In Table 3, when the features combined VFs with VIs and/or LiDAR data, it showed inferior performance in almost all the combinations, not only for 768 + avg_NDVI, compared to the results displayed in [7].

Table 3. Performance results using DINOv2 feature extractor with different dimensionality reduction and feature combinations.

Features	Dimensionality reduction	min R^2	max R^2
768	PCA	0.29	0.54
768	UMAP	0.16	0.12
max_NDVI	PCA	**0.38**	**0.60**
768 + max_NDVI	PCA	0.29 ↓	0.54 ↓
avg_NDVI	PCA	0.40	0.60
768 + avg_NDVI	PCA	**0.47** ↑	**0.71** ↑
Multispec_PCs	PCA	**0.62**	**0.81**
768 + Multispec_PCs	PCA	0.55 ↓	0.71 ↓
LiDAR_PCs	PCA	**0.50**	**0.80**
768 + LiDAR_PCs	PCA	0.49 ↓	0.58 ↓
Multispec_PCs + LiDAR_PCs	PCA	**0.62**	**0.88**
768 + Multispec_PCs + LiDAR_PCs	PCA	0.62	0.71 ↓

7 Conclusions

This work provides a comparative study evaluating dimensionality reduction in conjunction with image descriptors obtained by VFs. It shows that when the VFs are used with VIs and LiDAR data features, they can improve the regression results of the latter. The minor values are likely related to the limited number of observations in the dataset. When images and features extracted from those images are employed in predictions, as the complexity of colors, textures, and forms they deal with is huge, the number of observations must be significantly larger. However, the occurrence of better outcomes could not be verified using this dataset, possibly due to the small sample size of observations. Accessing datasets with sufficient observations to make better predictions is challenging in sugarcane cultivation due to the absence of an organization responsible for collecting and sharing this data, often without proper agreements with farmers.

In future work, accessing datasets collected in a larger number of surveys from producers is an essential improvement to model predictions, leading to better outcomes.

Acknowledgements. We thank the authors of [1] for making the sugarcane crop dataset available, as this type of data is difficult to find when one wants to study its productivity due to producers not disclosing it. This study was financed in part by the Coordenação de Aperfeiçoamento de Pessoal de Nível Superior - Brasil (CAPES). The authors are grateful to São Paulo Research Foundation - FAPESP (grant #2024/04890-5) and Brazilian National Council for Scientific and Technological Development - CNPq (grants #313193/2023-1), and the Coordination for the Improvement of Higher Education Personnel - CAPES (grant #88887.899737/2023-00).

Disclosure of Interests. Brazilian Agricultural Research Corporation-Embrapa employed author Luiz Antonio Falaguasta Barbosa at the time of the study; however, the company was not involved in the design, execution, or funding of the research, nor did it have any influence on the manuscript. The research was conducted independently as part of the author's doctoral studies at São Paulo State University. The remaining authors declare that the research was conducted in the absence of any commercial or financial relationships that could be construed as a potential conflict of interest.

References

1. Shendryk, Y., Sofonia, J., Garrard, R., Rist, Y., Skocaj, D., Thorburn, P.: Fine-scale prediction of biomass and leaf nitrogen content in sugarcane using uav lidar and multispectral imaging. Int. J. Appl. Earth Observat. Geoinf. **92**, 102177 (2020). https://doi.org/10.1016/j.jag.2020.102177
2. Corley, I., Robinson, C., Dodhia, R., Ferres, J.M.L., Najafirad, P.: Revisiting pre-trained remote sensing model benchmarks: resizing and normalization matters. In: Proceedings of the IEEE/CVF Conference on Computer Vision and Pattern Recognition, pp. 3162–3172 (2024)
3. Wang, Y., Braham, N.A.A., Xiong, Z., Liu, C., Albrecht, C.M., Zhu, X.X.: SSL4EO-S12: a large-scale multimodal, multitemporal dataset for self-supervised learning in earth observation [software and data sets]. IEEE Geosci. Remote Sens. Mag. **11**(3), 98–106 (2023). https://doi.org/10.1109/MGRS.2023.3281651
4. Bu, Y., et al.: Resnet incorporating the fusion data of rgb & hyperspectral images improves classification accuracy of vegetable soybean freshness. Sci. Rep. **14**(1), 2568 (2024). https://doi.org/10.1038/s41598-024-51668-6
5. He, K., Zhang, X., Ren, S., Sun, J.: Deep residual learning for image recognition. In: Proceedings of the IEEE Conference on Computer Vision and Pattern Recognition (CVPR) (2016)
6. Oquab, M., et al.: Dinov2: learning robust visual features without supervision (2024)
7. Barbosa, L.A.F., Pedronette, D.C.G., Guilherme, I.R., Tisseyre, B.: Meta-features extracted from use of kNN regressor to improve sugarcane crop yield prediction. Remote Sens. **17**(11), 1846 (2025). https://doi.org/10.3390/rs17111846
8. Barbosa, L.A.F.; Pedronette, D.C.G.; Guilherme, I.R. A meta-feature model for exploiting different regressors to estimate sugarcane crop yield. In: Proceedings of the IGARSS 2023 - 2023 IEEE International Geoscience and Remote Sensing Symposium, pp. 2030–2033 (2023). https://doi.org/10.1109/IGARSS52108.2023.10283309
9. Ippolito, P.P.: Hyperparameter tuning: the art of fine-tuning machine and deep learning models to improve metric results. In: Applied Data Science in Tourism: Interdisciplinary Approaches, Methodologies, and Applications, pp. 231–251. Springer, Heidelberg (2022). https://doi.org/10.1007/978-3-030-88389-8_12
10. Abdullahi, M., et al.: A systematic literature review of visual feature learning: deep learning techniques, applications, challenges and future directions. Multimedia Tools Appl. 1–58 (2024). https://doi.org/10.1007/s11042-024-19823-3

11. Shui, J., Ding, S., Li, M., Ma, Y.: Entity semantic feature fusion network for remote sensing image-text retrieval. In: Asia-Pacific Web (APWeb) and Web-Age Information Management (WAIM) Joint International Conference on Web and Big Data, pp. 130–145. Springer, Heidelberg (2024). https://doi.org/10.1007/978-981-97-7244-5_9
12. Dosovitskiy, A., et al.: An image is worth 16×16 words: transformers for image recognition at scale (2021)
13. Jia, W., Sun, M., Lian, J., Hou, S.: Feature dimensionality reduction: a review. Complex Intell. Syst. **8**(3), 2663–2693 (2022). https://doi.org/10.1007/s40747-021-00637-x
14. Moon, K.R., et al.: Visualizing structure and transitions in high-dimensional biological data. Nat. Biotechnol. **37**(12), 1482–1492 (2019). https://doi.org/10.1038/s41587-019-0336-3
15. Coifman, R.R., Lafon, S.: Diffusion maps: applied and computational harmonic analysis. Appl. Comput. Harmon. Anal. **21**(1), 5–14 (2004). https://doi.org/10.1016/j.acha.2006.04.006
16. Tenenbaum, J.B., de Silva, V., Langford, J.C.: A global geometric framework for nonlinear dimensionality reduction. Science **290**(5500), 2319–2323 (2000). arXiv:1910.00204. https://doi.org/10.1126/science.290.5500.2319
17. Amid, E., Warmuth, M.K.: TriMAP: large-scale dimensionality reduction using triplets (2019). https://arxiv.org/abs/1910.00204
18. Wang, Y., et al.: PaCMAP: dimension reduction via pairwise controlled manifold approximation. MLJ (2021)
19. Grill, J.-B., et al.: Bootstrap your own latent-a new approach to self-supervised learning". Adv. Neural. Inf. Process. Syst. **33**, 21271–21284 (2020)
20. Chhikara, P., Jain, N., Tekchandani, R., Kumar, N.: Data dimensionality reduction techniques for industry 4.0: research results, challenges, and future research directions. Softw. Pract. Exp. **52**(3), 658–688 (2022). https://doi.org/10.1002/spe.2876
21. Villares, M., Saunders, C.M., Fey, N.: Comparison of dimensionality reduction techniques for the visualisation of chemical space in organometallic catalysis. Artif. Intell. Chem. **2**(1), 100055 (2024). https://doi.org/10.1016/j.aichem.2024.100055
22. Johnstone, I.M.: On the distribution of the largest eigenvalue in principal components analysis. Ann. Stat. **29**(2), 295–327 (2001). https://doi.org/10.1214/aos/1009210544
23. Gewers, F.L., et al.: Principal component analysis: a natural approach to data exploration. ACM Comput. Surv. **54**(4) (2021). https://doi.org/10.1145/3447755
24. Fei, Y., Liu, Y., Jia, C., Li, Z., Wei, X., Chen, M.: A survey of geometric optimization for deep learning: from Euclidean space to Riemannian manifold. ACM Comput. Surv. **57**(5) (2025). https://doi.org/10.1145/3708498

ForestSAM: A Novel Integration of DeepForest and SAM2 for Oil Palm Crown Segmentation in Aerial Imagery

Juan Suarez[1(✉)], David Melo[2], Iván Ramírez[1,4], and Ariolfo Camacho[3]

[1] Vision Computer, Conexalab SAS BIC, Bucaramanga 680001, Santander, Colombia
juansuarez@ico2.co, iramirez@conexalab.com
[2] Coordination of Computer Sciences, INAOE, 72840 Tonantzintla, Puebla, Mexico
jdom1824@inaoep.mx
[3] Remote Sensing, Agrosavia, Bogotá 110110, Cundinamarca, Colombia
acamachov@agrosavia.co
[4] Coordination of Industrial Engineer, Universidad Politécnica de Valencia, 46022 Valencia, Spain

Abstract. The oil palm industry in Colombia faces significant challenges in large-scale plantation management, particularly in the early detection of diseases that cause morphological changes in the crown of palm trees, severely impacting crop health and productivity. Early identification of these symptoms is typically performed manually through structural examination of the tree crowns, a process that is time-consuming, labor-intensive, and prone to human error. Although recent studies have made progress using deep learning techniques for tree counting, detection, and crown classification, the application of these methods for detailed crown segmentation remains largely unexplored. In this study, we propose ForestSAM, the first integration of DeepForest for oil palm tree detection with Segment Anything Model 2 (SAM2) for precise crown segmentation using drone and aerial imagery. While this study does not directly validate disease detection, the proposed method achieves a Dice coefficient (DSC) of 0.826 and a Jaccard index (IoU) of 0.712, showing its effectiveness in accurately segmenting oil palm crowns. These results establish a foundation upon which future methods for early disease detection such as bud rot monitoring, carbon capture metrics, and optimized resource allocation.

Keywords: Oil Palm · Tree Crown Segmentation · Drone Imagery · Aircraft Imagery · DeepForest · SAM2

1 Introduction

Oil palm farming is a key economic activity in Colombia. Large-scale plantations still depend on manual or semi-automated monitoring practices to maintain

D. Melo, I. Ramírez and A. Camacho—These authors contributed equally to this work.

D. Chaves et al. (Eds.): CIARP 2025, LNCS 16528, pp. 30–43, 2026.
https://doi.org/10.1007/978-3-032-23161-1_3

productivity, highlighting the growing need for precise, efficient, and scalable monitoring solutions. One of the most critical challenges is the late detection of diseases such as bud rot, caused by the pathogen *Phytophthora palmivora*, which can devastate entire plantations [1]. In Santander, this disease has affected approximately 23,000 hectares, threatening both the local economy and the sustainability of the sector [2,3].

Early detection of bud rot and other diseases can be achieved by analyzing morphological changes in the leaves and crown structure, where the first symptoms become visible. From an aerial perspective, such symptoms often manifest as chlorosis on the adaxial surface of the leaflets and the collapse of one or several spear leaves. Although the present study does not provide direct evidence of detecting these anomalies, accurate crown delineation is essential, as it constitutes a necessary prerequisite for any automated system aiming to identify early indicators of infection. The use of drones and high-resolution image analysis, with tools such as i-CO_2 developed by CONEXALAB SAS BIC, has become a strategic supporter in precision agriculture to address these challenges in Colombia [4].

Recently, several techniques have been explored for the automatic counting and detection of oil palms, using methods such as YOLOv8, Faster R-CNN [5,6], and semi-supervised methods [7], as well as deep learning models for crown delineation such as DeepForest [8]. However, the specific integration of detection and segmentation for detailed crown delineation in oil palm plantations remains an underexplored area, particularly in the context of high-resolution aerial imagery.

To address this gap, we propose ForestSAM, a novel method that combines DeepForest for tree detection with SAM2 [6] for crown segmentation. This method is designed to improve the accuracy of palm crown delineation, which is critical for early disease detection and plantation monitoring.

Our main contributions are:

1. A new method that integrates DeepForest and SAM2 to perform detection and segmentation of oil palm crowns from aerial RGB imagery.
2. A manually annotated dataset with crown-level segmentation masks for oil palms in both homogeneous and heterogeneous plantations.
3. A quantitative evaluation showing that ForestSAM outperforms each individual model (DeepForest and SAM2) in both tree counting and crown segmentation accuracy.

The proposed method achieves a DSC of 0.826 and an IoU of 0.712, demonstrating its efficacy in accurately segmenting oil palm crowns. This approach is designed to facilitate the timely detection of critical health issues affecting oil palm trees, paving the way for future applications, including biomass estimation, carbon capture monitoring, and optimized resource management, ultimately contributing to the sustainability of the oil palm industry.

2 Background

This section describes the fundamentals needed to understand this work. We introduce DeepForest, a model designed for tree detection, followed by SAM2, a general purpose segmentation model.

2.1 DeepForest

DeepForest is a deep learning model built on the Faster R-CNN architecture, designed for the detection of tree crowns in high-resolution RGB images. It is commonly used in ecological and forestry applications to automatically identify and count individual trees. The model is pre-trained on millions of tree crown annotations generated from LiDAR data and can be fine tuned with user provided data to adapt to different environments and tree species [8].

2.2 Segment Anything Model 2 (SAM2)

The Segment Anything Model 2 (SAM2) is a general-purpose image segmentation model capable of generating object masks from user-defined prompts, such as points, bounding boxes, or existing masks. It is based on a transformer architecture and has been trained on a large and diverse dataset comprising millions of segmentation samples. SAM2 has been successfully applied across a wide range of domains including medical imaging, autonomous driving, and remote sensing—for precise object delineation [6].

3 Related Work

This section reviews related work, summarized in Table 1, and classified into three categories: (1) automatic counting methods using models such as YOLO and Faster R-CNN, focused on speed and efficiency; (2) detection approaches using semi-supervised learning for better generalization; and (3) crown delineation methods based on deep learning, aimed at precise segmentation. Recent hybrid models also integrate segmentation and classification for multi-species detection.

Korom et al. (2012) [9] propose a method for the individual delineation of oil palms. The approach combines pixel based masking and object-based segmentation, utilizing the watershed technique for segmentation. Vegetation indices are employed to enhance the segmentation. The algorithm achieves a good segmentation rate, reaching up to 100% in the evaluated trees. However, the study acknowledges limitations related to the imperfection of pixel masking, which can lead to under or oversegmentation.

Table 1. Summary of related studies on oil palm tree detection, delineation, and classification

Authors	Approach	Method	Data	Resolution
Korom et al. (2012 [9])	Delineation	Pixel-based masking + object-based segmentation	WorldView02	~0.5 m
Weinstein et al. (2019 [7])	Detection and Counting	Semi-supervised deep learning	Aerial RGB images	Not specified
Weinstein et al. (2020 [8])	Delineation	Deep learning (Faster R-CNN)	Aerial RGB images	Not specified
Zhorif et al. (2024 [10])	Counting	YOLOv8 + SAHI method	Aerial images (Indonesia)	~0.3 m
Teo et al. (2025 [11])	Detection and Classification	SAM + EfficientNetB3	Drone images (Malaysia)	Not specified
This work (2025)	Counting and Delineation	DeepForest + SAM2	Aerial RGB images	Not specified

Weinstein et al. (2019) [7], present a semi-supervised deep learning approach for individual tree crown detection in RGB images, addressing the challenge of limited labeled training data. The method uses unsupervised LIDAR-based delineation to generate initial, albeit noisy, training data. This data is used to train an initial RGB crown detection model. The model is then refined using several hand-annotated RGB images to improve accuracy. Results show that this method can produce accurate predictions in natural landscapes, achieving a tree crown recall of 0.69 and a precision of 0.61.

Weinstein et al. (2020) [8] introduce DeepForest: a Python package that leverages deep learning for tree crown delineation in high-resolution RGB imagery. Based on Faster R-CNN, the model is pretrained on over 30 million LiDAR-generated crowns and fine tuned with 10,000 manually labeled crowns. DeepForest enables users to apply the pretrained model, fine-tune it with new data, or train custom models, demonstrating effectiveness across diverse forest types and urban environments. Its performance varies with data resolution and forest structure, highlighting the need for site specific training.

The article by Zhorif et al. (2024) [10] presents a methodology for counting oil palm trees using high-resolution aerial images and the YOLOv8 object detection model. To improve accuracy, the Slicing Aided Hyper Inference (SAHI) method was implemented, achieving a Mean Absolute Percentage Error (MAPE) of 0.01758 over an area of 73.2 hectares, with a detection time of five minutes and 45 s. This approach overcomes the limitations of manual counting, which is labor-intensive and costly. The results demonstrate the effectiveness of combining YOLOv8 and SAHI for automated mapping and counting in oil palm plantations, contributing to a more efficient and accurate resource management.

Rejeki et al. (2024) [5] compared the effectiveness of Faster R-CNN models with ResNet50 and VGG16 backbones for detecting and counting oil palm trees using drone-captured images. The results showed that VGG16 achieves higher accuracy (97–99%) compared to ResNet50 (not exceeding 95%) due to its simpler architecture that better adapts to the dataset used. The study notes that

ResNet50 requires precise parameter tuning or a dataset to optimize its performance. In addition, both models can be prone to overfitting, which affects their ability to generalize to unknown data.

Teo et al. (2025) [11] proposed an end-to-end model for the detection and classification of oil palm, coconut, and banana trees from drone imagery. The method integrates the Segment Anything Model (SAM) for instance segmentation and EfficientNetB3 for species classification. This approach addresses the challenge of distinguishing morphologically similar tree species in complex agricultural environments. Their system achieved F1-scores of 97%, 89%, and 91% for detection, and 92%, 88%, and 91% for classification of oil palm, coconut, and banana trees, respectively.

Despite advances in palm species detection, counting and classification, to our knowledge, no studies have combined palm detection with segmentation models for oil palm delineation. In particular, the integration of DeepForest with models such as SAM2 remains an underexplored area, but has significant potential for agro-industrial application. This paper addresses this research gap by proposing a hybrid approach for accurate delineation in drone imagery.

4 Materials, Experimental Setup and Method

This section outlines the experimental setup used to develop and assess the ForestSAM approach. It includes a description of the dataset and image acquisition method, the hardware setup for implementation, and the evaluation metrics used to quantify segmentation performance. These details are crucial for ensuring reproducibility and providing context for interpreting the obtained results.

4.1 Data Source

For this study, a set of high-resolution images was selected from two different locations: Colombia and Saudi Arabia, as shown in Fig. 1. These locations represent distinct plantation environments, allowing the evaluation of the method under varying field conditions.

Figure 2 presents a row of three images used in this study, captured using different sensors and under varying geographical and spectral conditions. These images illustrate the visual diversity of oil palm plantations across different acquisition settings.

Subfigure 2a is sourced from a public dataset collected in Kharj, Saudi Arabia. It was acquired using a DJI Phantom 4 Pro drone equipped with a DJI FC6310 camera, providing a resolution of 4864 $\times$ 3648 pixels [12,13].

Subfigures 2b and 2c are cropped patches extracted from a high-resolution hyperspectral image originally sized at 26396 $\times$ 31897 pixels. The image was captured in the VNIR spectral range using a HySpex VNIR-1600 camera (Norsk Elektro Optikk AS, Norway) [14]. Each cropped image has a resolution of 662 $\times$ 812 pixels. These images were acquired in 2017 over oil palm plantations in the department of Caquetá, Colombia. The study area covers approximately 130 hectares, and the data collection was performed from a flight altitude of 430 m.

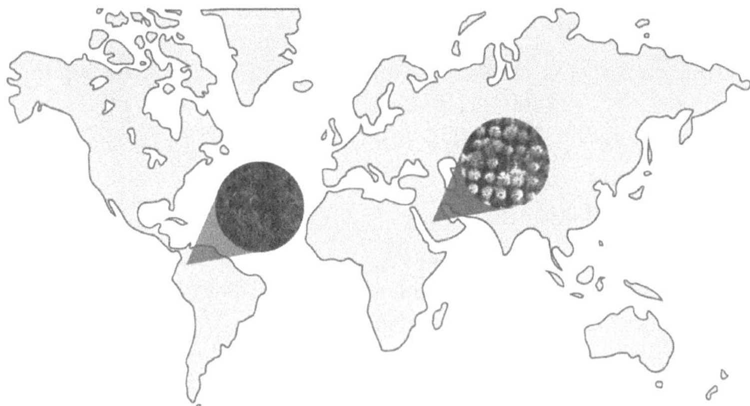

Fig. 1. Geographical location of the input images: Colombia and Kharj, Saudi Arabia.

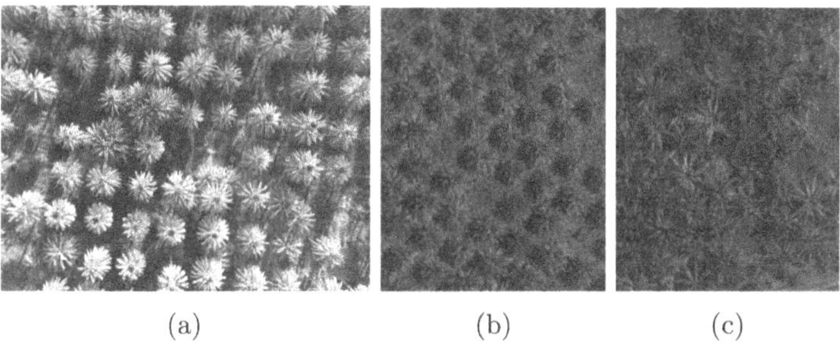

(a) (b) (c)

Fig. 2. Example row of three image columns used in this study: (a) aerial image from Saudi Arabia captured by a DJI Phantom 4 Pro drone; (b) and (c) aerial image crops from oil palm plantations in Caquetá, Colombia.

4.2 Ground-Truth Masks

Since no publicly available segmented dataset exists, the Roboflow platform is used to manually annotate palm crowns [15]. These annotations serve as the ground truth for the following comparisons.

To evaluate the segmentation accuracy of SAM2 and ForestSAM, ground-truth masks are created through a detailed, tree-by-tree segmentation on each selected RGB image. After delineating all crowns, the individual masks are merged to obtain a single reference mask for each image. This procedure is indispensable because, to the best of our knowledge, no public databases provide palm-crown segmentations with the resolution and accuracy required for a rigorous comparison.

To ensure the reproducibility of the experiment, the annotated masks have been released in the repository https://doi.org/10.7910/DVN/WDJZO1, where they can be downloaded together with the evaluation scripts.

4.3 Hardware Setup

The experiments are conducted in a computational environment equipped with an Intel® Core™ i5-11400 (11th Gen) processor, 64 GB of RAM, and an NVIDIA GeForce RTX 3060 GPU.

4.4 Experimental Setup

The proposed method, illustrated in Fig. 3, integrates DeepForest and SAM2 to achieve precise detection, automated counting, and detailed segmentation of oil palm crowns from aerial imagery. Initially, a manual dataset is created using images of palm trees to perform manual counting and segmentation, establishing a ground truth dataset necessary for comparative analysis and validation, as no publicly available detailed segmentation dataset for oil palm trees currently exists.

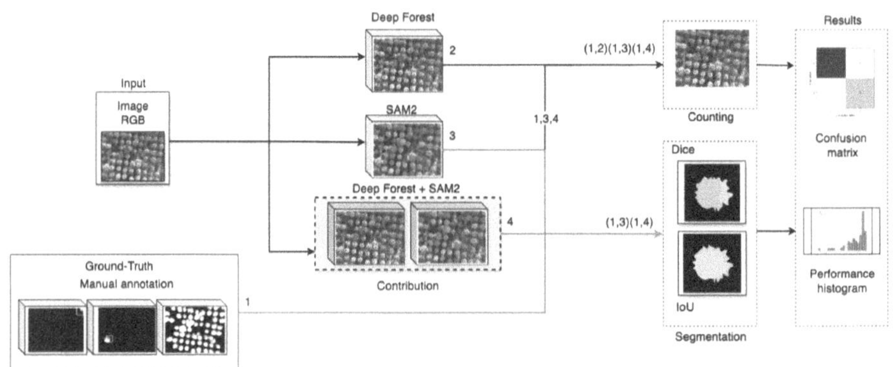

Fig. 3. Overview of the ForestSAM method: RGB input, palm detection with DeepForest, segmentation with SAM2, and evaluation using DSC and IoU.

Next, previously selected RGB images are individually processed with DeepForest and SAM2, then combined into our primary contribution, which integrates these two methods to enhance counting precision and crown segmentation accuracy.

In the first stage of experimentation, a counting comparison is conducted: first contrasting ground truth with DeepForest results (1,2). Next, we compare the ground truth with the count derived from SAM2 segmentation (1,3). Finally, we assess our integrated method, which aims to surpass previous methods by combining both approaches (1,4).

Finally, we specifically evaluate our method's segmentation capabilities against manual segmentation. To achieve this, we analyze the variation between ground truth and SAM2 results, as well as between manual segmentation and DeepForest results. This analysis can be observed in the green lines in the figure, specifically sections (1,3) and (1,4).

4.5 ForestSAM: Proposed Detection and Segmentation Approach

ForestSAM is introduced as a hybrid method that integrates the detection and segmentation of oil palm crowns in aerial images. The DeepForest neural network is used to detect and localize the palms within the image, generating bounding boxes whose values correspond to pixel coordinates in the original image. For this process, a patch size of 2500 pixels and a trust threshold of 40% are used for prediction filtering; these parameters were selected after several manual iterative tests, as they provided the best results on the evaluated dataset. DeepForest's overlapping strategy is to use sliding windows with overlap to capture edge trees, then apply Non-Max Suppression to remove duplicate detections. It is important to note that DeepForest is specifically designed for detection tasks, so its output is limited to the localization of palms through bounding boxes, without providing a precise segmentation of the crowns. This is because detection models, such as DeepForest, identify the presence and position of objects, but are not trained to delineate their exact contours.

To overcome this limitation, the SAM2 model is used, which takes as input the boxes generated by DeepForest and refines the segmentation, producing masks that more accurately represent the shape and extent of the palm crowns within each box. At this stage, SAM2 is used with its default parameters. In this way, SAM2 leverages the spatial information provided by DeepForest to focus the segmentation exclusively on the relevant regions of the image, improving accuracy and reducing the segmentation of irrelevant areas or noise.

Thus, ForestSAM constitutes a hybrid approach that not only enables the quantification of the number of palms present in the image, but also allows for detailed segmentation of their crowns, combining the strengths of both models to obtain more robust and precise results.

4.6 Performance Measure

The performance of the proposed algorithm is evaluated using two standard metrics in image segmentation: the Dice Similarity Coefficient (DSC) and the Jaccard Index (Intersection over Union, IoU), both of which were introduced in the background section.

4.6.1 Dice Similarity Coefficient (DSC)

The DSC is a widely adopted metric for assessing the similarity between two sets, particularly in the context of image segmentation tasks [16]. It measures the degree of overlap between the predicted regions and the ground truth, producing a value between 0 and 1, where 1 denotes complete agreement. The DSC is useful in scenarios with class imbalance because it emphasizes correctly predicted overlapping areas. Its simplicity and interpretability make it a standard choice in medical image analysis and other segmentation applications.

The DSC measures the similarity between the predicted segmentation (A) and the ground truth (B), and is defined as:

$$DSC = \frac{2\,A \cap B|}{|A| + |B|} \tag{1}$$

This metric ranges from 0 (no overlap) to 1 (perfect overlap), and is particularly effective in evaluating segmentation performance when the target regions are small or class imbalance is present.

4.6.2 Jaccard Index (IoU)

The Jaccard Index, also known as Intersection over Union (IoU), is another commonly used metric for evaluating the similarity between predicted and ground truth regions in segmentation tasks [17]. It assesses how much the two sets overlap relative to their combined size, with values ranging from 0 to 1—where higher values indicate better alignment. Due to its intuitive interpretation and strong theoretical foundation, the Jaccard Index is widely used in computer vision, particularly for benchmarking segmentation algorithms.

The IoU quantifies the overlap between the predicted and actual segments relative to their union, and is given by:

$$IoU = \frac{|A \cap B|}{|A \cup B|} \tag{2}$$

Compared to the DSC, the IoU is more conservative, as it penalizes false positives and false negatives more strictly.

Together, these metrics provide a complementary and robust evaluation framework for assessing the accuracy and reliability of segmentation results.

5 Results

To evaluate the performance of the methods, we compute the true positives (TP), false positives (FP), and false negatives (FN) by comparing the predicted masks from each model against the manually generated ground truth masks. In this context, the analytical process follows two stages: first, the identification and counting of individual trees; and second, the precise segmentation of tree crowns. A TP corresponds to each predicted mask that matches a ground truth mask with an IoU of 0.5 or higher. Predicted masks that do not match any ground truth mask count as FP, while ground truth masks not detected by any model count as FN.

Figure 4 provides a visual comparison of the results obtained for the three evaluated images using the different tree counting and crown segmentation methods. The first row shows the original RGB images, followed by the second row with the manually generated ground truth masks. The third row corresponds to the crown detection results using DeepForest, where the predicted bounding boxes are displayed over the original images, allowing for the assessment of each method's counting capability. The fourth row presents the segmented masks obtained with the SAM2 method, while the fifth row shows the results of the combined ForestSAM approach, which integrates crown localization and refined segmentation.

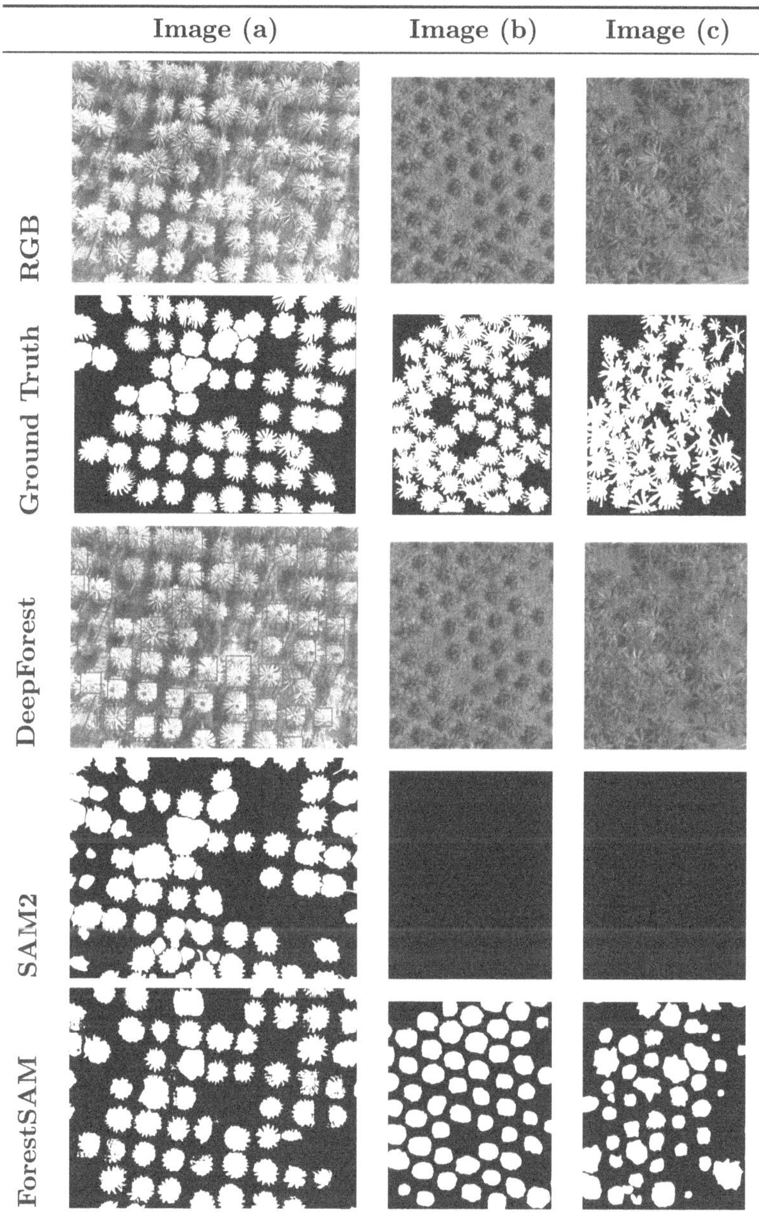

Fig. 4. Visual comparison of the segmentation and counting results for the three evaluated images.

Table 2 summarizes the accumulated quantitative results for the evaluated methods across the three test images. For each method and image, the number of TP, FP, and FN was computed, along with the corresponding precision, recall, mean DSC, and mean IoU.

Table 2. Summary of quantitative results for each evaluated method and image, including mean and standard deviation (SD) for DSC and IoU. TP: true positives, FP: false positives, FN: false negatives.

Image	Method	TP	FP	FN	Precision	Recall	Mean DSC	SD DSC	Mean IoU	SD IoU
Img (a)	SAM2	48	16	49	0.75	0.49	0.94	0.02	0.89	0.04
	ForestSAM	53	10	11	0.84	0.82	0.89	0.07	0.80	0.10
	Ground Truth	63								
Img (b)	SAM2	0	53	0	0	0	0	–	0	–
	ForestSAM	45	11	8	0.80	0.84	0.77	0.05	0.63	0.06
	Ground Truth	53								
Img (c)	SAM2	0	46	0	0	0	0	–	0	–
	ForestSAM	15	36	31	0.29	0.32	0.75	0.05	0.61	0.07
	Ground Truth	46								
Acc. SAM2		48	122	49	0.28	0.49	0.94	0.026	0.891	0.046
Acc. ForestSAM		**113**	**57**	**50**	**0.66**	**0.69**	**0.82**	**0.08**	**0.71**	**0.15**

This visualization allows for a qualitative comparison of each method's ability to not only accurately identify and count palm tree crowns, but also to precisely segment them under varying image conditions. The combined ForestSAM method demonstrates more precise and consistent segmentation across all three images. In contrast, SAM2 shows limitations in images (b) and (c), where it fails to detect any crowns. DeepForest, meanwhile, provides adequate crown localization for counting purposes but does not perform instance segmentation. ForestSAM achieved higher precision in the Saudi dataset, characterized by more uniform tree arrangements, while Colombian imagery presented more variability and challenges due to ecological heterogeneity.

ForestSAM achieved a superior balance between the number of detected crowns and segmentation quality, as evidenced by the accumulated values of precision 0.665, recall 0.693, mean DSC 0.826, and mean IoU 0.712, along with moderate standard deviation, indicating consistent results. The integration of both models leveraged the strength of DeepForest in localization and SAM2's ability to accurately delineate crown contours, minimizing the segmentation of irrelevant areas and noise.

It is important to highlight that, although the accumulated mean DSC and mean IoU for SAM2 are high, these metrics only reflect the quality of the few matches achieved by the method. In cases where SAM2 failed to detect any crowns (resulting in zero true positives), these images do not contribute to the average, potentially leading to an overestimation of segmentation quality. Therefore, while mean DSC and mean IoU provide insight into the accuracy of the detected instances, they should be interpreted alongside precision and recall, which account for the overall detection performance. The low precision and recall values for SAM2 indicate that despite the high segmentation quality in a limited number of cases, the method was generally ineffective for the data set as a whole.

The results confirm that this hybrid approach delineates oil palm crowns more effectively and automatically than previous methods that focus only on detection, counting, or classification. This advance supports early disease detection, such as bud rot, and lays the basis for applications in carbon capture monitoring and agronomic resource optimization in large-scale plantations.

6 Limitations

A key limitation of this study lies in the restricted dataset, which included only three images and a limited set of manually annotated crowns. This small sample constrains the robustness, generalizability, and statistical significance of the reported results, as performance may vary under different ecological, environmental, or sensor conditions. Additionally, no specific strategy was implemented to mitigate annotation bias during the creation of ground-truth masks. The dataset was generated by a single annotator, without procedures such as multiple annotators, validation protocols, or error analysis in complex cases. Such cases include overlapping crowns or trees partially obscured by shadows and surrounding vegetation, where segmentation can be more ambiguous.

7 Conclusions

This study shows the efficacy of integrating the DeepForest and SAM2 models for detailed segmentation of oil palm crowns using aerial imagery. The proposed method achieves strong performance, with a DSC of 0.826 and an IoU of 0.712, highlighting its ability to generate highly accurate segmentation masks. In addition, the system detects (53, 45, 15) palms compared to (63, 53, 46) manually annotated ones, indicating high sensitivity in contexts with natural variations in tree density and species. The results confirm that this hybrid approach delineates oil palm crowns more effectively and automatically than previous methods that focus only on detection, counting, or classification. However, the direct utility for detecting diseases such as bud rot depends on integrating segmentation output with health classification models and datasets that include infected samples. ForestSAM should be understood as a technical basis for future diagnostic methods that supports early detection of diseases, such as bud rot, and lays the basis for applications in carbon capture monitoring and agronomic resource optimization in large plantations.

8 Future Work

With palm crowns already segmented automatically, future research is focused on developing classification models, such as convolutional neural networks or vision transformers, to categorize each palm tree's health status (e.g., healthy, suspected, infected). This approach enables large-scale tree-level health monitoring and facilitates targeted intervention strategies.

For the early detection of bud rot disease, it is essential to extract physiological indicators that are not visible in the RGB spectrum. Future work may explore the integration of crown masks generated by DeepForest+SAM2 with multispectral or hyperspectral imagery, focusing on the identification of early-stage physiological changes associated with infection. We anticipate that the overall ForestSAM workflow will remain compatible and will require only minor adjustments such as spectral normalization and preprocessing, without fundamental algorithmic changes.

Given the limitation of the current dataset, for future research will be its expansion in both scale and diversity. The construction of larger, multi-source repositories that include both healthy and diseased palms across diverse geographic, ecological, and sensor conditions will allow for a more comprehensive validation of the proposed approach and will support the development of fully integrated methods for sustainable plantation monitoring. Incorporating these datasets into pilot studies will further provide valuable insight into how crown delineation can be effectively integrated into operational methods for the early detection of diseases such as bud rot.

Author contributions. All authors contributed equally to the conception, design, implementation, analysis, and writing of the manuscript.

Funding Information. CONEXALAB SAS BIC.

Data Availibility Statement. The annotated ground-truth masks used in this study are available at: https://doi.org/10.7910/DVN/WDJZO1.

Conflict of Interest. The authors declare no conflict of interest.

Ethics Approval and Consent to Participate. Not applicable.

Consent for Publication. Not applicable.

References

1. García-Gaona, M., Botero-Rozo, D., Araque, L., Romero, H.M.: The dynamic interaction between oil palm and phytophthora palmivora in bud rot disease: insights from transcriptomic analysis and network modelling. J. Fungi **10**(3), 164 (2024). https://doi.org/10.3390/jof10030164
2. Ruiz Herrera, A.: La pudrición del Cogollo afecta cerca de 23.000 hectáreas de palma en Santander. La República (2012). https://www.larepublica.co/economia/la-pudricion-del-cogollo-afecta-cerca-de-23-000-hectareas-de-palma-en-santander-2026290
3. Radio, R.: Dramática situación atraviesan palmicultores de Santander por la pudrición del cogollo. RCN Radio. rcnradio.com

4. ConexaLab: i-CO_2 Carbon Capture. ConexaLab (2025). https://www.conexalab.com/i-co2-carbon-capture. Accessed 8 May 2025
5. Rejeki, D.S., Irwansyah, E., Golda, B.: Analysis of comparison faster R-CNN backbone for oil palm tree counting using drone imagery data. In: International Conference on Information Technology and Computing (ICITCOM), pp. 230–235 (2024)
6. Ravi, N., et al.: SAM 2: segment anything in images and videos. arXiv preprint arXiv:2408.00714 (2024)
7. Weinstein, B.G., Marconi, S., Bohlman, S., Zare, A., White, E.: Individual tree-crown detection in RGB imagery using semi-supervised deep learning neural networks. Remote Sens. **11**(11), 1309 (2019). https://doi.org/10.3390/rs11111309
8. Weinstein, B.G., Marconi, S., Aubry-Kientz, M., Vincent, G., Senyondo, H., White, E.P.: DeepForest: a python package for RGB deep learning tree crown delineation. Methods Ecol. Evol. **11**(12), 1743–1751 (2020). https://doi.org/10.1111/2041-210X.13472
9. Korom, A., Phua, M., Hirata, Y., Matsuura, T.: Individual delineation of oil palm tree using worldview02 satellite image. In: 33rd Asian Conference on Remote Sensing 2012, ACRS 2012, vol. 2 (2012)
10. Zhorif, N.N., Anandyto, R.K., Rusyadi, A.U., Irwansyah, E.: Implementation of slicing aided hyper inference (SAHI) in yolov8 to counting oil palm trees using high-resolution aerial imagery data. Int. J. Adv. Comput. Sci. Appl. **15**(7) (2024). https://doi.org/10.14569/IJACSA.2024.0150786
11. Teo, K.K., Noor, N.F.B.M., Palaiahnakote, S., Ayub, M.N.B.: An EfficientNet-based model for classification of oil palm, coconut and banana trees in drone images. Smart Agric. Technol. **10**, 100748 (2025). https://doi.org/10.1016/j.atech.2024.100748
12. Ammar, A., Koubaa, A.: Aerial images of palm trees. Kaggle (2023). https://doi.org/10.34740/KAGGLE/DSV/6382990, https://www.kaggle.com/dsv/6382990
13. Ammar, A., Koubaa, A., Benjdira, B.: Deep-learning-based automated palm tree counting and geolocation in large farms from aerial geotagged images. Agronomy **11**(8), 1458 (2021). https://doi.org/10.3390/agronomy11081458
14. Camacho, A., Vargas, E., Arguello, H.: Hyperspectral and multispectral image fusion addressing spectral variability by an augmented linear mixing model. Int. J. Remote Sens. **43**(5), 1577–1608 (2022). https://doi.org/10.1080/01431161.2022.2041762
15. Roboflow: Roboflow: Organize, Annotate, and Train Computer Vision Models. https://roboflow.com/. Accessed 25 July 2025
16. Dice, L.R.: Measures of the amount of ecologic association between species. Ecology **26**(3), 297–302 (1945). https://doi.org/10.2307/1932409
17. Jaccard, P.: Étude de la distribution florale dans une portion des alpes et du jura. Bull. Soc. Vaudoise Sci. Naturelles **37**, 547–579 (1901)

Geo@gro: A Framework for Crop Monitoring in Colombia. Case Study Potato Mapping Using Time Series

Ariolfo Camacho Velasco[1(✉)], Cristhian C. Rey Rueda[2], William Alfonso Leon Rueda[3], Hector Miguel Vargas García[4], and César A. Vargas García[2]

[1] Corporación Colombiana de Investigación Agropecuaria (AGROSAVIA), CI La Suiza, Bucaramanga, Colombia
acamachov@agrosavia.co
[2] Corporación Colombiana de Investigación Agropecuaria (AGROSAVIA), CI Tibaitatá, Bogotá, Colombia
[3] Federación Colombiana de Productores de Papa - Fondo Nacional de Fomento de la Papa, Bogotá, Colombia
[4] Universidad Manuela Beltrán, Bogotá, Colombia
http://www.agroavia.co

Abstract. Modern precision agriculture requires automated monitoring systems capable of analyzing spatiotemporal vegetation patterns influenced by climatic variability and landscape heterogeneity. This study presents Geo@gro, an open source Python-based framework designed for automated crop monitoring in Colombia. The system integrates high-resolution Planet NICFI satellite imagery with meteorological data to analyze crop phenology through time series vegetation indices. Its modular pipeline includes automated image acquisition, preprocessing, vegetation index calculation (NDVI, EVI), statistical time series analysis, clustering of homogeneous zones, and integration of local weather data. A case study on potato crops in Boyacá and Cundinamarca, conducted with the Colombian Federation of Potato Producers (FEDEPAPA), showed that clustering results aligned with expert-reported phenological stages. The framework also detected stress events such as pest or disease outbreaks and subsequent recovery. Geo@gro's modular, open source design using freely available data provides a scalable, cost-effective solution adaptable to other crops and regions, enabling early stress detection and data-driven agricultural decisions in developing countries.

Keywords: Precision agriculture · Time series analysis · Clustering · NDVI · Potato crops · Colombian agriculture

1 Introduction

Modern precision agriculture increasingly emphasizes the identification of spatio-temporal patterns in vegetation dynamics as influenced by climatic variability (e.g. temperature, precipitation) and landscape heterogeneity [13]. Geo@gro is designed to automate this process by ingestion of high-resolution satellite

D. Chaves et al. (Eds.): CIARP 2025, LNCS 16528, pp. 44–59, 2026.
https://doi.org/10.1007/978-3-032-23161-1_4

mosaics, extraction of vegetation indices over time, and correlating their trajectories with local weather records to reveal phenological and stress response patterns [1,18].

Existing open source frameworks provide valuable tools for geospatial data management and visualization, but they are limited in scope: most do not offer integrated workflows for automated time series decomposition, clustering of vegetation dynamics, and direct coupling with freely available meteorological data. As a result, there is a critical gap in open source solutions capable of delivering end-to-end, reproducible workflows for crop monitoring in data limited contexts such as Colombia.

The Geo@gro framework addresses this gap by combining statistical decomposition and clustering techniques to isolate seasonal cycles, long-term trends, and transient anomalies, thus delineating management zones based both on temporal signatures and spatial context [26]. By integrating freely available meteorological APIs, Geo@gro aligns index-derived clusters with temperature and precipitation series, facilitating the interpretation of crop vigor regarding weather events and climatic stressors. These integrated components provide a cohesive pipeline for detecting and mapping vegetation patterns in Colombian potato fields, improving both the scalability and reproducibility of precision agriculture analyses [9,31].

1.1 Overview of Software for Crop Monitoring

A wide range of software tools has been developed for crop monitoring, spanning commercial platforms and open source initiatives. Commercial solutions such as Climate FieldView [4], Xarvio® [30], Granular [8], DroneDeploy [6], and Pix4Dfields [24] offer integrated services that aggregate field data and deliver real-time insights to growers. Their strengths lie in usability and advanced analytics, but they are subscription based, rely on proprietary architectures, and are rarely adapted to the conditions of smallholder agriculture. On the other hand, open source tools like FarmOS [7] and QGIS PAT [22] provide accessible environments for data management and geospatial analysis. While valuable for community driven innovation, they typically function as isolated modules and lack automated workflows for spatiotemporal vegetation analysis.

Geo@gro distinguishes itself from these approaches by offering a modular, Python-based framework that integrates high resolution satellite imagery, vegetation index time series decomposition, clustering, and freely available meteorological data in a single reproducible pipeline. Unlike commercial platforms, it is open source and cost free; unlike general purpose open source GIS tools, it delivers an end-to-end methodology explicitly tailored to crop monitoring in heterogeneous landscapes under climatic variability.

1.2 Background: Agriculture in Colombia

Agriculture remains crucial to Colombia's rural economy, supporting 23.3% of the population, though less than half of arable land is used for intensive agriculture [23]. Precision Agriculture (PA) offers a strategic solution by optimizing crop management through decision support tools that address spatial and temporal variability. However, PA adoption in Colombia remains limited due to factors including farm size, producer education, income levels, and access to digital technologies [23].

Digital and geospatial technologies are transforming Colombian agriculture. Hyperspectral imaging (HSI) provides high resolution spectral data for crop characterization, with studies showing the Adaptive Coherence Estimator (ACE) algorithm achieving over 90% detection probability across crops like oil palm, rubber, and sugarcane [3]. However, accuracy decreases in complex environments such as diseased crops or mountainous areas with fragmented plots, highlighting the need for context aware approaches in Colombia's diverse geography. Software platforms processing these datasets enable early detection of crop stress, pests, and nutrient deficiencies through spectral analysis, supporting the transition toward efficient, data driven agricultural practices essential for Colombia's sustainable development.

1.3 Satellite-Based Remote Sensing for Crop Monitoring

Earth monitoring using satellites is a fundamental tool for obtaining valuable information on land use and land cover (LULC) changes, essential for environmental management [16,25]. Remote sensing offers a viable and cost-effective alternative for generating timely information on cultivated areas, particularly when compared to expensive large scale field surveys [16]. Furthermore, satellite data are crucial for monitoring critical hydrometeorological variables for agriculture, such as precipitation [15].

Various satellite data sources are used for crop monitoring:

- The Landsat family has provided unique data continuity since the 1970s, enabling long-term time series analyses to map LULC and cash crop dynamics [25].
- Sentinel-1 (SAR) and Sentinel-2 (optical) sensors provide useful data for crop classification due to their spatial and temporal resolution [16]. However, cloud cover limits optical sensors like Sentinel-2, particularly in Colombia's potato producing regions characterized by persistent cloud cover [16,25].
- Radar images (Sentinel-1) are unaffected by clouds and sensitive to crop identification at different phenological stages, complementing optical images and improving mapping accuracy when combined [16,25].
- Image fusion techniques combine optical and radar data to generate enhanced information that overcomes individual sensor limitations, such as fusing hyperspectral and multispectral images for improved spatial and spectral resolution in agricultural applications [2].

1.4 Tools and APIs for Downloading Free Weather Data

Crop monitoring requires open source tools and free access to weather data via APIs. QGIS [20], an open source GIS, enables visualization, analysis and management of geospatial data for agricultural applications. Open-Meteo [28] provides free meteorological data including forecasts, historical records, and real time observations without API keys. Similarly, Colombia's DHIME system [19] offers access to national hydrological and weather data through IDEAM's platform. These tools form the foundation for comprehensive crop monitoring frameworks in precision agriculture.

This study develops a crop monitoring framework for Colombia using open source tools, particularly Python, based on time series analysis of vegetation indices to characterize crop phenological behavior. A case study with Fedepapa analyzed potato crop cycles in Toca, Boyacá and Villapinzón, Cundinamarca, with patterns showing strong correspondence to expert reported phenological stages. The Geo@gro framework was validated by agronomists and Fedepapa experts, confirming its practical utility for agricultural decision making.

2 Materials and Methods

2.1 Geo@gro Software Environment and Hardware Configuration

Geo@gro was implemented in Python within a Conda managed environment, using open source libraries to ensure reproducibility and accessibility. Core raster and vector operations were performed with GDAL, Rasterio, and GeoPandas; time series analysis relied on Statsmodels for stationarity testing and decomposition; and tslearn was used for K-means clustering on normalized series. Numerical computations were supported by NumPy and SciPy, while visualization was enabled through Matplotlib and Plotly. Parallel execution routines were incorporated to improve scalability when processing multiple polygons. To demonstrate operational feasibility, the pipeline was executed on a workstation equipped with an NVIDIA RTX 4500 GPU. In this configuration, the full pipeline from raw tile extraction through final spatial clustering for 28 field polygons was completed in approximately 11 min of wall-clock time for 9 generated clusters.

Figure 1 presents an overview of the Geo@gro framework pipeline, illustrating the complete workflow from satellite image acquisition to spatial cluster generation.

2.2 Data Download and Preprocessing

Satellite imagery is automatically acquired using a Python script that interfaces with the Planet API [12]. Triggered manually or via the Windows Task Scheduler, the script searches for scenes using geospatial inputs and previous metadata, places orders through the Order API, and organizes downloaded images into folders by PSScene ID. This pipeline streamlines satellite image management with full logging of both successful and failed operations.

The integration of satellite imagery into agricultural decision making remains challenging due to limitations in robust data refinement and interpretation, particularly in the spatiotemporal analysis of crop fields. While optical satellite

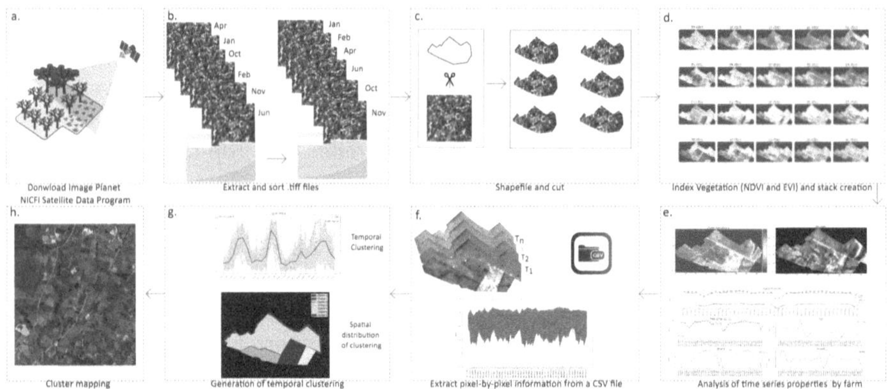

Fig. 1. Geo@gro pipeline overview. (a) Download Planet NICFI mosaics; (b) extract and date-sort TIFFs; (c) clip scenes by field boundaries; (d) compute NDVI/EVI stacks; (e) decompose mean time series into trend, seasonal, and residual components; (f) vectorize per-pixel indices into CSV; (g) apply K-means clustering and plot cluster signatures; (h) map cluster classes back onto base imagery.

sensors capture surface reflectance data, they do not directly measure key biophysical properties such as the Normalized Difference Vegetation Index (NDVI) [29], Enhanced Vegetation Index (EVI) [10], or crop height. Consequently, raw satellite imagery must undergo specialized processing to convert surface observations into meaningful crop metrics. For the Geo@gro framework, all raw Planet NICFI mosaic tiles are initially transformed into analysis-ready per-field image stacks of vegetation indices. First, we ingest the imagery by recursively scanning the source directory; for each match, the embedded year and month are extracted via regular expression and every TIFF within the folder is moved into a flat directory and renamed. Next, to focus on individual potato parcels, each renamed tile is clipped to the extent of each field polygon: the shapefile of study fields is loaded in GeoPandas, its geometries and unique ID values are iterated, and Rasterio's mask function is applied to produce, for each field, a set of cropped TIFFs. Finally, within each field folder, two key spectral indices are calculated: NDVI and EVI. These indices are calculated according to the following equations:

$$\mathrm{NDVI} = \frac{\mathrm{NIR} - \mathrm{Red}}{\mathrm{NIR} + \mathrm{Red}} \tag{1}$$

$$\mathrm{EVI} = 2.5 \times \frac{\mathrm{NIR} - \mathrm{Red}}{\mathrm{NIR} + 6 \cdot \mathrm{Red} - 7.5 \cdot \mathrm{Blue} + 1} \tag{2}$$

Subsequently, the calculated vegetation index values are then normalized to the range [0,1] per acquisition and temporally stacked along the band dimension to obtain georeferenced multiband files ready for further analysis. Figure 2 presents an example polygon showing the results of NDVI and EVI clipping, where each segment represents data extracted for a different time period.

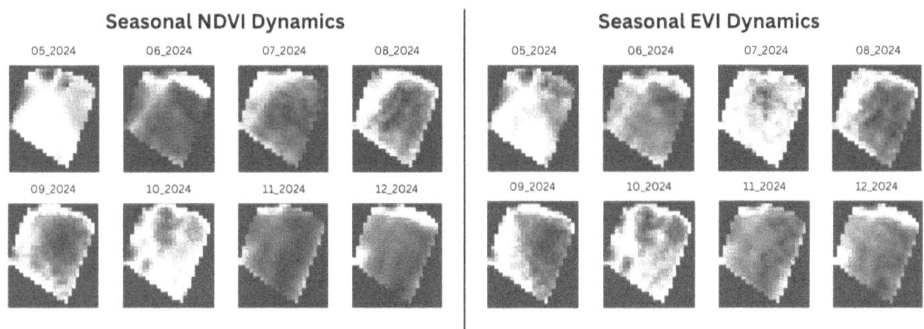

Fig. 2. Comparative temporal sequences of normalized vegetation indices for a single field polygon. Left panel shows monthly NDVI rasters from May to December 2024, illustrating the progression from peak greenness through senescence. Right panel presents the corresponding EVI rasters over the same period. Consistent color ramps (red = low index, green = high index) facilitate direct comparison of NDVI versus EVI dynamics across the growing season. (Color figure online)

2.3 Time Series Analysis

In the next phase of the framework, all per-pixel index stacks are converted into time series, cleaned of anomalies, statistically characterized, decomposed into constituent components, and visualized for exploratory analysis. First, each multiband TIFF is vectorized into a CSV of pixel coordinates and dates, then outliers are removed using a $1.5 \times IQR$ (Interquartile Range) filter. Next, stationarity is assessed on each cleaned series via the Augmented Dickey-Fuller (ADF) and Kwiatkowski Phillips Schmidt and Shin (KPSS) tests [11]. The ADF regression equation is:

$$\Delta Y_t = \alpha Y_{t-1} + \sum_{i=1}^{p-1} \phi_i \Delta Y_{t-i} + e_t \tag{3}$$

where Y_t is the value of the time series at time t, Y_{t-1} is the lagged level of the series, α is the parameter of interest, related to the presence of a unit root, ϕ_i are autoregressive coefficients for the lagged differences and e_t is the error term, assumed to be white noise. The null hypothesis $H_0 : \alpha = 0$ indicates the presence of a unit root (non-stationarity), while $H_1 : \alpha < 0$ implies stationarity. [17].

Conversely, the KPSS test decomposes a series as

$$Y_t = \delta t + r_t + e_t \tag{4}$$

where δt is the deterministic trend, r_t is the random walk, and e_t is the stationary error. Its null hypothesis assumes trend stationarity, while the alternative indicates a unit root [27]. Since KPSS reverses the null and alternative hypotheses compared to ADF, our framework requires both tests to confirm stationarity, ensuring robust time series characterization. The seasonal component of each series is then extracted with the seasonal decompose routine from `statsmodels` and recorded alongside spatial metadata. To capture field scale dynamics, the mean NDVI trajectory of each stack is computed and decomposed into trend, seasonal, and residual signals. Figure 3 displays the seasonal decomposition components for a temporal stack, including a representative clipped region and the corresponding average time series for the complete stack.

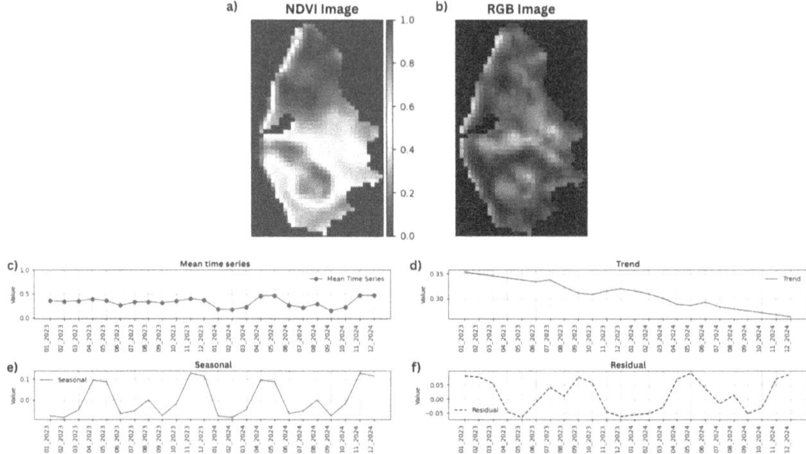

Fig. 3. Temporal decomposition of the NDVI time series for a representative field polygon. (a) Reference NDVI clip from the multiband stack (first date) with its corresponding (b) contrast enhanced RGB image (adaptive histogram equalization). The spatial mean of NDVI over each date in the stack produces the one dimensional series shown in (c), which is decomposed using an additive seasonal model into (d) trend, (e) seasonal, and (f) residual components via `statsmodels.tsa.seasonal_decompose`. The decomposition period is set automatically based on the number of acquisitions, revealing both long-term decline and repeating seasonal cycles in crop vigor.

2.4 Clustering and Integration with Meteorological Data

For the next phase of the Geo@gro framework, we segment per pixel vegetation index trajectories into homogeneous clusters and then contextualize those clusters with local meteorological records. To begin, we determine the optimal number of clusters (K) using the "elbow" method [5] on both the raw and normalized time series data. For each candidate k, we fit a K-Means algorithm [14] model and record its inertia; plotting inertia versus k reveals a point of diminishing returns, which we select as the elbow. The elbow method is a heuristic approach for determining the optimal number of clusters in K-Means clustering, particularly effective for relatively small values of K [5]. This method evaluates the trade-off between the number of clusters and the quality of clustering by analyzing the sum of squares within the cluster (WCSS) at different values of K.

At the optimal K value, each data point essentially becomes its own cluster center, resulting in a distance of zero between the point and its centroid, which represents the theoretical minimum of the WCSS function. Once K is identified, we apply K-Means algorithm directly to the raw band-value matrix to assign each pixel a `cluster_raw` label. In parallel, each series is scaled to zero mean and unit variance, projected into a low-dimensional subspace using principal components analysis (PCA, up to three components), and then clustered with the same algorithm to yield `cluster_norm` labels. This dual approach enables comparison of patterns emerging from absolute versus normalized spectral signatures. To enrich clusters with environmental context, we compute the geographic centroid of each study field (shapefile) and query the Open-Meteo archive API [21] for daily temperature, relative humidity, and precipitation over the study period. These daily values are aggregated to the chosen temporal resolution (daily or monthly) and aligned with the index acquisition dates. In the final phase, we transform the pixel-level cluster assignments into geospatially explicit, thematic maps suitable for visualization and further spatial analysis. We first load the CSV of individual pixel signatures (with fields `pixel_x`, `pixel_y` and `cluster_label`) for each study polygon. For each polygon, we open its corresponding reference raster (`stack_ndvi.tif`) to retrieve the height, width, geotransform and CRS. We then allocate an empty `uint8` mask array matching those dimensions and populate it by iterating over the CSV rows: each pixel at coordinates (x, y) is assigned an integer index based on its `cluster_label`. To enable consistent coloring across all polygons, we generate a global color palette from Matplotlib's "jet" colormap, reserving index 0 for background (no data) and mapping each cluster index to an RGBA tuple. Finally, we write the mask as a single-band, paletted GeoTIFF (photometric=PALETTE, nodata=0) using Rasterio's `write_colormap` functionality, exporting one thematic TIFF per polygon (e.g. `polygon_1897_clusters.tif`). These high-resolution, color-coded cluster maps succinctly convey spatial patterns in crop dynamics and can be overlaid on base imagery or incorporated into GIS workflows. Figure 4 presents the temporal-to-spatial clustering transformation: individual polygon time series, extracted cluster profiles with representative patterns, and the resulting spatial cluster distribution across a study area.

3 Case Study: Potato Mapping Using the Geo@gro Framework

We evaluated Geo@gro on potato crops (Solanum tuberosum L.) in two high Andean municipalities (Villapinzón, Cundinamarca; Toca, Boyacá). These regions are representative of Colombia's potato production and are characterized by frequent cloud cover, fragmented parcels, and marked intraannual rainfall regimes, conditions that pose a rigorous test for automated spatio-temporal monitoring (Fig. 5). The goal here is not to provide an exhaustive agronomic analysis, but rather to demonstrate that a pattern detection pipeline can recover typical phenological trajectories, identify stress induced deviations, and contextualize

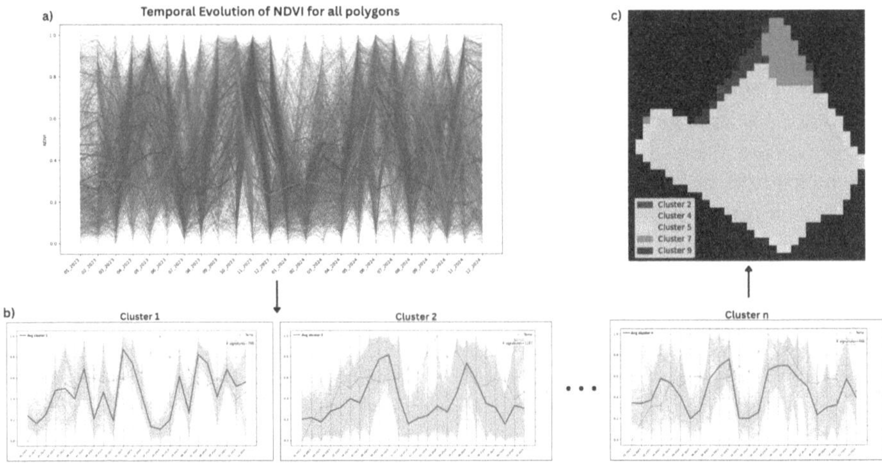

Fig. 4. Multi-panel analysis of NDVI time series clustering and spatial mapping. (a) Parallel-coordinates plot showing monthly NDVI trajectories for all pixel signatures across the study polygons. (b) Example cluster temporal signatures: pale blue lines are individual pixel NDVI profiles, the red line is the cluster-average NDVI, and orange markers denote corresponding monthly mean temperature; panels illustrate clusters 1, 2, ..., n. (c) Spatial distribution of cluster membership across the field, with each colour corresponding to a distinct cluster as in the legend. (Color figure online)

those deviations with field and weather information. It is important to note, however, that Geo@gro is not crop specific. To date, it has been applied to a wide range of crops, including sugarcane, rice, and cattle pastures. The phenological cycle captured through time series varies depending on the crop: for instance, the cycle of sugarcane and panela cane extends over 16 to 18 months, while the vegetative development cycle of potato ranges between 4 and 6 months. This highlights a key aspect of time series analysis: the shape and periodicity of the series differ across crops. Such variability allows the Geo@gro framework to be flexibly applied to diverse agricultural systems.

3.1 Study Area and Acquired Data

Data collection was carried out in the municipalities of Toca and Villapinzón, regions of great importance for potato production in Colombia. Both areas are located between 2,600 and 3,200 m above sea level, with average annual temperatures between 12 °C and 14 °C, and rainfall ranging between 900 and 1,200 mm per year, distributed in two main rainy periods (March-May and September-November). These agroclimatic conditions favor the establishment of potato crops in short cycles (approximately 120 days) and in intermediate or long cycles, depending on the varieties used and the production strategies of each farmer.

To structure a robust database that would allow the comprehensive characterization of potato production units, a methodology was developed to capture

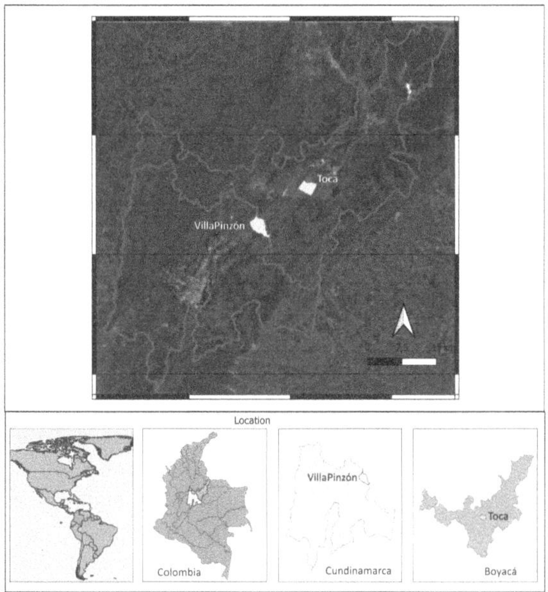

Fig. 5. Geographic location of the study sites in Villapinzón (Cundinamarca) and Toca (Boyacá), Colombia. The main panel shows parcel boundaries over satellite basemaps; insets locate sites within Colombia and their departments. High elevation ($\approx 2,600 - 3,200$ m a.s.l.) and frequent cloud cover make these areas well suited to evaluating time series monitoring pipelines.

data oriented to technical, sanitary, and agronomic management records at the lot level. Data collection was carried out in the field by extensionists from the National Potato Development Fund (FNFP for its acronym in spanish) administered by FEDEPAPA, who used a mobile application to capture the records. This strategy was applied in a total of 36 lots distributed in the municipalities of Toca and Villapinzón. The methodology was structured in three main components: the general characterization of the lot and the technical follow-up during the phenological cycle of the crop. These settings provide a rigorous test for time series vegetation analysis under persistent cloudiness and heterogeneous field geometry, motivating our use of dense mosaics and index based clustering.

3.2 Result Case Study

Initial Field Characterization. Each production unit was identified by a unique code, which allowed us to establish a relationship between the basic information of the plot and the events recorded throughout the production cycle. The initial characterization included variables associated with the location of the plot (including the polygon), the potato variety established, and the soil conditions, such as the texture class and topography of the soil. The history of land use was also recorded, including the previous crop, as well as the dates of the agricultural calendar, such as planting and harvesting (see Table 1).

Table 1. Experimental cultivation data by locality and variety. The harvest values are expressed in tons per hectare.

FID	Code	Place	Variety	Establishment	Harvest Date	Topography	Texture	Previous Cultivation	Harvest
258	1832	TOCA	Diacol Capiro	18/02/2023	17/08/2023	Semi-flat	Loam	Praderas	30.0
259	1833	TOCA	Diacol Capiro	18/02/2023	17/08/2023	Semi-flat	Loam	Praderas	53.7
260	1834	TOCA	Diacol Capiro	21/02/2023	20/08/2023	Semi-flat	Loam	Praderas	42.8
261	1835	TOCA	Diacol Capiro	18/02/2023	17/08/2023	Flat	Clay loam	Crop	26.0
811	793	VILLA	Rubi	29/04/2023	26/10/2023	Undulated	Loam	Praderas	30.0
812	794	VILLA	Diacol Capiro	11/05/2023	07/11/2023	Undulated	Loam	Praderas	27.0
817	799	VILLA	Ica única	17/05/2023	13/11/2023	Semi-flat	Loam	Praderas	33.0
818	800	VILLA	Superior	26/05/2023	22/11/2023	Undulated	Loam	Praderas	32.0

Technical Monitoring During the Crop Cycle. During crop development, technical visits were conducted for approximately 30 days to record field observations related to the phenological stage, the appearance of biotic limiting factors, and the management practices implemented. In each monitoring, the stage of crop development and the presence of diseases, pests, or weeds were documented. For each case, the causal agent identified and the level of severity observed were recorded on a categorical scale with five levels (AUSENTE, LEVE, MEDIO, FUERTE and SEVERO). In addition, phytosanitary and agronomic applications made in response to these incidences were recorded. Each record included information on the date of intervention, the product used, its functional category, active ingredient, dose applied, unit of measurement, and toxicological classification, according to current regulations (see Tables 2 and 3).

Data Standardization and Integration. All information was collected and stored in databases, which facilitated their validation and subsequent analysis. Quality control processes were implemented to verify the temporal consistency of records, standardize units and categories, and ensure the integrity of the linkage between lots and their associated events.

Table 2. Phytosanitary monitoring of Lot 260. Monitoring dates, applications, phenological status of the crop, type of problem identified, level of incidence and type of input applied are included.

Lot Name	Date/Time Monitoring	Application Date	Phenological State	Problem Type	Identified Problem	Incidence Level	Type of Input
260	24/02/2023 10:09	28/02/2023	Sowing	Diseases	Nematodos Sinfilidos	Leve	Fungicide
260	24/02/2023 10:09	28/02/2023	Sowing	Diseases	Others	Leve	Fungicide/Bactericide
260	15/03/2023 13:21	17/03/2023	Growth	Pests	*Epitrix cucumeris*	Medio	Insecticide
260	15/03/2023 13:21	17/03/2023	Growth	Pests	*Epitrix cucumeris*	Medio	Insecticide
260	15/03/2023 13:21	17/03/2023	Growth	Diseases	*Phytophthora infestans*	Medio	Fungicide
260	15/03/2023 13:21	17/03/2023	Growth	Diseases	*Phytophthora infestans*	Medio	Fungicide
260	05/04/2023 15:10	07/04/2023	Growth	Pests	*Bactericera cockerelli*	Ausente	Insecticide
260	25/04/2023 12:43	28/04/2023	Flowering	Fertilization	Foliar	Ausente	Fertilizer/Conditioner
260	30/05/2023 15:56	09/06/2023	Filling	Pests	*Tecia solanivora*	Ausente	Insecticide
260	30/05/2023 15:56	09/06/2023	Filling	Pests	*Tecia solanivora*	Ausente	Insecticide

Table 3. Phytosanitary monitoring of Lot 261. Monitoring dates, applications, phenological status of the crop, type of problem identified, level of incidence and type of input applied are included. The program covers from the preparation of the land to the senescence of the crop.

Lot Name	Date/Time Monitoring	Application Date	Phenological State	Problem Type	Identified Problem	Incidence Level	Type of Input
261	24/02/2023 11:11	28/02/2023	Preparation	Weeds	*Rumex crispus*	Medio	Herbicide
261	15/03/2023 14:09	17/03/2023	Sowing	Diseases	*Rhizoctonia solani*	Medio	Fungicide
261	15/03/2023 14:09	17/03/2023	Sowing	Diseases	Bacteria	Medio	Fungicide/Bactericide
261	15/03/2023 14:09	17/03/2023	Sowing	Diseases	Nematodos Sinfilidos	Medio	Nematicide
261	05/04/2023 15:44	10/04/2023	Emergence	Pests	*Epitrix cucumeris*	Medio	Insecticide
261	25/04/2023 13:39	28/04/2023	Growth	Diseases	*Phytophthora infestans*	Leve	Fungicide
261	30/05/2023 16:59	03/06/2023	Growth	Pests	*Premnotrypes vorax*	Leve	Insecticide
261	30/05/2023 16:59	03/06/2023	Growth	Pests	*Premnotrypes vorax*	Ausente	Insecticide
261	30/05/2023 16:59	03/06/2023	Growth	Nutritional Deficiency	Phosphorus	Ausente	Fertilizer/Conditioner
261	15/06/2023 16:39	19/06/2023	Filling	Fertilization	Foliar	Leve	Fertilizer/Conditioner
261	15/06/2023 16:39	19/06/2023	Filling	Fertilization	Foliar	Leve	Fertilizer/Conditioner
261	15/06/2023 16:39	19/06/2023	Filling	Diseases	*Phytophthora infestans*	Leve	Fungicide
261	15/06/2023 16:39	19/06/2023	Filling	Diseases	*Phytophthora infestans*	Leve	Fungicide
261	19/07/2023 18:40	22/07/2023	Senescence	Diseases	*Phytophthora infestans*	Ausente	Fungicide

The time series analysis revealed a strong correspondence between vegetation patterns and field observations. Plot 260 (Cluster 3), illustrated in Fig. 6 (left) and detailed in Table 2, spans the FebruaryAugust 2023 cycle. Its NDVI trajectory rose during the first month but showed an unusual slowdown and plateau in April, deviating from the expected phenological curve. Table 2 records a medium-level pest attack (*Epitrix cucumeris*) in March and a concurrent disease outbreak (*Phytophthora infestans*), both of which disrupted crop development. Following targeted insecticide and fungicide treatments, the crop required two months to recover before resuming its normal sequence of flowering, tuber filling, senescence, and harvest. In contrast, Plot 261 (Cluster 5), shown in Fig. 6 (right) and summarized in Table 3, exhibited the canonical temporal trajectory of potato phenology. NDVI increased steadily through emergence and vegetative growth, peaking around the third month, before descending through flowering, tuber filling, and senescence. No major stress events were recorded in Table 3, confirming that the spectral signal reflected a healthy crop cycle. These results demonstrate that Geo@gro is not only technically effective but also agronomically relevant. The early detection of stress in Plot 260, illustrates how time series clustering of NDVI can provide actionable warnings before yield losses accumulate. For Colombian potato farming, where smallholder producers often operate with limited resources and face high input costs, timely identification of anomalies supports more precise pesticide applications, reduces unnecessary treatments, and contributes to integrated pest and disease management. Beyond detecting anomalies, the framework's ability to confirm post-treatment recovery strengthens farmer confidence in interventions and guides decisions on subsequent applications. A distinctive and innovative factor of Geo@gro lies in its systematic use of time series analysis, which allows the characterization of phenological cycles and the detection of deviations specific to each crop and context. By leveraging the shape and periodicity of vegetation indices across time, the framework can capture crop dynamics more faithfully than static approaches, making it adaptable across diverse agricultural systems. Compared with commercial platforms such as Climate FieldView or Xarvio®, Geo@gro delivers comparable monitoring functions in an open source, cost free form adapted to local conditions. Unlike general purpose open source tools (e.g. FarmOS, QGIS plugins), it offers a fully integrated pipeline that automates image acquisition, statistical decomposition, clustering, and weather integration, making it reproducible and transferable to other crops and regions. By bridging methodological rigor with practical needs, Geo@gro is positioning itself as a scalable alternative for data driven decision making in precision agriculture.

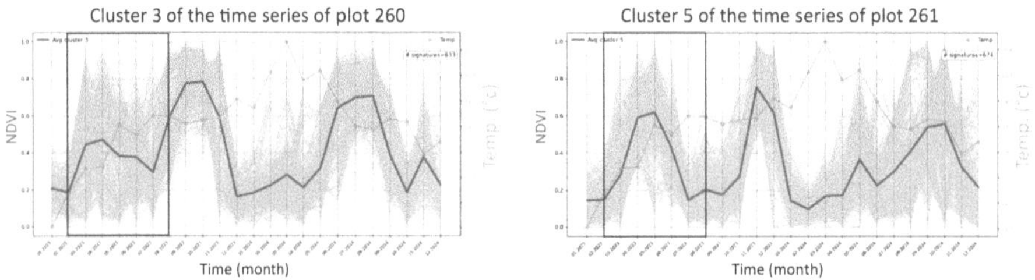

Fig. 6. Time series patterns for two representative parcels with field-reported phenology. In the left panel (Plot 260, Cluster 3), the NDVI trajectory exhibits an atypical slowdown and plateau in April, even though the blue box indicates the expected growth cycle. Field observations from MarchApril confirmed infestations of Epitrix cucumeris and Phytophthora infestans. Following treatment, the vegetation signal recovered and resumed its expected late season decline. In the right panel (Plot 261, Cluster 5), the trajectory follows a canonical pattern: a steady rise during emergence and vegetative growth, a peak around the third month, and a gradual decline through tuber filling and senescence, with no major disruptions recorded. In both cases, the red cluster-average curves condense pixel-level profiles into representative trends. (Color figure online)

4 Limitations

While Geo@gro provides a reproducible, open source framework for crop monitoring, certain limitations affect its scalability and scope. Regarding data acquisition and orchestration, the automated retrieval of satellite imagery through the Python script that interacts with the Planet API still requires manual or locally scheduled activation. This reliance on manual triggers poses challenges for fully operational, large scale deployments at regional or national levels, even though the pipeline incorporates parallel execution routines to enhance scalability when processing multiple polygons. The validation scope of this study is also bounded: experiments were limited to potato crops in two municipalities of Colombia (Villapinzón, Cundinamarca; Toca, Boyacá). A broader assessment across diverse crops, regions, and climatic conditions would be necessary to fully evaluate the robustness and generalizability of the framework. Finally, with respect to data analysis, the current implementation relies primarily on vegetation indices such as NDVI and EVI. While these have proven effective, the framework could be further strengthened by integrating additional biophysical variables (e.g., soil moisture, canopy temperature) to improve sensitivity to specific stressors and capture a wider range of crop responses.

5 Conclusions

Geo@gro demonstrates that open-source, modular pipelines can make advanced crop monitoring approaches accessible to researchers, practitioners, and agricultural organizations in developing countries. By integrating satellite imagery, vegetation index time series, clustering, and meteorological data, the framework provides a reproducible and cost-effective methodology for detecting crop stress and interpreting phenological dynamics at the field scale. In practice, this enables more precise agronomic recommendations, early detection of stress-induced anomalies, and evidence-based decision making for organizations such as FEDEPAPA.

At the same time, the present study highlights opportunities for extending the framework. While Geo@gro currently identifies anomalies in vegetation trajectories, it does not yet implement automated early warning mechanisms. Future work will investigate the use of artificial intelligence and machine learning algorithms such as recurrent and transformer based sequence models, anomaly detection with autoencoders, and hybrid approaches that integrate weather covariates to anticipate anomalous events before they are fully expressed in the field. Such predictive capabilities would strengthen the framework's role as an operational early warning system for pest and disease outbreaks, nutrient deficiencies, or extreme weather stress. Further development will also focus on scaling Geo@gro to multi-crop and multi-year datasets, and on incorporating additional biophysical indicators such as soil moisture and canopy temperature to improve sensitivity to specific stressors. A key priority is the operational deployment of the framework into regional or national decision support platforms, enabling real time, cost-effective monitoring that strengthens the resilience of smallholder agriculture under increasing climatic variability.

Acknowledgments. The authors acknowledge the Minciencias under Grant 934 - Convocatoria de estancias posdoctorales orientadas por misiones, titled: "Clasificación y detección de cambios en cultivos agrícolas usando imágenes satelitales ópticas y de radar como herramienta de apoyo para Agrosavia en los centros de investigación La Suiza, de Rionegro Santander y Norte de Santander, Centro de Investigación Nataima, Tolima y el distrito de conservación de suelos del Centro de Investigación Tibaitatá".

References

1. Brook, A., et al.: A smart multiple spatial and temporal resolution system to support precision agriculture from satellite images: proof of concept on Aglianico vineyard. Remote Sens. Environ. 240, 111679 (2020). https://doi.org/10.1016/j.rse.2020.111679, https://www.sciencedirect.com/science/article/pii/S0034425720300481. ISSN: 0034-4257
2. Camacho, A., Vargas, E., Arguello, H.: Hyperspectral and multispectral image fusion addressing spectral variability by an augmented linear mixing model. Int. J. Remote Sens. **43**(5), 1577–1608 (2022)
3. Velasco, A.C., García, C.A.V., Fuentes, H.A.: A comparative study of target detection algorithms in hyperspectral imagery applied to agricultural crops in Colombia. Tecnura **20**(49), 86–99 (2016)
4. Climate FieldView: All-in-One Digital Farming Solution. Climate Corporation (2025). URL: https://climate.com/. Accessed 18 June 2025
5. Cui, M.: Introduction to the K-means clustering algorithm based on the elbow method. Geosci. Remote Sens. **3**, 9–16 (2020). https://doi.org/10.23977/geors.2020.030102
6. DroneDeploy Agriculture: Drones for Farming. DroneDeploy, Inc. (2025). https://www.dronedeploy.com/solutions/agriculture. Accessed 18 June 2025
7. farmOS: Farm Management and Record Keeping Platform. farmOS Community (2025). https://farmos.org/. Accessed 18 June 2025
8. Granular Insights: Digital Solutions for Precision Agriculture. Corteva Agriscience (2025). https://www.corteva.us/products-and-solutions/digital-solutions/granular-insights.html. Accessed 18 June 2025
9. Haque, S., Hossain, S., Billah, M.: Precision agriculture through remote sensing and GIS: advancing sustainable farming and climate resilience. Am. J. Smart Technol. Solut. **4**, 30–36 (2025). https://doi.org/10.54536/ajsts.v4i1.4418
10. Huete, A.R., et al.: A comparison of vegetation indices over a global set of TM images for EOS–MODIS. Remote Sens. Environ. **59**(3), pp. 440–451 (1997). https://doi.org/10.1016/S0034-4257(96)00112-5, https://www.sciencedirect.com/science/article/pii/S0034425796001125. ISSN 0034-4257
11. Kębłowski, P., Welfe, A.: The ADF–KPSS test of the joint confirmation hypothesis of unit autoregressive root. Econ. Lett. **85**(2), 257–263 (2004). https://doi.org/10.1016/j.econlet.2004.04.013, https://www.sciencedirect.com/science/article/pii/S0165176504001880. ISSN 0165-1765
12. Kim, J.Y.: Open-source software for satellite-based crop health monitoring. J. Biosyst. Eng. **49**, 419–433 (2024). https://doi.org/10.1007/s42853-024-00242-z
13. Kooistra, L., et al.: Reviews and syntheses: remotely sensed optical time series for monitoring vegetation productivity. Biogeosciences **21**(2), 473–511 (2024). https://doi.org/10.5194/bg-21-473-2024, https://bg.copernicus.org/articles/21/473/2024/
14. Lin, J., Vlachos, M., Keogh, E., Gunopulos, D.: Iterative incremental clustering of time series. In: Bertino, E., et al. (eds.) EDBT 2004. LNCS, vol. 2992, pp. 106–122. Springer, Heidelberg (2004). https://doi.org/10.1007/978-3-540-24741-8_8
15. López, P.L., et al.: Spatial downscaling of satellite-based precipitation and its impact on discharge simulations in the Magdalena River basin in Colombia. Front. Earth Sci. **6**, 68 (2018)
16. Luque, N., Martinez, L.J., Monsalve, O.: Classification of potato crops in the high Andean zone through sentinel-1/2 image fusion (2024). https://ssrn.com/abstract=4848415, https://doi.org/10.2139/ssrn.4848415

17. Lütkepohl, H., Krätzig, M. (eds.): Applied Time Series Econometrics. Cambridge University Press, Cambridge (2004)
18. Melgar-García, L., et al.: Discovering spatio-temporal patterns in precision agriculture based on triclustering, pp. 226–236 (2021). https://doi.org/10.1007/978-3-030-57802-2_22. ISBN 978-3-030-57801-5
19. Moreno-Cueva, L.A., Peña-Cortés, C.A., Henriquez-Miranda, C.N.: Planeación estratégica aplicada a la administración de cultivos de alverja empleando herramientas de inteligencia artificial basada en datos hidrometeorológicos y económicos. Respuestas **29**(3), 6–14 (2024)
20. Moyroud, N., Portet, F.: Introduction to QGIS. QGIS Generic Tools **1**, 1–17 (2018)
21. Open-Meteo. Open-Meteo API (2025). https://open-meteo.com/. Accessed 18 June 2025
22. PAT – Precision Agriculture Tools Plugin for QGIS. QGIS Python Plugins Repository (2025). https://plugins.qgis.org/plugins/pat/. Accessed 18 June 2025
23. Pérez-Ortega, D.J., Bolaños-Alomia, F.A., da Silva, A.M.: Variables que influyen en la aplicación de la agricultura de precisión en Colombia: revisión de estudios. Ciencia y Tecnolog í a Agropecuaria **23**(1) (2022)
24. PIX4Dfields: Drone Software for Agriculture Mapping. Pix4D SA (2025). https://www.pix4d.com/product/pix4dfields. Accessed 18 June 2025
25. Salomão, C., et al.: Mapping hydropower expansion and cash crop dynamics in Colombia using Landsat time series. Geocarto Int. **39**(1), 2322064 (2024)
26. Sishodia, R.P., Ray, R.L., Singh, S.K.: Applications of remote sensing in precision agriculture: a review. Remote Sens. **12**(19) (2020). https://www.mdpi.com/2072-4292/12/19/3136. ISSN 2072-4292
27. Sjösten, L.: A comparative study of the KPSS and ADF tests in terms of size and power (2022)
28. Sodadasi, S.: Weather assistant (2024)
29. Tucker, C.J., et al.: Satellite remote sensing of total herbaceous biomass production in the Senegalese Sahel: 1980–1984. Remote Sens. Environ. **17**(3), 233–249 (1985). https://doi.org/10.1016/0034-4257(85)90097-5, https://www.sciencedirect.com/science/article/pii/0034425785900975. ISSN 0034-4257
30. xarvio ® Digital Farming Solutions. xarvio GmbH (2025). https://www.xarvio.com/. Accessed 18 June 2025
31. Zhang, D., et al.: Precision agriculture: temporal and spatial modeling of wheat canopy spectral characteristics. Agriculture **15**(3) (2025). https://www.mdpi.com/2077-0472/15/3/326. ISSN 2077-0472

Vis-NIR Spectroscopy and Machine Learning for Estimating Soil Nutrient Quality Index and Chemical Properties in Remote Cacao-Growing Regions of Colombia

Eliana Romero[1], Diego Leonardo Cortés[2], Eliana Martínez Pachón[3], Andrea Montenegro[3], Paula H. Reyes-Herrera[3], and Francisco Calderón[1](✉)

[1] Pontificia Universidad Javeriana, Bogotá, Colombia
calderonf@javeriana.edu.co

[2] Corporación Colombiana de Investigación Agropecuaria AGROSAVIA, CI Obonuco, Pasto, Nariño, Colombia

[3] Corporación Colombiana de Investigación Agropecuaria AGROSAVIA, CI Tibaitatá, Bogotá, Colombia

Abstract. This study presents an innovative system integrating Vis-NIR spectroscopy and Machine Learning regressors to estimate soil chemical properties and a soil quality index in remote cacao-growing regions of Colombia. By incorporating expert evaluations, the developed soil nutrient quality index is practical and actionable, offering farmers a cost-effective and environmentally friendly alternative to traditional soil analysis. This study analyzes 138 soil samples from Nariño and 284 from Norte de Santander—two cacao-producing, hard-to-access regions in Colombia. The data were evaluated separately for each region and also in combination.

Key chemical soil properties —such as pH, organic matter, calcium, magnesium, iron, cation exchange capacity, and the composite soil nutrient quality index—were predicted using regression models including Partial Least Squares Regression, Support Vector Regression, Least Absolute Shrinkage and Selection Operator, and Cubist. Model performance was optimized using cross-validation and hyperparameter tuning. Results highlight the effectiveness of Cubist model, with R^2 values as high as 0.912 for calcium in Nariño and 0.831 in the combined dataset. The study also estimated the soil nutrient quality index (SNQI) with an R^2 of 0.583. This approach demonstrates great potential for improving decision-making in sustainable cacao cultivation.

Keywords: vis-NIRS · Precision Agriculture · Soil nutrient properties · SVR · Cubist · LASSO · PLSR · cocoa · SNQI

Supplementary Information The online version contains supplementary material available at https://doi.org/10.1007/978-3-032-23161-1_5.

D. Chaves et al. (Eds.): CIARP 2025, LNCS 16528, pp. 60–74, 2026.
https://doi.org/10.1007/978-3-032-23161-1_5

1 Introduction

Traditionally, soil analysis has been conducted in laboratories using chemical methods which, although effective, are costly, time-consuming, and require highly specialized personnel. These methods also have a significant environmental impact due to the extensive use of chemical reagents. As agricultural demands shift toward real-time decision-making and sustainable practices, Visible-Near Infrared spectroscopy (vis-NIRS) has emerged as a promising technique for precision agriculture. This non-destructive method allows for rapid and cost-effective analysis of soil chemical properties without the use of chemical agents, offering an efficient and environmentally friendly alternative [14]. Vis-NIR takes advantage of light absorption by different materials in the visible and near-infrared regions of the electromagnetic spectrum, enabling the inference of multiple soil components from a single spectrum.

The integration of Vis–NIR spectroscopy with machine learning (ML) methodologies began around 2010, but its application expanded rapidly after 2020, primarily to improve the estimation of soil chemical attributes [23]. Since then, ML has become central to advancing soil analysis with Vis–NIR data. Partial Least Squares Regression (PLSR) remains the traditional baseline; however, non-linear algorithms are increasingly achieving superior predictive performance. For instance, in Colombian sugarcane soils, Cubist and Support Vector Regression (SVR) improved external-validation accuracy for exchangeable K from $R^2 = 0.55$ to 0.71 compared with PLSR [5]. Beyond single learners, ensemble methods have further enhanced prediction. Tavakoli et al. [21], for example, combined Random Forests, Gradient Boosting, and Elastic Net into a meta-learner, increasing soil organic carbon (SOC) prediction from $R^2 = 0.80$ (best single model) to 0.88.

At the same time, deep learning (DL) approaches are gaining traction as larger datasets become available. A 1-D convolutional neural network (CNN) surpassed Cubist models when training samples exceeded 1500 [15], while hybrid Transformer–CNN architectures have recently achieved state-of-the-art performance ($R^2 = 0.92$ for SOC; $R^2 = 0.88$ for total N) [21]. In data-limited contexts, transfer learning and federated learning are emerging alternatives. Gallios et al. [7] demonstrated that decentralized federated CNNs trained across multiple soil datasets achieved accuracy comparable to centralized models while preserving data privacy. Finally, sensor fusion has opened new opportunities: combining Vis–NIR with portable X-ray fluorescence (pXRF) increased external-validation R^2 for SOC from 0.52 to 0.90 [22], highlighting the value of multimodal approaches for field-ready soil fertility diagnostics.

These methodological advances build on substantial evidence that Vis–NIR spectroscopy effectively predicts a wide range of soil properties. Laboratory Vis–NIR reflectance (350–2500 nm) has repeatedly outperformed mid-infrared spectroscopy for attributes such as pH, SOC, and total nitrogen (TN). In a 12,000-sample Brazilian spectral library, Ng et al. [15] showed that Savitzky–Golay smoothing and standard-normal-variate (SNV) correction supported $R^2 = 0.87$ for SOC and 0.82 for TN under PLSR. Portable sensors, despite narrower ranges (320–1100 nm), have also delivered actionable accuracies: Jiang et al. [8] reported

$R^2_{\text{ext}} = 0.77$ and RPIQ = 3.0 for pH in alpine-pasture soils, while SOC predictions achieved RMSE = 0.22%. Previous work in Colombia confirmed this potential, as Delgadillo-Durán et al. [5] reported $R^2 = 0.81$ for pH and 0.73 for exchangeable Ca, along with similarly robust predictions for Na, K, and Mg using portable Vis–NIR. Importantly, this approach reduced testing costs by 87%, making soil diagnostics more accessible to smallholder farmers.

Nonetheless, prediction of macronutrients remains more challenging due to their indirect spectral signatures. Even so, promising advances have been reported. Serrano et al. [17] achieved $R^2 = 0.76$ for total phosphorus and 0.85 for soil organic matter in Mediterranean agroforestry soils using NIR (1000–2500 nm), while Zhang et al. [25] improved TN prediction in mine-affected calcareous soils from $R^2 = 0.81$ with PLSR to 0.91 with SVR after recursive-feature elimination. Collectively, these studies demonstrate that Vis–NIR spectra contain sufficient information to predict a broad suite of soil chemical parameters, especially when coupled with advanced machine learning methods.

Beyond single attributes, composite soil-nutrient quality indices (SNQI) integrate chemical nutrient related properties into a holistic score. Conceptual frameworks by [2,9] emphasise pH, organic matter, nutrients, cation-exchange capacity (CEC) and salinity as core indicators. In cacao plantations of Tolima, [13] introduced a fuzzy-logic SNQI that mapped pH, CEC, base saturation and micronutrients into linguistic membership functions; the resulting index correlated ($r = 0.78$) with expert soil-fertility rankings. The Soil Nutrient Quality Index (SNQI) proposed by [4] standardises five fertility variables—%OM, available P, K, Mg, Ca, plus Al saturation—against cacao requirements and aggregates them by weighted arithmetic sum.

Spectroscopy offers rapid input for such indices. A Latin-American meta-analysis [16] concludes that portable spectra-driven soil quality index can halve analytical costs and enable near-real-time advisory services for smallholders. However, robust local calibration, moisture correction and socio-economic acceptance remain open issues.

Three gaps persist. **(i) Geographic bias:** most models are trained on temperate or single-crop datasets; transferability to highly weathered tropical soils is uncertain [1]. **(ii) Limited field validation:** only a handful of studies [6,26] evaluate low-cost handheld sensors under *in-situ* moisture and roughness conditions. **(iii) Under-representation of agro-forest systems:** cacao soils, with their distinctive organic inputs and shading micro-climate, remain sparsely represented in spectral libraries [18,19].

This gap is particularly relevant given cacao's strategic role. Between 2005 and 2021, Colombian production grew by 122%, now covering approximately 194,000 ha and supporting 65,000 families [12]. Many plantations are located in regions recovering from armed conflict, where sustainable income alternatives are vital. Providing farmers with rapid soil diagnostics therefore has both agronomic and socio-economic significance.

Building on Vis–NIR applications in sugarcane soils [5] and guided by established SNQI frameworks [2,9], the objective of this study is to develop a Vis–NIR spectroscopy system, coupled with machine learning, to predict key soil chemical properties and generate a soil nutrient quality index tailored to cacao crops.

Using samples from the departments of Nariño and Norte de Santander, we evaluate regression models—Partial Least Squares Regression (PLSR), Support Vector Regression (SVR), LASSO, and Cubist—for the prediction of soil properties including pH, organic matter (OM), exchangeable calcium (Ca), magnesium (Mg), iron (Fe), and CEC. These same models were also applied to predict a soil nutrient quality index (SNQI), designed to offer farmers a comprehensive and user-friendly tool for evaluating soil fertility, with the goal of optimizing productivity and promoting the sustainable cultivation of cacao.

2 Materials and Methods

We developed a comprehensive workflow integrating field sampling, laboratory analysis, and machine learning to predict soil chemical properties in cacao cultivation systems (Fig. 1). Soil samples were collected from smallholder farms in two Colombian Andean departments, analyzed using certified methods, and scanned with Vis–NIR spectroscopy. Machine learning regressors were then trained on spectral data to predict key chemical properties and the Soil Nutrient Quality Index (SNQI). This approach provides a rapid, scalable alternative to conventional soil testing.

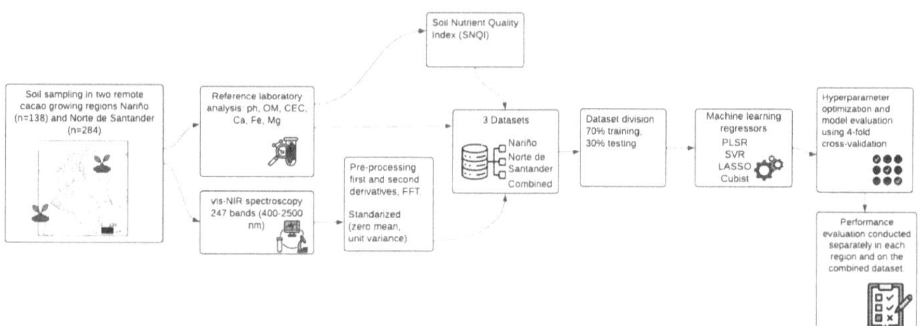

Fig. 1. Workflow for soil property prediction from Vis–NIR spectroscopy in cacao farms. Samples from two Colombian departments underwent chemical analysis (for SNQI calculation) and spectral scanning. Preprocessed spectral data were split for training and testing machine learning regressors, evaluated via cross-validation across three datasets (Nariño, Norte de Santander, and combined).

2.1 Study Area and Sampling Design

Soil samples were collected in 2023–2024 from smallholder cacao farms in two Andean departments of Colombia: Nariño (S 1°10'–1°45', W 77°10'–77°55'), located in the southern Andes near the Ecuadorian border; and Norte de Santander (N 7°00'–8°05', W 72°30'–73°10'), in the northeastern Andes near the Venezuelan border (Fig. 2D).

A stratified sampling design based on farm elevation and management history was used to collect 422 composite topsoil samples: 138 from Nariño (Fig. 2A–C) and 284 from Norte de Santander (Fig. 2 B–E). Each composite sample consisted of 5–7 subsamples taken along a 30-meter transect at a depth of 0–20 cm, after removing surface litter. Samples were air-dried, sieved through a 2 mm mesh, and stored in hermetic polyethylene bags until spectral and laboratory analysis.

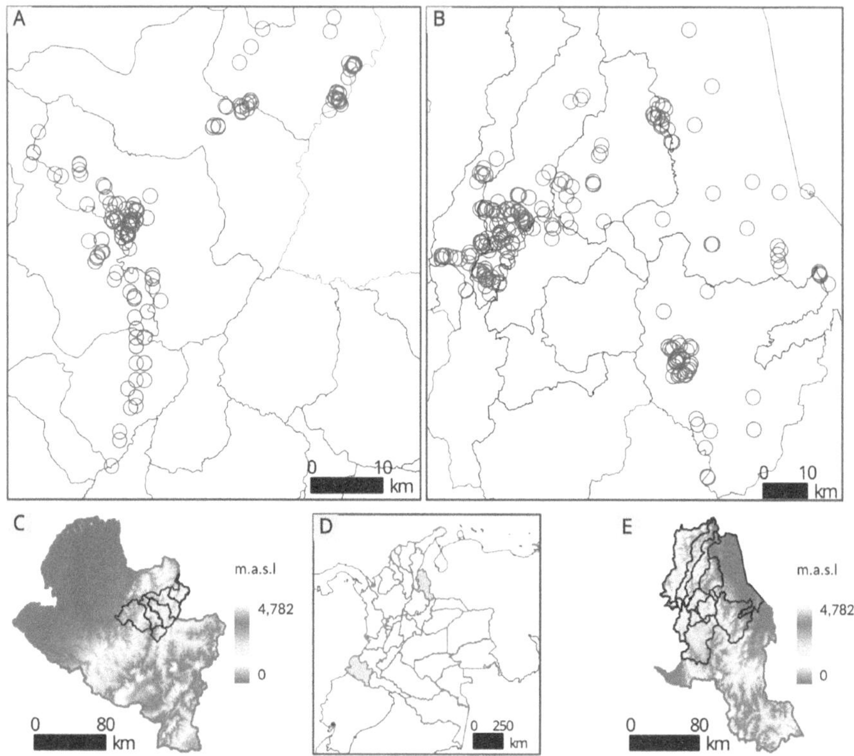

Fig. 2. A. Approximate location of soil samples from Norte de Santander. **B.** Approximate location of soil samples from Nariño. **C.** Terrain relief of Norte de Santander, with the collection site marked in black. **D.** Map of Colombia showing the departments of Nariño and Norte de Santander in grey. **E.** Terrain relief of Nariño, with the collection site marked in black.

2.2 Reference Laboratory Analyses.

Conventional wet chemistry tests were performed in the Agrosavia analytical chemistry laboratory, which uses ISO/IEC 17025:2017-certified techniques. The variables measured were: pH_{H_2O} (1:2.5 soil:water) (ICONTEC, 2008), and organic matter (OM) with Walkley-Black (1934) wet digestion methods. We used the NH_4OAc extraction method to measure exchangeable cations (Ca+2, Mg+2, K+, and Na+) by NTC 5349–2008 (Aguiar-Herrera, 2001). Available Fe was measured by NTC (5526:2007 Method D). The analytical precision ($\pm^{11}$ SD of duplicates) was 2% for the major cations and 0.03 pH units.

2.3 Vis–NIR Spectral Acquisition

The samples passed through a humidity homogenization process; samples were dried at 40 °C for 48–96 h, with drying time adjusted according to soil type. Samples were placed in a 50 mm diameter annular cup and scanned from 400 to 2500 nm using a NIR spectrophotometer (FOOS-DS2500).

2.4 Soil Nutrient Quality Index

To obtain the Soil Nutrient Quality Index (SNQI) for cacao cultivation, we used the index proposed in [4]. For this purpose, soil data were collected from cacao farms in each region, evaluating macro and micronutrients, pH, organic matter OM, and aluminum saturation. These variables were standardized using scoring functions based on the nutritional requirements of the cacao crop. The SNQI was then calculated as a linear combination of the standardized and weighted variables. Each variable I_i was assigned a relative weight W_i, determined according to the importance of each nutrient or soil attribute in supporting cacao growth and development, as detailed in Table 1.

Table 1. Weights and function types used in the Soil Nutrient Quality Index for cacao cultivation.

Variable I_i	Group	Primary Weight	Group Weight (%) W_i	Function Type
P	Macronutrient	0.4	0.2	More is better
K			0.2	More is better
Ca			0.2	Optimum
Mg			0.2	More is better
S			0.2	Optimum
Zn	Micronutrient	0.25	0.15	Optimum
B			0.3	Optimum
Mn			0.225	Optimum
Cu			0.225	Optimum
Fe			0.1	Optimum
pH	General Fertility	0.1	0.1	Optimum
Al Sat.	General Fertility	0.05	0.05	Less is better
OM	General Fertility	0.2	0.2	More is better

2.5 Machine Learning Regressors for Estimating Soil Chemical Properties and SNQI

The modeling process focused on six key chemical properties relevant to cacao cultivation: pH, organic matter (OM), cation exchange capacity (CEC), calcium (Ca), magnesium (Mg), and iron (Fe). The same models were also used to estimate the Soil Nutrient Quality Index (SNQI). The full dataset—comprising 422 soil samples from Nariño and Norte de Santander—was processed into three subsets: Nariño only, Norte de Santander only, and the consolidated set combining both.

Spectral features were derived from Vis–NIR reflectance data spanning 400–2491 nm at 10 nm resolution (247 bands), which were pre-processed like in [5]. Using first and second-order derivatives and the magnitude of the Fast Fourier Transform (FFT), also Features were then standardized (zero mean, unit variance)

The dataset was split into 70% for training and 30% for testing. Four regression algorithms were employed to predict soil chemical properties and SNQI from Vis-NIR spectral data:

- **Partial Least Squares Regression (PLSR)** – a baseline linear model that projects spectral features onto latent variables that maximize covariance with target soil properties.
- **Support Vector Regression (SVR)** – a kernel-based method tested with linear, polynomial, and radial basis function (RBF) kernels. Hyperparameters optimized included: regularization parameter $C \in [0.1, 100]$, epsilon-tube width $\varepsilon \in [0.01, 1]$, kernel coefficient $\gamma \in [0.001, 1]$, and polynomial degree $\in [2, 5]$.
- **Least Absolute Shrinkage and Selection Operator (LASSO) Regression** – applies L1 regularization to induce sparsity in wavelength selection. Tuned parameters included regularization strength $\alpha \in [0.0001, 10]$ and maximum iterations up to 5000.
- **Cubist** – a rule-based regression approach combining decision tree partitioning with local linear models at terminal nodes. Parameters optimized included the number of committees $\in [1, 50]$ and nearest neighbors $\in [0, 9]$ for instance-based adjustment.

Hyperparameters were optimized using exhaustive grid search within a 4-fold cross-validation framework applied to the training partition (70% of data). Model performance was evaluated using three metrics: coefficient of determination (R^2), mean squared error (MSE), and Pearson's correlation coefficient (ρ). Final model assessment was conducted on an independent test set (30% of data) that was withheld from all training and optimization procedures to ensure unbiased performance estimates and prevent data leakage.

3 Results

Descriptive statistics for soil properties in each region are summarized in Table S1 and Figure S1. Regional differences in concentration ranges are substantial:

Fe spans 5.0–209.94 mg/kg in Nariño versus 24.54–778.0 mg/kg in Norte de Santander, directly influencing MSE magnitudes, as this metric is scale-dependent. Consequently, scale-normalized metrics (R^2, ρ) provide more appropriate bases for cross-regional model comparison.

Model performance is summarized in terms of cross-validation R_2CV on the training set, and test set metrics including R^2, mean squared error MSE, and Pearson's correlation coefficient ρ. Results are presented for each property in Table 2 (Nariño), Table 3 (Norte de Santander), and Table 4 (combined dataset). Each table also includes the best hyperparameters set. For R_2CV, R^2, and ρ, cells are color-coded from red (lowest) to green (highest). For MSE, the colors are inverted: red indicates higher error, green indicates lower.

Table 2. Results of Regressor Models for Estimating Chemical Properties Using the Nariño Dataset.

Property	Model	Best_Params	R² CV	R²	MSE	ρ
Ca	PLSR	n_components: 2	0.544	0.746	11.251	0.865
	SVR	C: 0.2, gamma: scale, kernel: linear	0.694	0.833	7.405	0.939
	LASSO	alpha: 0.2	0.62	0.818	8.078	0.922
	Cubist	n_committees: 10, neighbors: 8	0.73	0.912	3.905	0.956
CEC	PLSR	n_components: 10	0.903	0.865	11.378	0.945
	SVR	C: 5, gamma: scale, kernel: linear	0.915	0.902	8.259	0.954
	LASSO	alpha: 0.02	0.896	0.907	7.846	0.954
	Cubist	n_committees: 10, neighbors: 2	0.844	0.869	11.01	0.949
pH	PLSR	n_components: 5	0.784	0.712	0.174	0.861
	SVR	C: 0.1, gamma: scale, kernel: linear	0.817	0.711	0.175	0.856
	LASSO	alpha: 0.01	0.796	0.742	0.156	0.864
	Cubist	n_committees: 10, neighbors: 7	0.805	0.636	0.22	0.826
Mg	PLSR	n_components: 10	0.8	0.785	2.07	0.892
	SVR	C: 1, gamma: scale, kernel: linear	0.797	0.786	2.057	0.889
	LASSO	alpha: 0.01	0.765	0.801	1.91	0.895
	Cubist	n_committees: 10, neighbors: 2	0.797	0.829	1.646	0.913
OM	PLSR	n_components: 5	0.605	0.197	0.289	0.739
	SVR	C: 0.2, gamma: scale, kernel: linear	0.649	0.348	0.235	0.771
	LASSO	alpha: 0.01	0.623	0.511	0.176	0.765
	Cubist	n_committees: 10, neighbors: 9	0.606	0.411	0.212	0.73
Fe	PLSR	n_components: 10	0.394	0.241	1794.318	0.658
	SVR	C: 10, gamma: scale, kernel: linear	0.447	0.497	1188.397	0.763
	LASSO	alpha: 0.2	0.458	0.364	1502.918	0.698
	Cubist	n_committees: 5, neighbors: 9	0.436	0.502	1176.968	0.719
SNQI	PLSR	n_components: 5	0.458	0.487	0.01	0.72
	SVR	C: 5, gamma: auto, kernel: rbf	0.464	0.583	0.008	0.766
	LASSO	alpha: 0.01	0.281	0.343	0.012	0.716
	Cubist	n_committees: 10, neighbors: 9	0.53	0.515	0.009	0.748

In the Nariño region, the best-performing soil property, based on R^2 on the test set, was calcium (Ca), with an R^2 of 0.912, followed by cation exchange capacity (CEC) with an R^2 of 0.907, and magnesium (Mg) with an R^2 of 0.829. In contrast, in Norte de Santander, the highest R^2 was observed for Ca (0.761), followed by pH ($R^2 = 0.721$). When datasets from both regions were combined, the models achieved R^2 values of 0.831 for Ca, 0.823 for CEC, 0.81 for Mg and

Table 3. Results of Regressor Models for Estimating Chemical Properties Using the Norte de Santander Dataset.

Property	Model	Best Parameters	R² CV	R²	MSE	ρ
Ca	PLSR	n_components: 10	0.489	0.734	4.751	0.867
	SVR	C: 2, gamma: scale, kernel: linear	0.568	0.761	4.262	0.881
	LASSO	alpha: 0.01	0.527	0.745	4.543	0.87
	Cubist	n_committees: 10, neighbors: 5	0.562	0.752	4.431	0.886
CEC	PLSR	n_components: 10	0.243	0.546	8.326	0.748
	SVR	C: 2, gamma: scale, kernel: linear	0.376	0.593	7.456	0.78
	LASSO	alpha: 0.05	0.206	0.574	7.812	0.761
	Cubist	n_committees: 10, neighbors: 7	0.285	0.567	7.94	0.77
pH	PLSR	n_components: 10	0.645	0.663	0.215	0.816
	SVR	C: 2, gamma: scale, kernel: linear	0.708	0.721	0.178	0.85
	LASSO	alpha: 0.01	0.6	0.615	0.246	0.832
	Cubist	n_committees: 10, neighbors: 9	0.675	0.711	0.184	0.852
Mg	PLSR	n_components: 2	0.006	0.017	1.036	0.337
	SVR	C: 5, gamma: scale, kernel: linear	0.383	0.431	0.599	0.7
	LASSO	alpha: 0.02	0.075	0.267	0.773	0.527
	Cubist	n_committees: 1, neighbors: 9	0.032	0.457	0.572	0.725
MO	PLSR	n_components: 10	0.558	0.494	0.341	0.742
	SVR	C: 0.5, gamma: scale, kernel: linear	0.583	0.535	0.313	0.75
	LASSO	alpha: 0.01	0.513	0.537	0.312	0.737
	Cubist	n_committees: 10, neighbors: 4	0.527	0.538	0.312	0.742
Fe	PLSR	n_components: 5	0.283	0.502	6299.153	0.71
	SVR	C: 10, gamma: scale, kernel: linear	0.271	0.512	6162.472	0.768
	LASSO	alpha: 1	0.305	0.567	5472.521	0.758
	Cubist	n_committees: 10, neighbors: 4	0.389	0.527	5980.949	0.73
SNQI	PLSR	n_components: 2	0.191	0.351	0.012	0.612
	SVR	C: 0.01, gamma: scale, kernel: linear	0.182	0.421	0.011	0.675
	LASSO	alpha: 0.01	0.002	0.017	0.018	0.455
	Cubist	n_committees: 10, neighbors: 5	0.217	0.367	0.012	0.613

0.776 for pH. These results are consistent with findings reported in previous studies [5].

Iron (Fe) was the most difficult property to predict, exhibiting relatively low R^2 values: 0.502 in Nariño, 0.567 in Norte de Santander, and 0.494 in the combined dataset. Although the MSE for Fe was relatively high, this is attributed to the inherent challenges of estimating Fe using NIRS and the substantially larger range of this property compared to others (Supplementary Fig. 1). Organic matter showed a similar trend, with R^2 values of 0.511 in Nariño, 0.380 in Norte de Santander, and 0.567 in the combined dataset.

In the case of magnesium (Mg), combining both datasets proved beneficial. While the models performed well on the Nariño dataset, achieving an R^2 of 0.829, their performance was considerably lower on the Norte de Santander dataset, with an R^2 of 0.457. However, when the datasets were combined, the model performance improved substantially, reaching an R^2 of 0.81.

Table 4. Results of Regressor Models for Estimating Chemical Properties Using Nariño and Norte de Santander joint Dataset

Property	Model	Best Parameters	R^2 CV	R^2	MSE	ρ
Ca	PLSR	n_components: 10	0.794	0.761	6.31	0.884
	SVR	C: 10, gamma: scale, kernel: linear	0.839	0.781	5.793	0.893
	LASSO	alpha: 0.02	0.808	0.771	6.056	0.879
	Cubist	n_committees: 10, neighbors: 9	0.843	0.831	4.46	0.924
CEC	PLSR	n_components: 10	0.838	0.771	7.928	0.882
	SVR	C: 10, gamma: scale, kernel: linear	0.858	0.797	7.033	0.894
	LASSO	alpha: 0.01	0.852	0.783	7.535	0.886
	Cubist	n_committees: 10, neighbors: 8	0.845	0.823	6.124	0.908
pH	PLSR	n_components: 10	0.762	0.736	0.146	0.871
	SVR	C: 0.5, gamma: scale, kernel: linear	0.787	0.77	0.127	0.881
	LASSO	alpha: 0.01	0.686	0.676	0.179	0.825
	Cubist	n_committees: 10, neighbors: 9	0.77	0.776	0.124	0.881
Mg	PLSR	n_components: 5	0.577	0.675	1.173	0.828
	SVR	C: 5, gamma: scale, kernel: linear	0.684	0.81	0.684	0.914
	LASSO	alpha: 0.02	0.623	0.739	0.94	0.865
	Cubist	n_committees: 10, neighbors: 8	0.512	0.167	3.002	0.682
MO	PLSR	n_components: 10	0.439	0.567	0.409	0.759
	SVR	C: 0.5, gamma: scale, kernel: linear	0.49	0.6	0.378	0.778
	LASSO	alpha: 0.01	0.379	0.506	0.467	0.725
	Cubist	n_committees: 10, neighbors: 9	0.385	0.529	0.446	0.727
Fe	PLSR	n_components: 10	0.439	0.358	6888.981	0.691
	SVR	C: 10, gamma: scale, kernel: linear	0.355	0.413	6294.152	0.655
	LASSO	alpha: 0.2	0.486	0.368	6783.042	0.688
	Cubist	n_committees: 10, neighbors: 2	0.356	0.494	5433.487	0.743
SNQI	PLSR	n_components: 10	0.415	0.292	0.013	0.59
	SVR	C: 0.05, gamma: scale, kernel: linear	0.415	0.333	0.012	0.594
	LASSO	alpha: 0.01	0.122	0.148	0.015	0.57
	Cubist	n_committees: 10, neighbors: 7	0.368	0.324	0.012	0.587

Moreover, the SNQI was successfully estimated using the SVR model, achieving an R^2 of 0.583, which was higher than the values obtained for the most difficult individual properties to predict. This result demonstrates the feasibility of estimating a composite index from VIS-NIR spectral data.

The correlation between the estimated values and laboratory measurements on the test set indicates moderate to strong associations. In Nariño, the highest correlations ranged from 0.763 to 0.956; in Norte de Santander, from 0.725 to 0.887; and in the combined dataset, from 0.743 to 0.924. These results indicate that the estimated values for all soil properties were correlated with the corresponding measured values.

The performance of the model varied according to the properties of the soil and the region. In the Nariño dataset, the Cubist and LASSO models achieved the best results, while the SVR outperformed other models in the Norte de Santander dataset. For the combined dataset, Cubist and SVR demonstrated the highest predictive performance. Despite regional variability, the overall performance patterns remained consistent across models, with similar trends in the most accurately predicted soil properties. Notably, Partial Least Squares Regression (PLSR), one of the most commonly used regressors in the literature, did not rank as the best-performing model in any scenario, highlighting the importance of evaluating alternative regression approaches.

4 Discussion

Regional differences in model performance reflect soil compositional heterogeneity across the study areas. The volcanic soils of Nariño exhibit more consistent spectral-chemical relationships due to their relatively uniform pedogenic origin, whereas the alluvial deposits of Norte de Santander's mid-Magdalena valley introduce greater structural and mineralogical variability that complicates spectroscopic prediction.

When analyzing model performance by region, significant differences emerged based on the dataset used. In Nariño, the LASSO and Cubist models consistently outperformed others, likely due to specific characteristics of the local soil samples that enabled these algorithms to more effectively capture variable relationships. In contrast, in Norte de Santander, the SVR and Cubist models showed better performance. These results underscore the importance of considering regional soil characteristics when selecting and applying predictive models. Adapting modeling approaches to the specific conditions of each region not only enhances prediction accuracy but also contributes to a deeper understanding of soil dynamics across diverse geographic contexts.

A key finding of this study is that the diverse distribution of soil samples in Nariño was particularly beneficial for modeling properties such as cation exchange capacity (CEC), calcium (Ca), and magnesium (Mg). This diversity provided a wide range of low, medium, and high values for each property, resulting in a more comprehensive representation of soil conditions and improved model performance. A total of 138 samples were collected in Nariño and 284 in Norte de Santander, reflecting a 55.65% greater sample size in the latter. The hypothesis was confirmed through the coefficient of determination (R^2) values, calculated for each region and for the combined dataset. For CEC, the R^2 was 0.907 in Nariño, 0.593 in Norte de Santander, and 0.823 in the combined set. For Ca, the R^2 values were 0.912, 0.761, and 0.831, respectively. For Mg, the R^2 was 0.829 in Nariño, 0.457 in Norte de Santander, and 0.81 when combining both datasets.

Iron (Fe) was the most challenging nutrients to estimate using Vis-NIR spectroscopy. This difficulty arises from multiple factors. First, iron exhibits variable spectroscopic behavior depending on regional lithology and climate. Second, in the cacao soils of both Nariño and Norte de Santander, Fe predominantly occurs as oxides such as hematite and goethite, which produce weak and poorly defined spectral features in the visible and near-infrared regions [3]. Furthermore, spectral interference from organic matter, clays, and quartz diminishes sensitivity and limits the models' ability to detect subtle variations in Fe content [11,24]. The challenge of predicting Fe is further compounded by the substantially broader concentration ranges observed in Norte de Santander (24.54–778.0 mg/kg) compared to Nariño (5.0–209.94 mg/kg; Table S1), which contributes to higher absolute MSE values in the former region. Consequently, iron predictions using Vis-NIR spectroscopy alone remain suboptimal. Consistent with previous research, we recommend complementing optical spectroscopy with techniques such as X-ray fluorescence (XRF), which enables direct elemental quantification and circumvents the limitations of the optical spectrum [10]. These findings underscore the importance of employing scale-normalized performance metrics (R^2, ρ) when comparing models across regions with markedly different property distributions.

Regression techniques were employed to estimate the Soil Nutrient Quality Index (SNQI) [4] from Vis-NIR spectral data, with SVR yielding the best performance across all three datasets. For Nariño, the optimal model achieved $R^2 = 0.583$ and $\rho = 0.766$, while Norte de Santander showed $R^2 = 0.421$ and $\rho = 0.67$. Unexpectedly, the combined dataset produced inferior results ($R^2 = 0.333$, $\rho = 0.59$), indicating that merging regional datasets does not improve predictive accuracy for this index. Although R^2 values for SNQI were lower than those obtained for the best-predicted individual soil properties, the index warrants further investigation due to its potential as a more intuitive and actionable metric for farmers. This composite indicator may offer greater practical value for sustainable soil management decision-making despite its reduced prediction accuracy.

Although applications of Vis-NIR spectroscopy in cacao systems remain limited, a few studies demonstrate its potential. Singh et al. [19] developed a spectral library from 507 soils collected from smallholder cacao farms in Papua New Guinea and showed that near-infrared diffuse reflectance spectroscopy can accurately estimate most chemical properties, achieving concordance correlation coefficients > 0.8 for soil organic carbon, Mg, DTPA-Fe, and exchangeable Ca. More recently, Silue et al. [18] applied Vis-NIR models to assess soil organic carbon in 20-year-old cacao agroforestry systems in Côte d'Ivoire, revealing that shade tree species strongly influence SOC stocks. Their NIRS-based predictions of SOC achieved an R^2 of 0.79 in the validation set. Collectively, these studies, together with our findings, confirm the feasibility of Vis-NIR spectroscopy for characterizing cacao soils while highlighting the continued under-representation of such systems in global spectral libraries.

Additional Insights and Practical Implications

Support-Vector Regression with a linear kernel and the rule-based Cubist obtained the highest scores in more than 60 % of the region–property combinations. SVR benefits from a relatively small hyper-parameter space that limits over-fitting in high-dimensional spectra, while Cubist partitions the spectral manifold and captures local interactions—behaviour that aligns with recent ensemble studies on geoscience data [20]. PLSR, although introduced several decades ago, achieved competitive results for Ca and CEC under strong multicollinearity and still provides interpretable loadings linked to specific chemical vibrations. LASSO mainly acted as a wavelength selector, revealing sparse sets of bands that agree with earlier laboratory analyses and that can guide the design of simpler field instruments.

Accurate predictions of Ca and CEC allow rapid lime-requirement assessments, while reliable pH and Mg estimates support balanced fertilization in Andean cacao fields. Even the composite Soil Nutrient Quality Index (SNQI) reached R^2 values up to 0.583, showing that a single high-dimensional spectrum can provide enough information for on-site decision support. These findings are relevant to soft-computing techniques for field-deployable sensing.

5 Conclusions

In conclusion, vis-NIR spectroscopy combined with machine learning methods has proven to be a valuable tool for diagnosing the Soil Nutritional Quality Index and key chemical properties in Colombia. The results support informed decision-making by farmers regarding crop management and associated practices. A quality index ranging from 0 to 1 facilitates the identification of optimal conditions for cacao cultivation, where a higher index indicates better soil conditions. Additionally, accurate estimation of chemical properties enables more appropriate and tailored fertilization recommendations.

Vis–NIR combined with machine learning explained between 71% and 91 % of the variance in Ca, CEC, Mg and pH, and achieved an R^2 of 0.583 for the SNQI—this is, to our knowledge, the first report of this index derived from NIR data for Andean cacao soils. This approach drastically reduces diagnostic turnaround compared to conventional laboratory analysis and can translate into significant cost savings per sample. Moreover, the consolidated model trained on all regions outperformed region-specific calibrations for Mg and OM, demonstrating effective transferability when spectral heterogeneity increases. A portable implementation of this algorithm within handheld spectrometers can empower cacao technicians to make precise, sustainable fertilization decisions in the field.

Limitations remain: spectral saturation at high Fe or OM concentrations, instrument-specific calibration transfer, and moisture-related variability under field conditions, all of which limited the prediction of SNQI. Future work will address these constraints through domain-invariant feature learning, hybrid sensor fusion, and targeted sampling of underrepresented concentration ranges—particularly low-Mg and high-Fe soils in Norte de Santander. Tackling these challenges will enable more robust and scalable Vis–NIR and machine learning solutions for field-deployable soil monitoring, aligned with the evolving requirements of pattern recognition applications

Acknowledgments. The authors gratefully acknowledge the Corporación Colombiana de Investigación Agropecuaria (AGROSAVIA) for providing the infrastructure and institutional support necessary to carry out this research. Funding was provided by the Environmental Fund of Portugal through the project "Improving Environmental Sustainability through the Implementation of Technologies within the Framework of the National Strategy for Cocoa Cultivation" (Agreement 2035—AGROSAVIA

1002031), as well as by Fundación Ayuda en Acción for the project "Development of a Low-Cadmium Cocoa Production Model for the Nariño Mountain Range Sub-region".

References

1. Benedet, L., et al.: Soil texture prediction using portable x-ray fluorescence spectrometry and visible near-infrared diffuse reflectance spectroscopy. Geoderma **376**, 114553 (2020)
2. Castillo-Valdez, X., Etchevers-Barra, J., Hidalgo-Moreno, C., Aguirre, A.: Evaluación de la calidad de suelo: generación e interpretación de indicadores. Rev. Terra Latinoamericana **39** (2021). https://doi.org/10.28940/terra.v39i0.698
3. Cornell, R.M., Schwertmann, U., et al.: The Iron Oxides: Structure, Properties, Reactions, Occurrences, and Uses, vol. 664. Wiley-vch Weinheim (2003)
4. Cortés Delgadillo, D.L., Lerma Lasso, J.L., López Rendon, J.F., Meneses Buitrago, D.H., Martínez Pachón, E.: Índice de calidad nutricional en suelos cacaoteros de la subregión cordillera de nariño. Siembra **11**(1), e6027 (2024). https://doi.org/10.29166/siembra.v11i1.6027, https://revistadigital.uce.edu.ec/index.php/SIEMBRA/article/view/6027
5. Delgadillo-Duran, D.A., Vargas-García, C.A., Varón-Ramírez, V.M., Calderón, F., Montenegro, A.C., Reyes-Herrera, P.H.: Vis-NIR spectroscopy and machine learning methods to diagnose chemical properties in Colombian sugarcane soils. Geoderma Reg. **31**, e00588 (2022). https://doi.org/10.1016/j.geodrs.2022.e00588
6. Ewing, P.M., TerAvest, D., Tu, X., Snapp, S.S.: Accessible, affordable, fine-scale estimates of soil carbon for sustainable management in sub-Saharan Africa. Soil Sci. Soc. Am. J. **85**(5), 1814–1826 (2021). https://doi.org/10.1002/saj2.20263, https://acsess.onlinelibrary.wiley.com/doi/abs/10.1002/saj2.20263
7. Gallios, G., Tsakiridis, N., Tziolas, N.: Federated learning applications in soil spectroscopy. Geoderma **456**, 117259 (2025). https://doi.org/10.1016/j.geoderma.2025.117259
8. Jiang, C., Zhao, J., Li, G.: Integration of Vis–NIR spectroscopy and machine learning techniques to predict eight soil parameters in alpine regions. Agronomy **13**(11) (2023). https://doi.org/10.3390/agronomy13112816
9. Karlen, D.L., Stott, D.E.: A Framework for Evaluating Physical and Chemical Indicators of Soil Quality, chap. 4, pp. 53–72. Wiley (1994). https://doi.org/10.2136/sssaspecpub35.c4, https://acsess.onlinelibrary.wiley.com/doi/abs/10.2136/sssaspecpub35.c4
10. Kirsten, M., et al.: Iron oxides and aluminous clays selectively control soil carbon storage and stability in the humid tropics. Sci. Rep. **11**(1), 5076 (2021)
11. Krivoshein, P.K., Volkov, D.S., Rogova, O.B., Proskurnin, M.A.: FTIR photoacoustic and ATR spectroscopies of soils with aggregate size fractionation by dry sieving. ACS Omega **7**(2), 2177–2197 (2022)
12. Meneses Buitrago, D.H., Lerma Lasso, J.L., Guerrero Díaz, G.F., Cortés Delgadillo, D.L., Martínez Pachón, E., Montenegro, A.C.: Estado del cultivo de cacao en la cordillera nariñense con énfasis en la concentración de cadmio (2023)
13. Mora-Herrera, D.Y., Guillaume, S., Snoeck, D., Zúñiga Escobar, O.: A fuzzy logic based soil chemical quality index for cacao. Comput. Electron. Agric. **177**, 1–30 (2020). https://doi.org/10.1016/j.compag.2020.105624

14. Nanni, M.R., Demattê, J.A.M.: Spectral reflectance methodology in comparison to traditional soil analysis. Soil Sci. Soc. Am. J. **70**(2), 393–407 (2006). https://doi.org/10.2136/sssaj2003.0285
15. Ng, W., Minasny, B., Mendes, W.D.S., Demattê, J.A.M.: The influence of training sample size on the accuracy of deep learning models for the prediction of soil properties with near-infrared spectroscopy data. SOIL **6**(2), 565–578 (2020). https://doi.org/10.5194/soil-6-565-2020, https://soil.copernicus.org/articles/6/565/2020/
16. Poppiel, R.R., Cherubin, M.R., Novais, J.J.M., Demattê, J.A.M.: Soil health in Latin America and the Caribbean. Commun. Earth Environ. **6**, 141 (2025). https://doi.org/10.1038/s43247-025-02021-w
17. Serrano, J., et al.: Evaluation of near infrared spectroscopy (NIRS) for estimating soil organic matter and phosphorus in Mediterranean Montado ecosystem. Sustainability **13**(5) (2021). https://doi.org/10.3390/su13052734, https://www.mdpi.com/2071-1050/13/5/2734
18. Silue, B.K., Koné, A.W., Masse, D., Moulin-Esmard, P., Kotaix, A.J., Chapuis-Lardy, L.: Contrasted effects of shade tree legumes on soil organic carbon stock and carbon balance in 20-year cacao agroforestry, ivory coast. Geoderma Reg. **37**, e00807 (2024). https://doi.org/10.1016/j.geodrs.2024.e00807, https://www.sciencedirect.com/science/article/pii/S2352009424000543
19. Singh, K., et al.: Near infrared diffuse reflectance spectroscopy for rapid and comprehensive soil condition assessment in smallholder cacao farming systems of Papua New Guinea. CATENA **183**, 104185 (2019). https://doi.org/10.1016/j.catena.2019.104185, https://www.sciencedirect.com/science/article/pii/S0341816219303273
20. Tavakoli, H., Correa, J., Sabetizade, M., Vogel, S.: Predicting key soil properties from Vis-NIR spectra by applying dual-wavelength indices transformations and stacking machine learning approaches. Soil Tillage Res. **229**, 105684 (2023). https://doi.org/10.1016/j.still.2023.105684
21. Tavakoli, H., Correa Reyes, J., Sabetizade, M., Vogel, S.: Predicting key soil properties from Vis-NIR spectra using dual-wavelength indices and stacked machine learning. Soil Tillage Res. **229**, 105684 (2023). https://doi.org/10.1016/j.still.2023.105684
22. Tavares, T.R., Molin, J.P., Javadi, S.H., Carvalho, H.W.P.D., Mouazen, A.M.: Combined use of Vis-NIR and XRF sensors for tropical soil fertility analysis: assessing different data fusion approaches. Sensors **21**(1) (2021). https://doi.org/10.3390/s21010148
23. Vullaganti, N., Ram, B.G., Sun, X.: Precision agriculture technologies for soil site-specific nutrient management: a comprehensive review. Artif. Intell. Agric. (2025)
24. Wight, J.P., Ashworth, A.J., Allen, F.L.: Organic substrate, clay type, texture, and water influence on NIR carbon measurements. Geoderma **261**, 36–43 (2016)
25. Zhang, T., Li, Y., Wang, M.: Prediction of soil organic carbon and total nitrogen in calcareous mine soils using Vis-NIR spectroscopy and machine learning. Sci. Rep. **14**, 28014 (2024). https://doi.org/10.1038/s41598-024-73761-6
26. Zhou, P., et al.: Development of a mobile online detector for soil total nitrogen based on visible and short-wave near-infrared spectroscopy. Measurement **248**, 116933 (2025). https://doi.org/10.1016/j.measurement.2025.116933, https://www.sciencedirect.com/science/article/pii/S0263224125002921

Classifying Nutritional Deficiencies in Coffee Leaves with Transfer Learning

Sarath Mukundan Adavakkat[1], Suhara Shahul[1], Deisy Chaves[2](✉), and Claudia Mazo[1]

[1] School of Computing, Dublin City University, Dublin, Ireland
{sarath.mukundanadavakkat2,suhara.shahulhameed2,claudia.mazo}@dcu.ie
[2] School of Systems Engineering and Computing, Universidad del Valle, Cali, Colombia
deisy.chaves@correounivalle.edu.co

Abstract. Coffee plays a crucial role globally due to their cultural and economic significance. It serves as a vital agricultural commodity, sustaining millions of farmers, especially in underdeveloped nations, and contributing significantly to various countries' economies, playing a pivotal role in global trade. Coffee plants require vital elements such as nitrogen, phosphorus, potassium, magnesium, iron, and zinc for proper growth and bean formation. Nutrient deficiencies in crops can result in reduced yields, poor quality, and financial losses for farmers. The ability to discover these problems quickly and reliably using automated analysis represents a proactive approach to agricultural management. In this study is proposed a method for automatic detection of four nutritional deficiencies—potassium (K), boron (B), calcium (Ca) and iron (Fe)—in coffee leaf images using transfer learning techniques—MobileNetV2, InceptionV3, VGG19, EfficientNetV2 and ResNet50—and data augmentation strategies. The robustness and generalization capability of the proposed method was evaluated using the CoLeaf data set V2.0 with five scenarios: (E1) raw unbalanced dataset without data augmentation; (E2) data augmentation but preserving unequal class distribution; (E3) balanced dataset with oversampling; (E4) balanced dataset with undersampling; (E5) balanced dataset using data augmentation. Results shown that ResNet50 on E5 was the high-performing model with F1 scores of 0.96 (B), 0.93 (Ca), 0.98 (Fe) and 0.98 (K), which demonstrate its general efficiency and dependability to identify the four nutritional deficiencies when data augmentation and oversampling are used.

Keywords: Coffee Leaves · Nutritional Deficiencies Classification · Data Augmentation · Transfer Learning · CNNs

1 Introduction

Coffee plants play a crucial role worldwide, not only as source of a popular beverage but also because of their cultural and economic importance. For the

D. Chaves et al. (Eds.): CIARP 2025, LNCS 16528, pp. 75–89, 2026.
https://doi.org/10.1007/978-3-032-23161-1_6

2024/25 coffee year, global production is projected to reach 174.395 million 60-kg bags, representing an increase of about 3% compared with the 169.345 million coffee bags recorded in 2023/24 [2]. Besides, in countries with a large coffee production, such as Peru—consistently ranked among the world's top ten coffee producers—output is projected to reach 4.2 million 60-kg coffee bags in 2025/26, reflecting an approximate 8% rebound from the previous year [11]. These figures highlight the considerable scale of the coffee sector at both global and national levels, where robust export volumes strengthen the country's trade balance, while coffee cultivation acts as a crucial driver of rural income generation [4].

However, sustaining healthy growth of coffee plants requires an adequate supply of essential nutrients such as nitrogen, phosphorus, potassium, magnesium, iron, calcium, sulphur, zinc, and boron [10,19,21], and deficiencies in these nutrients are known to reduce crop productivity. Nutritional lack in coffee plants manifest in various forms, including leaf discolourations, stunted growth, limited fruiting, and twisted leaves and its manual detection requires pay close attention to leaf symptoms such as yellowing, shape deformation, discoloration, as well as abnormal growth patterns. This is a challenging and laborious task for farmers daily as its a time-consuming process and corrective actions can be required. Therefore, early identification of nutrient deficiencies gives farmers the foresight they need to make informed decisions, enhancing effective resource management and increasing overall output in coffee cultivation.

Recent studies on the classification of coffee leaf diseases, including nutritional deficiencies, have used the CoLeaf-DB dataset [19], which has evolved from a limited set of 335 leaf images in version V1.0 [10] to a more diverse collection of 1006 images in version V2.0. Early work on nutritional deficiency detection employed machine learning methods such as Random Forest [10] and 1-NN classifier [21], achieving modest accuracies (up to 67.5%), while recent approaches using deep Convolutional Neural Networks (CNN), like ResNet50, on CoLeaf V2.0 reported improved accuracy (87.75%) [20]. In parallel, transfer learning has shown high potential in classifying coffee leaf diseases, in particular, in the BRACOL dataset [7], where CNN models using InceptionV3 and MobileNetV2 achieved accuracies of 97.61% and 98.51%, respectively [3,8]. Broader studies on plant diseases in other crops confirm the effectiveness of transfer learning with models such as ResNet50, VGG19, and DenseNet121, especially when applied to large and diverse datasets [1,9,13,16], showing the potential of deep learning models for its application in the classification of coffee diseases.

In this work, we proposed a strategy for automatic classification of nutritional deficiencies on coffee leaves using transfer learning techniques and data augmentation on the CoLeaf dataset [19]. In particular, we focus on the detection of the most relevant nutritional deficiencies that coffee plants could have [21]: potassium, boron, calcium, and iron. This approach enhances the model's capacity to detect coffee nutritional deficiencies, thereby ensuring reliable results.

2 Related Works

2.1 Classification of Coffee Nutritional Deficiencies

Few studies in the literature using the publicly available CoLeaf-DB dataset [19] for the classification of nutritional deficiencies [10,20,21]. Initially, the first version (V1.0) of the CoLeaf-DB dataset consisted of 335 images across eight nutritional deficiency classes [10]. The image distribution per class was: boron (53), nitrogen (49), calcium (52), iron (36), magnesium (30), manganese (33), phosphorus (53), and potassium (29). In 2023, the dataset was updated to version V2.0, increasing to 1006 images comprising: healthy leaf (6), nitrogen (64), phosphorus (246), potassium (96), magnesium (79), boron (101), manganese (83), calcium (162), iron (65), and more than one deficiency (104) [19]. The limited size of V1.0 posed challenges for classification models generalization, but the update in V2.0 aimed to address these limitations by enhancing dataset size and diversity.

In [10], a machine learning approach using decision trees and Random Forest was applied to classify nutritional deficiencies, using local and global features extracted from leaves. Pre-processing involved Otsu thresholding, median filtering, and resizing. Local features included SIFT and Bag-of-Features with color descriptors, while global features comprised basic and shape measurements. Four Random Forest models were trained on the CoLeaf V1.0 dataset using different feature combinations, with global features achieving the highest accuracy of 67.5%. Despite reasonable accuracy, challenges such as poor vein representation and necrosis detection at the leaf tip were noted.

In [21], digital image processing and supervised classification algorithms were applied to a subset of CoLeaf V1.0 (269 images from four nutrients: Boron, Iron, Potassium, Calcium). Features were extracted using Blurred Shape Model (BSM) and Gray Level Co-occurrence Matrix (GLCM). Classifiers such as 1-NN, Naïve Bayes, and Neural Networks were evaluated. The Naïve Bayes classifier with BSM features achieved the best overall accuracy (65.05%), though accuracy across models remained modest. Each descriptor-classifier combination showed varied performance per nutrient, highlighting the importance of feature selection.

In [20], a deep CNN (ResNet50) was employed on CoLeaf V2.0 to detect nutritional deficiencies using shape and color features, achieving an overall accuracy of 87.75%. However, the study did not report class-specific metrics, limiting insights into model performance for individual deficiencies.

2.2 Transfer Learning for Classifying Plant Diseases

Few studies have used transfer learning for coffee disease classification [3,8,12]. In [8], a CNN model based on InceptionV3 with transfer learning, data augmentation, and Mini-Batch Gradient Descent (MBGD) was applied to the BRACOL dataset [7], achieving an accuracy of 97.61%. The BRACOL dataset includes 1747 images categorized into five classes: healthy, leaf miner, leaf rust, phoma, and cercospora spots.

Similarly, in [3], a CNN using MobileNetV2 with transfer learning and data augmentation achieved an accuracy of 98.51% and a loss of 0.2482 on the BRACOL dataset. Despite high accuracy, the study emphasized the importance of considering additional metrics and context, such as class imbalance, to fully assess model performance.

In [12], a deep learning model was developed for identifying and categorizing coffee leaf disease using 1120 images collected from Wolaita Sodo and Kaggle datasets [6]. Data augmentation, such as rotation and flipping were used to address data imbalance and overfitting, increased the dataset to 3360 images across four classes: healthy, rust, wilt, and brown eye spot. Both CNN trained from scratch and transfer learning models (ResNet50 and MobileNet) were compared. ResNet50 achieved the highest accuracy of 99.89%, confirming the effectiveness of transfer learning in plant disease classification.

Additionally, several studies have explored transfer learning for plant diseases in other crops. In [16], InceptionV3 achieved 92% accuracy on tomato diseases. In [9], ResNet50 showed strong performance on tomato and pepper bell disease. In [13], VGG19 achieved the highest accuracy among tested models on a plant disease dataset. In [1], DenseNet121 outperformed other models using a combined dataset of 54,000 images, illustrating the potential of dataset pooling and pre-trained models for robust plant disease classification.

3 Proposed Approach

Figure 1 depicts the workflow of the proposed strategy to classify nutrient deficiencies in coffee leaves using data augmentation and transfer learning techniques with CNN models. The strategy is a five-stage forward: 1) Creation of a training dataset composed by coffee leaf images of four nutrient deficiencies: boron, potassium, calcium, and iron. 2) Data augmentation applying methods such as rotation, scaling, and shearing to add diversity to the collection. 3) Transfer Learning model for effective training of classifiers of nutritional deficits. 4) Evaluation of the trained model using an independent testing dataset. 5) Classification of leaves identifying lacking nutrients. This methodology improves the model's ability to recognize and categorize nutritional deficits in coffee leaves, guaranteeing accurate outcomes.

3.1 Dataset of Coffee Nutritional Deficiencies

The publicly available CoLeaf dataset V2.0 [19] was choosen for training and testing our classification models. CoLeaf dataset V2.0 includes 1006 images of the leaves of coffee plants that have been grouped according to different nutritional deficits, including 10 classes: healthy leaf (6), nitrogen (64), phosphorus (246), potassium (96), magnesium (79), boron (101), manganese (83), calcium (162), iron (65) and images with more than one deficiency (104). The images in the dataset have a size of 4000×3000 pixels and are stored in JPEG format. In this work, we focused on the classification of the four most relevant nutritional

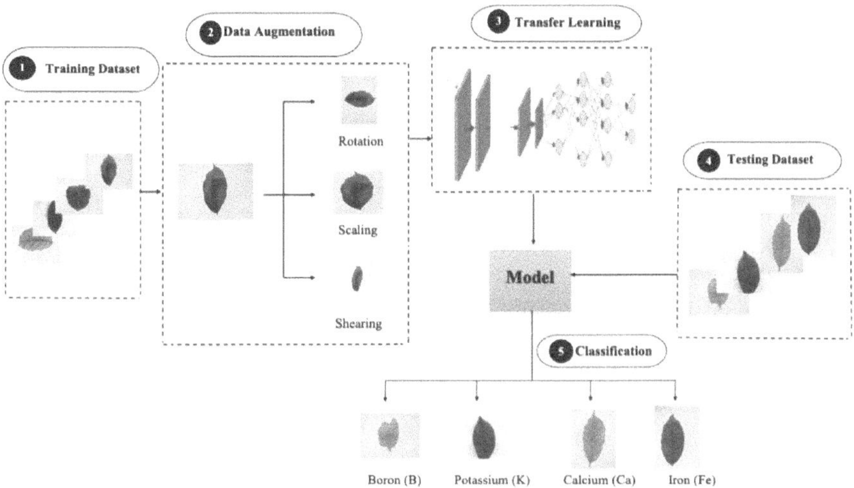

Fig. 1. Proposed approach for classifying nutritional deficiencies in coffee leaf images.

deficiencies that could affect coffee [21] (see Table 1): potassium (K), boron (B), calcium (Ca), and iron (Fe).

Table 1. Nutritional deficiencies and symptoms, adapted from [19,21].

Nutrition	Symptoms
Potassium (K)	Older leaves are impacted first; chlorosis along the leaf margins and tips, leading to necrosis
Boron (B)	Stunted development, interveinal chlorosis, necrosis of the leaf tips, and thick, brittle, malformed young leaves that exhibit the "brittle leaf" sign
Calcium (Ca)	Young leaves and growing tips are mainly impacted; new leaves are twisted, hooked, or have unusual forms. Leaf tips may show necrosis
Iron (Fe)	In young leaves, interveinal chlorosis causes the veins to stay green while turning the surrounding tissue yellow. Severe cases can result in total leaf yellowing and decreased growth

3.2 Data Augmentation

Data augmentation is performed on the nutritional deficiencies dataset to increase the number of coffee leaf images. This allows the models to use additional training data, particularly for uncommon conditions where source data variances are absent. Hence, our classifiers can acquire strong features and perform better in identifying and categorizing coffee nutritional deficits.

In particular, three methods to improve the dataset's diversity and resilience were used [3]: rotation, scaling and shearing. Rotation is applied to the leaf images to replicate orientations frequently found in real-world situations. This ensures that our model is effective from various viewpoints and makes it invariant to changes in leaf orientation. Moreover, scaling is performed, which entails resizing the leaf images to various sizes. This enables us to replicate differences in leaf size, which can improve the model's ability to generalize to different-sized leaves. Furthermore, shearing is carried out, which is a method that moves the pixels along one axis while maintaining the other axis stationary, thereby distorting the leaf images. Shearing can improve the model's capacity to identify and categorize leaves under various deformations by introducing geometric distortions akin to those brought on by elements like wind or physical damage to leaves.

3.3 Transfer Learning

CNN architectures combined with transfer learning were employed to build models for identifying and categorizing nutritional deficiencies of K, B, Ca, and Fe, which are among the most critical in coffee leaves [21]. Although CNNs have achieved state-of-the-art performance across various computer vision tasks, including the classification of plant diseases on images [3,16], their effective training from scratch requires large volumes of labeled data. To address this limitation, transfer learning has become a widely adopted approach, allowing the adaptation of models pre-trained on large-scale datasets to specific domain applications.

In this work, five well-known CNN's architectures were fine-tuned to classify the coffee nutritional deficiencies: MobileNetV2 [14], InceptionV3 [17], VGG19 [15], EfficientNetV2 [18], and ResNet50 [5]. These networks have been chosen for their proven effectiveness in image classification tasks, as well as the public availability of their pre-trained weights obtained from training on large-scale natural image datasets.

4 Experimental Set-up

We assessed the performance of the proposed classifier of coffee nutritional deficiencies using five experimental cases that allow us to analyze the effect of data augmentation with/without balancing the number of deficiency's samples on the dataset. Table 2 summarizes the experimental cases evaluated. Besides, Table 3 shows the number of coffee leaf images used by evaluation case.

In the first experiment (E1), the dataset used to build models is the original one without any data augmentation applied. This means that the dataset remains in its raw form as initially collected. Consequently, the class distribution in the dataset is imbalanced. The second experiment (E2) involves using the original dataset with the data augmentation techniques applied: rotation, shearing, and rescaling. All coffee leaf images were resized into 224×224 pixels. Despite the augmentation in this experimental case, the class distribution remains imbalanced. In the third experiment (E3), the original dataset is balanced through the process of oversampling. Oversampling involves duplicating samples from the minority class to ensure that all classes have a similar number of samples. Balancing the dataset in this way helps mitigate the bias toward the majority class, promoting fairer training across all classes. The fourth experiment (E4) also aims to balance the dataset, but it uses undersampling. Undersampling reduces the number of samples in the majority class to match the minority class, resulting in a balanced dataset. In the fifth experiment (E5), oversampling with data augmentation is applied to achieve a balanced dataset to ensure that each class has an equal number of samples, leading to a balanced class distribution.

Table 2. Summary of evaluated experimental cases.

Experimental case	Balanced/Imbalanced	Data augmentation
E1: Original Dataset	Imbalanced	No
E2: Original Dataset with Data Augmentation	Imbalanced	Yes
E3: Oversampling Original Dataset	Balanced	No
E4: Undersampling Original Dataset	Balanced	No
E5: Oversampling Data Augmented Dataset	Balanced	Yes

Table 3. Distribution of nutritional deficiencies on each evaluation case's dataset

Experimental case	Total leaf images	K	B	Ca	Fe
E1: Original Dataset	424	96	101	162	65
E2: Original Dataset with Data Augmentation	1275	288	306	486	195
E3: Oversampling Original Dataset	648	162	162	162	162
E4: Undersampling Original Dataset	260	65	65	65	65
E5: Oversampling Data Augmented Dataset	1944	486	486	486	486

Moreover, the transfer learning models fine-tuned for classifying coffee nutritional deficiencies were initialized with weights pre-trained on the ImageNet dataset, which contains over 14 million high-resolution images categorized into thousands of classes following the WordNet hierarchy. During training, the convolutional layers of the base models were kept frozen to preserve their learned feature representations, while the fully connected layers were modified to address the target task by adjusting the output vector to four classes (i.e., K, B, Ca, and Fe). After these modifications, model training was conducted using a 70:30 ratio for training and testing, respectively. Table 4 summarizes the implementation details of the five CNN architectures analyzed in this work: MobileNetV2 [14], InceptionV3 [17], VGG19 [15], EfficientNetV2 [18], and ResNet50 [5].

Table 4. Details of CNN architectures used for transfer learning in this study.

Item	MobileNetV2	InceptionV3	VGG19	EfficientNetV2	ResNet50
General					
Parameters	32.4M	23.3M	20.4M	7.9M	26.7M
Channels	3	3	3	3	3
Input size	224×224	299 × 299	224 × 224	224 × 224	224 × 224
Number of layers					
Convolutional	53	94	16	55	53
Max pooling	1	4	5	1	1
Fully connected	1	1	3	2	1
Presence of modules					
Batch normalization	yes	yes	no	yes	yes
Residual connections	no	no	no	no	yes

Furthermore, four metrics (i.e., Accuracy, F1-score, Precision, and Recall) obtained from the confusion matrix were used to quantify the performance of built models in classifying the lack of K, B, Ca, and Fe on coffee leaves. These measures are defined mathematically as follows (Eq. 1–4):

$$\text{Accuracy} = \frac{TP + TN}{TP + TN + FP + FN}, \tag{1}$$

$$\text{Precision} = \frac{TP}{TP + FP}, \tag{2}$$

$$\text{Recall} = \frac{TP}{TP + FN}, \tag{3}$$

$$\text{F1-Score} = 2 \cdot \frac{\text{Precision} \cdot \text{Recall}}{\text{Precision} + \text{Recall}}, \tag{4}$$

where TP is True Positive, FP is False positive, TN is True Negative, and FN is False Negative.

5 Results and Discussion

We analyzed the performance of the proposed classifier of coffee nutritional deficiencies in three stages: (i) comparison of the overall performance of the five fine-tuned CNNs across the five evaluation scenarios (see Sect. 4) with and without dataset balancing through data augmentation, (ii) comparison of the best-performing classifiers for each specific nutritional deficiency, (iii) comparison of the obtained results against the state-of-the-art approaches for classifying nutritional deficiencies in coffee leaves.

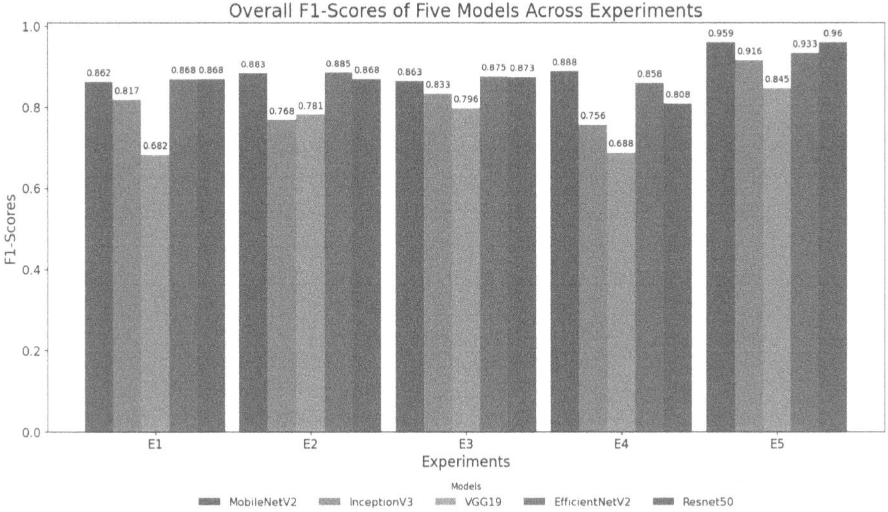

Fig. 2. Overall F1-Score of classification models per evaluation case

5.1 Comparison of Classification Models Across the Evaluation Scenarios

Table 5 presents the accuracy and Fig. 2 shows the F1-score of the trained classification models using transfer learning (MobileNetV2, InceptionV3, VGG19, EfficientNetV2, and Resnet50) with the five experimental cases, see Sect. 4.

Table 5. Overall accuracy of the built classification models across the five evaluation cases. The best accuracy values per model are displayed in bold

Model	E1	E2	E3	E4	E5
MobileNetV2	0.862	0.882	0.862	0.888	**0.959**
InceptionV3	0.815	0.772	0.832	0.762	**0.916**
VGG19	0.692	0.780	0.796	0.688	**0.842**
EfficientNetV2	0.869	0.885	0.872	0.862	**0.933**
Resnet50	0.869	0.869	0.872	0.812	**0.961**

Among all the evaluated scenarios, the best performance is achieved on E5, as evidenced by the overall accuracies in Table 5 and the F1-scores in Fig. 2 reported by CNN model. In particular, MobileNetV2, InceptionV3, VGG19, EfficientNetV2, and ResNet50 obtained overall accuracies of 0.959, 0.916, 0.842, 0.933, and 0.961, respectively, with corresponding F1-scores of 0.959, 0.916, 0.845, 0.933, and 0.960. These consistently high performance values highlight the superior effectiveness of E5. The integration of data augmentation and oversampling techniques in this test scenario likely enhanced the models' generalization ability to discriminate the coffee nutritional deficiencies (K, B, Ca, and Fe), thereby improving the classifiers robustness across diverse datasets. In particular, ResNet50 has the best performance in terms of accuracy and F1-scores.

In contrast, E4 yielded the weakest performance due to its reliance on undersampling, which reduced the number of training samples. This limitation particularly affected the performance of InceptionV3 and VGG19 models, which achieved accuracy scores of 0.762 and 0.688, respectively. The F1-scores also illustrate this trend with values of 0.756 and 0.688, respectively. This confirm E4 as the least effective configuration among all evaluated cases.

5.2 Comparison of Classification Performance by Nutritional Deficiency

Table 6 reports the precision and the recall by nutritional deficiency (K, B, Ca, and, Fe) of the five classification models built in E5, the test scenario that yielded the best overall performance, see Sect. 5.1. In addition, Fig. 3 presents their corresponding F1-Scores.

Table 6. Precision (Prec.) and Recall (Rec.) of the classification models in E5 by nutritional deficiency. The best performance values are highlighted in bold

Class	MobileNetV2		InceptionV3		VGG19		EfficientNetV2		ResNet50	
	Prec.	Rec.	Prec.	Rec.	Prec.	Rec.	Prec.	Rec.	Prec.	Rec.
B	0.96	**0.95**	0.96	0.88	0.82	0.87	**0.98**	0.88	0.96	**0.95**
Ca	**0.98**	0.91	0.91	0.86	0.71	0.92	0.91	**0.94**	0.94	0.92
Fe	0.95	0.99	0.90	0.97	**0.98**	0.77	0.87	**1.00**	0.97	0.98
K	0.95	**0.99**	0.89	0.95	0.94	0.82	**0.99**	0.91	0.97	**0.99**
Overall	**0.96**	**0.96**	0.91	0.91	0.86	0.84	0.94	0.93	**0.96**	**0.96**

Results shown that for detecting B deficiency, EfficientNetV2 achieved the highest precision (0.98) with a recall of 0.88, while MobileNetV2 and ResNet50 slightly surpassed it in recall (0.95). In relation to Ca deficiency, MobileNetV2 obtained the best precision (0.98), whereas EfficientNetV2 yield the highest recall (0.94). Concerning Fe deficiency, VGG19 recorded the highest precision (0.98), while EfficientNetV2 obtained a perfect recall of 1.00. Finally, with respect to K deficiency, EfficientNetV2 achieved the best precision (0.99), whereas MobileNetV2 and ResNet50 matched the highest recall (0.99).

Besides, among all evaluated models, ResNet50 achieved the most consistent performance across the four coffee nutritional deficiencies. In terms of precision, it closely followed the top-performing models, trailing by 0.02 for B and K (relative to EfficientNetV2), 0.04 for Ca (relative to MobileNetV2), and 0.01

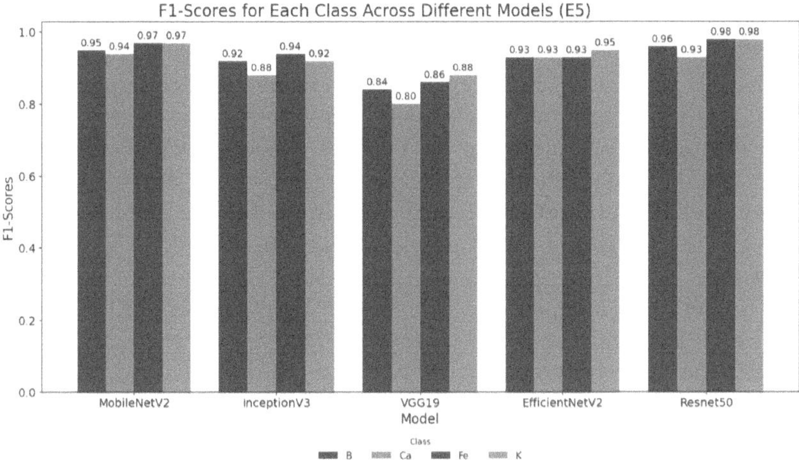

Fig. 3. F1-Score of the classification models in E5 by nutritional deficiency

for Fe (relative to VGG19). Likewise, in terms of recall, ResNet50 tied with MobileNetV2 for the highest values in detecting the lack of B (0.95) and K (0.99), and lagged EfficientNetV2 by 0.02 for the nutritional deficiencies of Ca and Fe. In particular, ResNet50 achieved precision scores of 0.96, 0.94, 0.97, and 0.97, and recall scores of 0.95, 0.92, 0.98, and 0.99 for B, Ca, Fe, and K, respectively. As shown in Fig. 3, its F1-scores were 0.96, 0.93, 0.98, and 0.98 for the same classes. Notably, deficiencies of Fe and K achieved the highest F1-scores (0.98), while Ca was the lowest at 0.93, though differences among classes remained relatively small.

In contrast, VGG19 was the least effective model, with substantially lower performance across all performance metrics. Its F1-scores were 0.84, 0.80, 0.86, and 0.88 for the deficiencies of B, Ca, Fe, and K, respectively. Precision scores were 0.82, 0.71, 0.98, and 0.94, while recall values were 0.87, 0.92, 0.77, and 0.82. These results confirm VGG19 as the weakest classifier among the five evaluated architectures.

5.3 Comparison of the Proposed Approach Against the State-of-Art

A limited number of studies in the literature [10,20,21] have used the publicly available CoLeaf-DB dataset [19]. In [21], an automatic approach for detecting nutritional deficiencies in coffee leaves was proposed using a subset of the CoLeaf dataset V1.0, consisting of 269 images distributed across four classes: B (69), Ca (74), Fe (70), K (54). Their method combined handcrafted descriptors, mainly the Blurred Shape Model (BSM) and the Gray-Level Co-occurrence Matrix (GLCM), with classical classifiers such as KNN, Naïve Bayes, and Neural Networks. The best results were obtained with the Naïve Bayes classifier using the BSM descriptor.

In contrast, our approach employed a subset of the larger CoLeaf dataset V2.0, with 424 images across the same four nutrient deficiencies: B (101), Ca (162), Fe (65), and K (96). However, instead of handcrafted descriptors and conventional classifiers, we apply transfer learning with state-of-the-art CNN architectures (MobileNetV2, InceptionV3, VGG19, EfficientNetV2, and ResNet50) together with data augmentation to balance the number of samples per class and increase model's generalization. In particular, we obtained the best results fine-tunning ResNet50 with an oversampling augmented dataset (E5), see Sect. 5.1 and Sect. 5.2.

Table 7 and Fig. 4 depict the comparison between our best classification model (ResNet50) and the proposed method (Naïve Bayes) from [21]. Results shown that ResNet50 outperformed Naïve Bayes across the detection of the four nutrient classes. In particular, ResNet50 obtained precision values of 0.96, 0.94, 0.97, and 0.97 for B, Ca, Fe, and K deficiencies, respectively, compared to 0.67, 0.64, 0.68, and 0.59 reported for Naïve Bayes [21]. Regarding to recall, score values improved from 0.69, 0.56, 0.70, and 0.64 with Naïve Bayes to 0.95, 0.92, 0.98, and 0.99 with ResNet50. Overall, ResNet50 reached precision, recall, and F1-score values of 0.96, representing improvements between 26% and 38% over Naïve Bayes (overall precision of 0.64, recall of 0.64 and F1-Score of 0,64). These findings demonstrates that our approach (ResNet50) outperforms Naïve Bayes in identifying nutritional deficits in coffee leaves, showing the advantage of transfer learning over traditional handcrafted feature-based approaches, particularly when combined with data augmentation to improve dataset variability.

Table 7. Precision (Prec.) and Recall (Rec.) comparison between state-of-art (Naïve Bayes [21]) and our model (ResNet50). The best performance values are displayed in bold

Class	Naïve Bayes		ResNet50 (our)	
	Prec.	Rec.	Prec.	Rec.
B	0.67	0.69	**0.96**	**0.95**
Ca	0.64	0.56	**0.94**	**0.92**
Fe	0.68	0.70	**0.97**	**0.98**
K	0.59	0.64	**0.97**	**0.99**
Overall	0.64	0.65	**0.96**	**0.96**

Other works have also explored the CoLeaf dataset [10,20]. In [10], the full CoLeaf dataset V1.0 was used to detect eight nutritional deficiencies employing Random Forest classifiers using local texture features extracted with the SIFT, along with global features based on shape and color. An experimental evaluation with 335 images of coffee leaves showed that global features performed better than local features, achieving an accuracy of 67.5%. Meanwhile, in [20], the full CoLeaf dataset V2.0 was used to identify 10 nutritional deficiencies. They trained a ResNet50 classifier that obtained an overall accuracy of 87.75%.

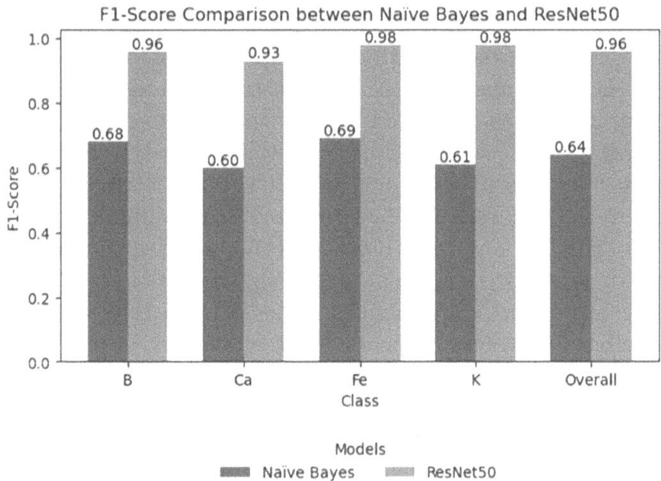

Fig. 4. Comparison of F1-Scores between State-of-Art (Naive Bayes [21]) and our model (ResNet50)

However, both [20] and [10] only reported overall accuracy without providing per-class performance metrics, which limits a detailed comparison with this work. In particular, our optimized ResNet50 model under scenario E5 achieved an overall accuracy of 96%, along with consistent per-class performance, thereby demonstrating superior effectiveness in detecting coffee nutritional deficiencies and greater robustness across nutrient classes.

6 Conclusions

In this work, we applied transfer learning and data augmentation approaches to classify of the most relevant nutritional deficiencies in coffee leaves: B, Ca, Fe, and K. The publicly available CoLeaf dataset, which contains images of coffee leaves exhibiting different nutritional deficits, was used for training and evaluating our proposal. In particular, by applying data augmentation techniques, such as rotation, scaling, and shearing, model performance was improved and the dataset's diversity was significantly expanded.

Among five experimental configurations, the E5 scenario the highest accuracy and F1-scores across all models, ranging from 0.86 to 0.96. Besides, among the evaluated architectures (MobileNetV2, InceptionV3, VGG19, EfficientNetV2, and ResNet50), the fine-tuned ResNet50 model achieved the best performance, obtaining an overall F1-score of 0.96. These findings highlight the importance of balanced datasets and data augmentation in improving the deep learning models' capacity for generalization. In addition to providing a reliable method for identifying nutritional deficits (B, Ca, Fe, and K) in coffee plants, this research offers insightful information about the application of transfer learning and data augmentation for coffee plants. This information may enable timely interventions that prevent yield losses and contribute to improvements in coffee bean quality.

Moreover, the performance of our best classification model (ResNet50 under the E5 configuration) was compared against three state-of-the-art studies that used the CoLeaf dataset, which reported overall accuracies ranging from 67.5% and 87.5%. Results shown that our approach outperformed these methods, achieving an overall accuracy of 96%, thereby establishing a new performance benchmark for this task. Furthermore, beyond accuracy improvements, this advancement in automated nutrient deficiency detection reduces reliance on labor-intensive and error-prone manual inspections in coffee plantations, promoting more efficient resource utilization and cost-effective coffee crop management.

As future work, we plan to explore more advanced data augmentation strategies, such as synthetic image generation using Generative Adversarial Networks (GANs), as well as ensemble approaches that combine multiple transfer learning models to leverage their complementary strengths.

References

1. Aarizou, A., Merah, M.: Transfer learning for plant disease detection on complex images. In: 2022 7th International Conference on Image and Signal Processing and their Applications (ISPA), pp. 1–6 (2022). https://doi.org/10.1109/ISPA54004.2022.9786306
2. Comunicaffè: USDA biannual report cuts estimate of world production for 2024/25 to 174.4 million. Comunicaffè (2025). https://www.comunicaffe.com/usda-biannual-report-cuts-estimate-of-world-production-for-2024-25-to-174-4-million/
3. Dutta, L., Rana, A.K.: Disease detection using transfer learning in coffee plants. In: 2021 2nd Global Conference for Advancement in Technology (GCAT), pp. 1–4 (2021). https://doi.org/10.1109/GCAT52182.2021.9587602
4. ElevaFinca: Peruvian coffee: Economic importance, sustainable practices, and market demand (2023). https://www.elevafinca.com/post/peruvian-coffee-economic-importance-sustainable-practices-and-market-demand. Accessed 21 Nov 2024
5. He, K., Zhang, X., Ren, S., Sun, J.: Deep residual learning for image recognition (2015). https://arxiv.org/abs/1512.03385
6. Jtaglione: Coffee leaf diseases (2022). https://www.kaggle.com/code/jtaglione/coffee-leaf-diseases/data
7. Krohling, R.A.: Bracol - a Brazilian arabica coffee leaf images dataset to identification and quantification of coffee diseases and pests (2019). https://data.mendeley.com/datasets/yy2k5y8mxg/1
8. Kumar, M., Gupta, P., Madhav, P., Sachin: Disease detection in coffee plants using convolutional neural network. In: 2020 5th International Conference on Communication and Electronics Systems (ICCES), pp. 755–760 (2020). https://doi.org/10.1109/ICCES48766.2020.9138000
9. Lakshmanarao, A., Supriya, N., Arulmurugan, A.: Plant disease prediction using transfer learning techniques. In: 2022 Second International Conference on Advances in Electrical, Computing, Communication and Sustainable Technologies (ICAECT), pp. 1–5 (2022). https://doi.org/10.1109/ICAECT54875.2022.9807956

10. Monsalve, D., Trujillo, M., Chaves, D.: Automatic classification of nutritional deficiencies in coffee plants. In: 6th Latin-American Conference on Networked and Electronic Media (LACNEM 2015), pp. 1–6 (2015). https://doi.org/10.1049/ic.2015.0317
11. Nolte, G., U.S. Department of Agriculture, F.A.S.: Coffee annual: Peru (report no. pe2025-0018). Technical report, U.S. Department of Agriculture, Foreign Agricultural Service (2025). https://apps.fas.usda.gov/newgainapi/api/Report/DownloadReportByFileName?fileName=Coffee+Annual_Lima_Peru_PE2025-0018.pdf
12. Paulos, E.B., Woldeyohannis, M.M.: Detection and classification of coffee leaf disease using deep learning. In: 2022 International Conference on Information and Communication Technology for Development for Africa (ICT4DA), pp. 1–6 (2022). https://doi.org/10.1109/ICT4DA56482.2022.9971300
13. Pendhari, H., Virkar, R., Edakkalathur, B., Jadhav, A.: A comparative study on algorithms for plant disease detection using transfer learning. In: 2023 5th International Conference on Inventive Research in Computing Applications (ICIRCA), pp. 528–533 (2023). https://doi.org/10.1109/ICIRCA57980.2023.10220597
14. Sandler, M., Howard, A., Zhu, M., Zhmoginov, A., Chen, L.C.: MobileNetv2: inverted residuals and linear bottlenecks (2019). https://arxiv.org/abs/1801.04381
15. Simonyan, K., Zisserman, A.: Very deep convolutional networks for large-scale image recognition (2015). https://arxiv.org/abs/1409.1556
16. Singh, H., Tewari, U., Ushasukhanya, S.: Tomato crop disease classification using convolution neural network and transfer learning. In: 2023 International Conference on Networking and Communications (ICNWC), pp. 1–6 (2023). https://doi.org/10.1109/ICNWC57852.2023.10127284
17. Szegedy, C., Vanhoucke, V., Ioffe, S., Shlens, J., Wojna, Z.: Rethinking the inception architecture for computer vision (2015). https://arxiv.org/abs/1512.00567
18. Tan, M., Le, Q.V.: EfficientNetv2: smaller models and faster training (2021). https://arxiv.org/abs/2104.00298
19. Tuesta-Monteza, V.A.: CoLeaf dataset (2023). https://data.mendeley.com/datasets/brfgw46wzb/1
20. Tuesta-Monteza, V.A., Mejia-Cabrera, H.I., Arcila-Diaz, J.: CoLeaf-DB: peruvian coffee leaf images dataset for coffee leaf nutritional deficiencies detection and classification. Data Brief **48**, 109226 (2023). https://doi.org/10.1016/j.dib.2023.109226, https://www.sciencedirect.com/science/article/pii/S2352340923003451
21. Vassallo-Barco, M., Vives-Garnique, L., Tuesta-Monteza, V.A., Mejía-Cabrera, H.I., de Sipán, S.: Automatic detection of nutritional deficiencies in coffee tree leaves through shape and texture descriptors (2017). https://api.semanticscholar.org/CorpusID:202707906

Parking Space Detection System: YOLOv12 Architecture Comparison and Deployment in a University Campus

Victor Ariza[1(✉)], Vladimir Riffo[2], and John W. Castro[3]

[1] Departamento de Ingeniería Informática y Ciencias de la Computación, Universidad de Atacama, 1537122 Copiapó, Chile
victor.ariza.24@alumnos.uda.cl

[2] Departamento de Ingeniería Informática y Ciencias de la Computación, Universidad de Atacama, 1537122 Copiapó, Chile
vladimir.riffo@uda.cl

[3] Departamento de Ingeniería Informática y Ciencias de la Computación, Universidad de Atacama, 1537122 Copiapó, Chile
john.castro@uda.cl

Abstract. Parking availability remains a persistent challenge in urban and institutional environments, particularly on university campuses with limited space and high demand. This work presents a vision-based detection system using YOLOv12 object detection models to identify vacant parking spots. Two model variants, YOLOv12-Medium (YOLOv12m) and YOLOv12-Large (YOLOv12l), were trained on a class-balanced dataset of 18,698 images, including diverse lighting conditions. A comparative evaluation revealed that YOLOv12m achieved a lower average detection time per image (4.86 ms) and a higher $F1$–score (0.984), whereas YOLOv12l provided marginal robustness under low-light scenarios. Based on this analysis, YOLOv12m was integrated into a modular web-based interface for continuous monitoring of parking occupancy. Campus deployment confirmed the system's usability, responsiveness, and potential for a scalable application. The proposed solution addresses key requirements for intelligent parking systems, including generalization, ease of deployment, and user-friendly interaction.

Keywords: Deep learning · Object detection · YOLOv12 · Parking space detection · Smart campus · Web deployment

1 Introduction

The growth of urban environments and the increasing number of vehicles have turned parking availability into a common yet persistent operational challenge. This issue is particularly evident on university campuses, where space is limited and vehicle turnover is high [1]. Conventional parking systems typically rely on physical sensors, which are expensive to install and maintain, and often fail to adapt to changing environmental conditions [2,3]. These limitations have encouraged the adoption of vision-based solutions, leveraging deep learning and video analytics to monitor parking occupancy more flexibly [4].

D. Chaves et al. (Eds.): CIARP 2025, LNCS 16528, pp. 90–101, 2026.
https://doi.org/10.1007/978-3-032-23161-1_7

Vision-based systems offer promising advantages in terms of flexibility and scalability. However, they must overcome critical challenges such as maintaining detection accuracy under variable lighting conditions, managing computational load for responsive performance, and integrating outputs into usable interfaces for end-users [5,6]. Moreover, many existing systems do not meet the requirements of large-scale deployment, often lacking modularity and user-oriented design features [7].

To address these limitations, this work proposes a fully functional parking space detection system based on the YOLOv12 object detection architecture [8,9]. YOLOv12 was chosen over previous versions because it introduces architectural optimizations that enhance both accuracy and efficiency. In particular, Tian et al. [9] report that YOLOv12 consistently outperforms YOLOv8, YOLOv10, and YOLOv11 across multiple scales, achieving higher mean Average Precision (mAP) while maintaining comparable or lower computational costs. For example, YOLOv12-M attains a mAP of 52.5% with 67.5 GFLOPs and 20.2M parameters, surpassing YOLOv8-M (50.3%) and YOLOv11-M (51.5%) with reduced inference latency (4.86 ms/image). These improvements make YOLOv12 a suitable architecture for real-time parking detection in campus environments. Two model variants, YOLOv12m and YOLOv12l, were trained and evaluated on a diverse dataset that reflects real-world conditions. After a comparative analysis, YOLOv12m was selected and integrated into a web-based interface for occupancy monitoring on a university campus.

The main contributions of this paper are as follows: *i*) an empirical comparison of YOLOv12m and YOLOv12l in terms of detection accuracy, average detection time per image, and robustness; *ii*) a scalable deployment framework that enables continuous monitoring and user interaction through a web interface; and *iii*) a demonstration of the system's applicability in a real campus scenario, satisfying the four key criteria for intelligent parking systems defined in Sect. 2 [10].

The remainder of this paper is organized as follows. Section 2 reviews prior work on parking detection using computer vision. Section 3 details the proposed system architecture, dataset composition, and model training procedures. Section 4 presents the experimental findings, including performance metrics and the visual and functional validation of the deployed web interface. Finally, Sect. 5 summarizes the contributions and discusses possible avenues for future research.

2 Related Work

Automated parking occupancy detection has become a key research focus in the development of intelligent transportation systems and smart city infrastructure [11,12]. Traditional systems have relied heavily on sensor-based technologies, which, although effective, tend to involve high deployment and maintenance costs [13]. Recent advances in computer vision and deep learning have led to a shift toward vision-based solutions that offer greater scalability and cost-efficiency [14]. In response to this shift, early methods explored lightweight Convolutional

Neural Networks (CNNs) for classifying occupancy status in individual parking slots. For instance, the work in [15] proposed an edge-based architecture that integrates camera networks and CNNs for distributed innovative parking management, highlighting the potential of real-time visual monitoring without dense sensor deployment.

More recent studies have adopted object detection frameworks such as YOLO for holistic scene analysis. In [16], the authors implemented a parking detection system using YOLOv3 combined with MobileNetv2 on Jetson TX2 hardware, achieving robust detection under various lighting conditions. Likewise, [17] utilized YOLOv3 to detect both parked vehicles and the physical layout of parking lots from video streams, demonstrating a low-cost, accurate solution applicable to scalable urban environments. A notable contribution in [18] integrated YOLOv4 with behavioral data from drivers to dynamically allocate parking spaces, offering a hybrid solution that adapts to real-time conditions and user patterns.

In terms of architecture comparison and refinement, [19] conducted an empirical evaluation between YOLO-Fastest V2 and a customized ResNet model. The authors reported limitations in the YOLO-Fastest variant and favored transfer learning for improved accuracy and generalization, reinforcing the need for more robust and flexible architectures in real-time parking applications.

To provide a structured comparison between prior works and the proposed approach, we define four technical criteria relevant to real-time deployment and user-oriented integration of parking detection systems: *i)* **real-time processing with synchronized statistical output**, where the system performs detection on continuous video input while updating occupancy indicators simultaneously; *ii)* **use of large-scale and class-balanced training datasets with more than 10,000 images**, which supports generalization across varying conditions; *iii)* **modular architecture with web-based deployment capability**, enabling separation of detection logic, user interface, and back-end services; and *iv)* **availability of an interactive prototype accessible via web**, supporting end-user interaction and visualization in operational environments. Table 1

Table 1. Comparison of related works based on four technical criteria. ✓ indicates that the work satisfies the criterion; ✗ indicates that it does not.

Work	Criterion 1	Criterion 2	Criterion 3	Criterion 4
Ahad et al. [18]	✓	✓	✗	✗
Amarasooriya et al. [17]	✓	✓	✗	✗
Bura et al. [15]	✓	✗	✗	✗
Chen et al. [16]	✓	✓	✗	✗
Li et al. [19]	✓	✗	✗	✗
This work	✓	✓	✓	✓

shows, for each related work, whether it satisfies or fails to satisfy each of the four criteria, defined above.

As observed in Table 1, although recent works have made substantial progress in video-based detection and data scale, they commonly lack scalable software architecture and deployable user interfaces. In contrast, our approach satisfies all four criteria by integrating a YOLOv12-based model into a modular, web-accessible system, validated in a real-world campus environment.

3 Methodology

This section presents the system architecture, dataset composition, model training procedure, and deployment strategy adopted in this work.

3.1 System Overview and Dataset

The proposed system is designed to detect available parking spaces using a deep learning-based object detection approach. It integrates an object recognition model based on the YOLOv12 architecture with a visualization interface that presents annotated outputs to the user. This configuration enables flexible deployment in university campuses and other parking environments where automated monitoring of space occupancy is necessary.

The input data is processed using OpenCV, resized to a resolution of 640×640, and then passed to a YOLOv12 model. The model, stored in the `best.pt` format was executed using the official Ultralytics package in Python. Postprocessing involves overlaying bounding boxes, with annotated results prepared for browser-based visualization.

The training dataset consisted of 18,698 labeled images compiled from three sources: publicly available parking datasets from Kaggle, curated collections from Roboflow, and original footage captured on-site at the Universidad de Atacama in Chile. Annotations were either manually generated using LabelImg or verified from pre-existing labels. All annotations follow the YOLO format and target the two classes mentioned earlier (*Space_ Empty* and *Space_ Occupied*). Images were selected to reflect diverse environmental conditions, including daylight, low-light, and nighttime scenarios, in both public and institutional parking areas. The dataset was class-balanced, and the image resolution was standardized to 640×640. Table 2 summarizes the dataset division used for model training and evaluation.

Table 2. Dataset split by usage and image count.

Subset	Percentage	Number of Images
Training	70%	13,088
Validation	20%	3,740
Testing	10%	1,870
Total	100%	18,698

Model evaluation involved controlled experiments conducted under various lighting conditions, utilizing diverse data sources. These tests confirmed the system's ability to maintain consistent detection performance across a range of real-world scenarios, as further detailed in Sect. 4.

3.2 Model Implementation and Training

Two YOLOv12 model variants were implemented: YOLOv12m and YOLOv12l. Both models were managed through the Ultralytics library and executed using the official `.pt` weight format, compatible with PyTorch and CUDA environments. The models were trained for 100 epochs on the curated parking dataset, using a resolution of 640 $\times$ 640. Training was performed on a system equipped with an NVIDIA RTX 4080 GPU (16 GB VRAM), an Intel Core i9-14900KF processor, and 32 GB of RAM. The environment was configured using Python 3.9, PyTorch, and CUDA 11.7, with all dependencies isolated through a Conda virtual environment to ensure reproducibility.

The default training parameters provided by the Ultralytics framework were used, including learning rate scheduling and data augmentation strategies optimized for object detection tasks. The same dataset split and annotations were used across both models to ensure a fair comparison. YOLOv12m was selected for deployment given its more favorable balance between average detection time per image and accuracy under the evaluated conditions. While YOLOv12l demonstrated slightly higher robustness in specific scenarios, its higher computational demands made it less suitable for scalable or responsive applications.

3.3 System Deployment

The trained YOLOv12m model was integrated into a functional detection system using a lightweight web application architecture. The deployment was built with Flask to serve the annotated detection results through a browser-accessible interface. OpenCV was used to handle input processing and render detection outputs, including bounding boxes and class labels. The system interface provides a visual representation of detected parking spaces, allowing users to monitor occupancy conditions through annotated frames and simple visual indicators. The interface was designed for ease of use, requiring minimal client-side resources and no installation beyond a web browser.

Operational tests were conducted using samples collected from both public datasets and proprietary sources, which were recorded at the Universidad de Atacama. These included sequences captured under various lighting conditions and viewpoints, simulating real-world deployment environments. The system demonstrated stable performance, validating its applicability to practical parking monitoring tasks without requiring specialized infrastructure.

4 Results and Analysis

This section presents the experimental results obtained from evaluating YOLOv12m and YOLOv12l on the parking space detection dataset. We focus on model performance, detection quality, and system behavior under deployment conditions.

4.1 Performance Comparison Between YOLOv12m and YOLOv12l

This subsection compares the computational characteristics of YOLOv12m and YOLOv12l. Both models were trained for 100 epochs under identical conditions and compared in terms of number of parameters, GFLOPs, and average detection time per image. The goal is to provide a clear overview of the trade-offs involved when choosing between the two YOLOv12 model variants.

Table 3 summarizes key aspects such as model size, number of parameters, computational load (GFLOPs), average detection time per image, and total training time. YOLOv12l exhibits higher computational demands across all metrics, requiring 6.77 ms to process each image and 175 h of training. In contrast, YOLOv12m is faster and lighter, processing each image in 4.86 ms and completing training in 36.7 h.

Table 3. Comparison of computational characteristics between YOLOv12m and YOLOv12l

Model	Model Size (MB)	Parameters (M)	GFLOPs	Detection Time (ms/image)	Training Time (h)
YOLOv12m	52.5	20.2	67.5	4.86	36.7
YOLOv12l	53.7	26.4	88.9	6.77	175

The results highlight a clear computational gap between the two models. While YOLOv12l may offer advantages in detection accuracy (analyzed in the following subsection), YOLOv12m provides a more efficient profile, making it a preferable choice for real-time or resource-constrained scenarios.

4.2 Evaluation Metrics and Quantitative Results

The evaluation of YOLOv12m and YOLOv12l followed standard object detection metrics, including *Precision*, *Recall*, *mean Average Precision* (mAP), and $F1 - score$. These indicators provide complementary insights into the quality and reliability of the predictions across multiple confidence thresholds and IoU levels.

Precision and *Recall* were computed according to Eq. 1, based on the number of *True Positives* (TP), *False Positives* (FP), and *False Negatives* (FN). These metrics respectively quantify the correctness and completeness of the detections:

$$Precision = \frac{TP}{TP + FP}, \quad Recall = \frac{TP}{TP + FN} \tag{1}$$

The $F1 - score$, defined in Eq. 2, provides a harmonic mean between *Precision* and *Recall*, offering a single metric that balances both aspects:

$$F1 - score = 2 \cdot \frac{Precision \cdot Recall}{Precision + Recall} \tag{2}$$

The mAP was used to quantify the overall detection quality. It is computed as the average area under the *Precision–Recall* curve (AP) for each class. In this study, both *mAP@0.5* and *mAP@0.5:0.95* were reported, following the COCO evaluation protocol.

The *Intersection over Union* (IoU) was used as the reference criterion to determine true positives. Equation 3 defines the IoU as the ratio between the overlapping area and the union area of predicted (BB_{dt}) and ground-truth (BB_{gt}) bounding boxes. A visual illustration is shown in Fig. 1.

$$\mathrm{IoU} = \frac{\mathrm{Area}(\mathrm{BB}_{dt} \cap \mathrm{BB}_{gt})}{\mathrm{Area}(\mathrm{BB}_{dt} \cup \mathrm{BB}_{gt})} \tag{3}$$

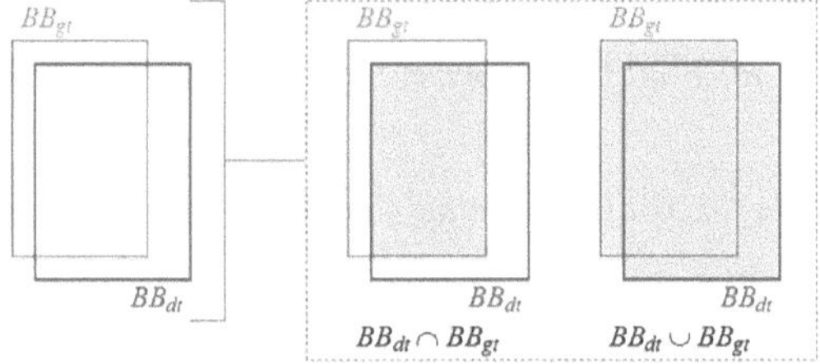

Fig. 1. Visual interpretation of bounding box overlap used in IoU computation. Green: ground truth (BB_{gt}), Red: detection (BB_{dt}). (Color figure online)

Table 4 summarizes the quantitative performance metrics for both models. YOLOv12m slightly outperformed YOLOv12l in overall $F1 - score$ (0.984 vs. 0.979) and mAP@0.5 (0.988 vs. 0.987), while YOLOv12l showed a marginal gain in mAP@0.5 : 0.95.

Table 4. Summary of mAP, *Precision*, *Recall*, and $F1 - score$ for YOLOv12m and YOLOv12l on the parking space detection task.

Model	mAP@0.5	mAP@0.5 : 0.95	*Precision*	*Recall*	$F1 - score$
YOLOv12m	0.988	0.972	0.969	0.999	0.984
YOLOv12l	0.987	0.974	0.967	0.991	0.979

To further strengthen the evaluation, we included a baseline comparison with previous YOLO versions, focusing on the medium scale (M) models. This choice is consistent with the deployment strategy described in Sect. 3, where YOLOv12m was selected as the most balanced configuration between accuracy and efficiency. Table 5 summarizes results reported in [9], showing that

YOLOv12m achieves higher mean Average Precision while maintaining competitive latency and computational cost compared to YOLOv8-M, YOLOv10-M, and YOLOv11-M. These findings confirm that YOLOv12m provides a superior trade-off, making it more suitable for real-time deployment in campus environments.

Table 5. Baseline comparison between YOLOv12m and prior YOLO versions (medium scale models).

Model	FLOPs (G)	Parameters (M)	mAP@0.5:0.95 (%)	Latency (ms)
YOLOv8-M	78.9	25.9	50.3	5.09
YOLOv10-M	59.1	15.4	51.1	4.74
YOLOv11-M	68.0	20.1	51.5	4.70
YOLOv12-M (Ours)	67.5	20.2	52.5	4.86

The *Precision–Recall* curves for YOLOv12m and YOLOv12l are presented in Figs. 2 and 3, respectively. Both models achieved high levels of detection *Precision* across the *Recall* spectrum, particularly for the two defined classes.

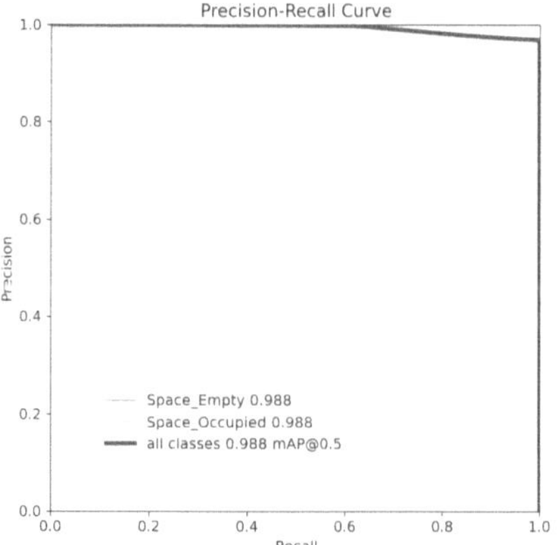

Fig. 2. *Precision–Recall* curve for YOLOv12m per class.

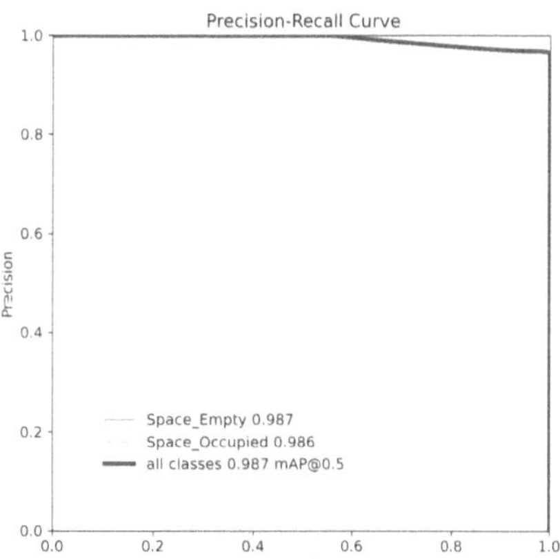

Fig. 3. *Precision–Recall* curve for YOLOv12l per class.

To complement these results, Figs. 4 and 5 show the $F1 - score$ evolution concerning the confidence threshold. YOLOv12m reached its optimal $F1 - score$ at a confidence of 0.626, while YOLOv12l peaked at 0.055, indicating different operating points for best performance.

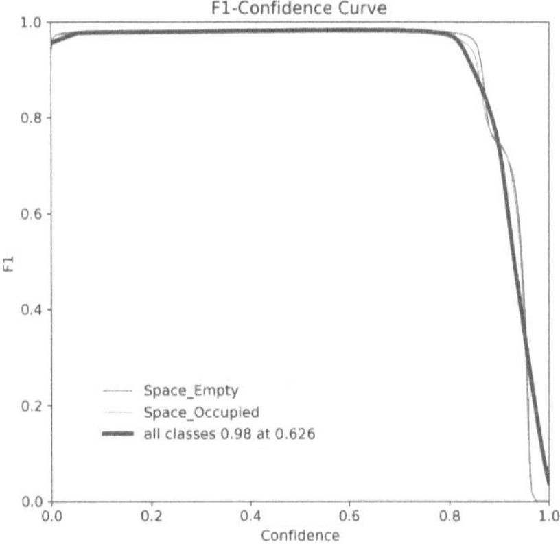

Fig. 4. $F1-score$ as a function of confidence threshold for YOLOv12m.

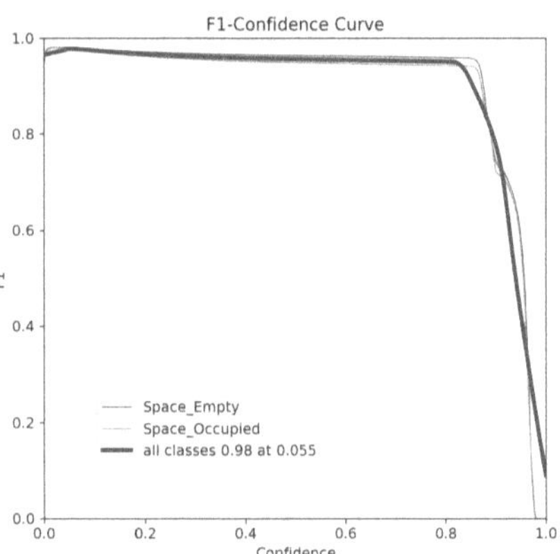

Fig. 5. $F1-score$ as a function of confidence threshold for YOLOv12l.

4.3 Visual Validation and Evaluation of the Functional Web Prototype

The proposed detection system was integrated into a functional web-based interface designed to support real-time visualization and usability testing. This interface displays annotated video streams and offers interactive controls for threshold adjustment and camera selection, enabling users to evaluate detection performance under operational conditions. To validate the visual effectiveness of the deployed solution, we collected qualitative evidence under three representative lighting scenarios: daylight, low-light, and nighttime. YOLOv12l was selected as the default backend due to its slightly higher robustness under varying illumination.

Figure 6 presents a composite view of the interface operating in these three conditions, demonstrating stable detection outputs and consistent bounding box rendering. In addition, Fig. 7 shows a live deployment instance of the system during a regular operation cycle. The visual interface presents annotated detections alongside a real-time chart of occupancy statistics, visible on the right panel. This dual-pane layout allows users to monitor the parking state while tracking temporal trends. The system was tested on-site at the Universidad de Atacama in Chile, using parking areas within the campus to refine detection thresholds, validate interface responsiveness, and assess overall deployment viability. These real-world tests ensured that the system was functionally validated in the same context described in the study and reflected in the title.

Both figures confirm the system's capacity to generalize across diverse lighting conditions and support real-world monitoring tasks through a front-end designed for accessibility and continuous operation. Beyond validating detection accuracy, the web interface demonstrates readiness for scalable deployment in practical environments. It marks the final step in translating the core model into

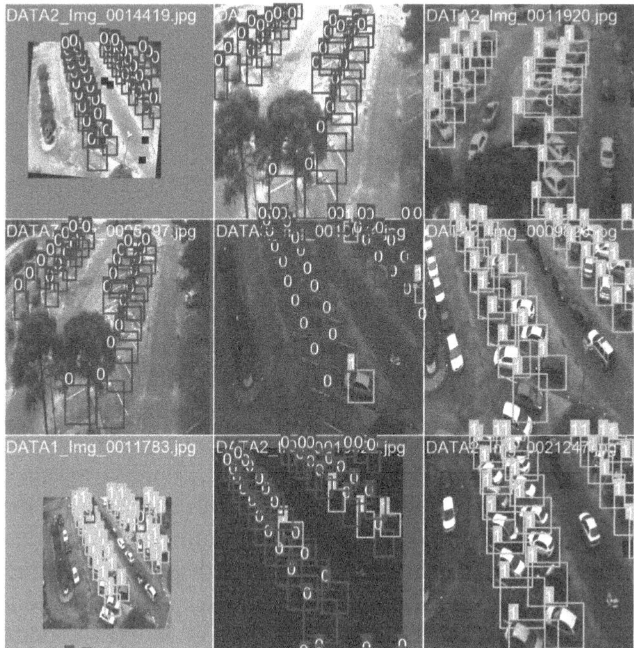

Fig. 6. Interface display under different lighting conditions (daylight, low-light, and night). Model: YOLOv12l.

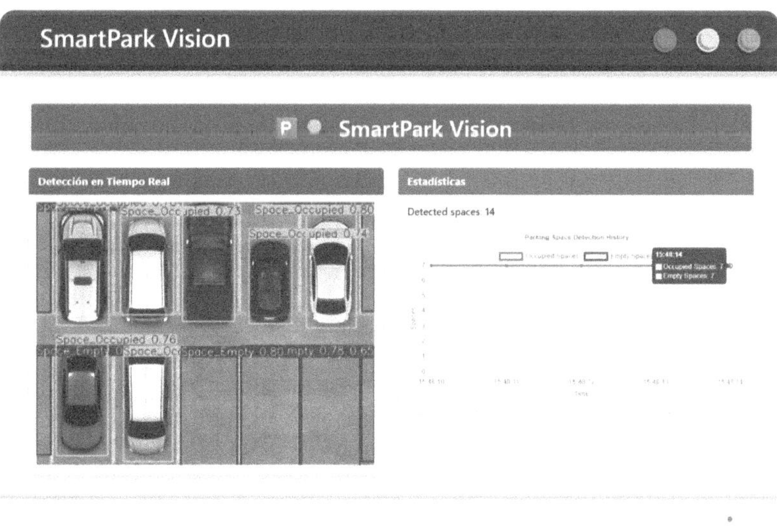

Fig. 7. Screenshot of the functional web-based interface showing real-time detections and occupancy statistics.

a usable tool, enabling sustained, remote supervision of parking areas without complex infrastructure.

5 Conclusions

This work presents a comparative evaluation of YOLOv12-based models for parking space detection under standardized training conditions and deployment scenarios. The goal was to assess the trade-offs in detection accuracy, computational load, and operational robustness for integration into a functional web-based monitoring tool within a university campus setting.

Experimental results showed that YOLOv12m achieved better inference speed and $F1-score$, making it well-suited for real-time detection in constrained environments. In contrast, YOLOv12l offered marginal gains in detection robustness under variable lighting conditions, supporting its use in scenarios where visual variability is a critical factor. Deployment through a web-based interface validated the system's practical viability. It confirmed the ability to provide continuous occupancy monitoring with minimal infrastructure.

Looking ahead, this work lays the groundwork for scalable parking supervision solutions in urban and institutional settings. The proposed system combines deep learning performance with a user-friendly interface, enabling broader implementations across larger campuses or smart cities. Ultimately, YOLOv12m stands out as the optimal balance between responsiveness and reliability for real-world parking detection applications.

References

1. Alkhudhayr, H.: Internet of things based parking slot detection and occupancy classification for smart city traffic management. Eng. Appl. Artif. Intell. **152**, 110802 (2025)
2. Farley, A., Ham, H., et al.: Real time IP camera parking occupancy detection using deep learning. Procedia Comput. Sci. **179**, 606–614 (2021)
3. Rafique, S., Gul, S., Jan, K., Khan, G.M.: Optimized real-time parking management framework using deep learning. Expert Syst. Appl. **220**, 119686 (2023)
4. Ling, X., Sheng, J., Baiocchi, O., Liu, X., Tolentino, M.E.: Identifying parking spaces & detecting occupancy using vision-based IoT devices. In: 2017 Global Internet of Things Summit (GIoTS), pp. 1–6. IEEE (2017)
5. Acharya, D., Yan, W., Khoshelham, K.: Real-time image-based parking occupancy detection using deep learning. Res.@ Locate **4**, 33–40 (2018)
6. Prabhavathi, K., Anusha, M.N., Manjunatha, G., Sowjanya, N.B., Naveen, K.B., et al.: Next-gen ai-embedded vision systems: Real-time optimization for surveillance, autonomous vehicles, and industry 4.0. In: 2024 International Conference on Distributed Systems, Computer Networks and Cybersecurity (ICDSCNC), pp. 1–8. IEEE (2024)
7. Padmasiri, H., Madurawe, R., Abeysinghe, C., Meedeniya, D.: Automated vehicle parking occupancy detection in real-time. In: 2020 Moratuwa engineering research conference (MERCon), pp. 1–6. IEEE (2020)

8. Redmon, J., Divvala, S., Girshick, R., Farhadi, A.: You only look once: unified, real-time object detection. In: Proceedings of the IEEE conference on computer vision and pattern recognition, pp. 779–788 (2016)
9. Tian, Y., Ye, Q., Doermann, D.: YOLOv12: attention-centric real-time object detectors. arXiv preprint arXiv:2502.12524 (2025)
10. Kim, J., Jeong, I., Jung, J., Cho, J.: Resource-efficient design and implementation of real-time parking monitoring system with edge device. Sens. (Basel Switz.) **25**(7), 2181 (2025)
11. Baroffio, L., Bondi, L., Cesana, M., Redondi, A.E., Tagliasacchi, M.: A visual sensor network for parking lot occupancy detection in smart cities. In: 2015 IEEE 2nd World Forum on Internet of Things (WF-IoT), pp. 745–750. IEEE (2015)
12. Singh, T., Khan, S.S., Chadokar, S.: A review on automatic parking space occupancy detection. In: 2018 International Conference on Advanced Computation and Telecommunication (ICACAT), pp. 1–5. IEEE (2018)
13. Szarka, R., Dakić, P., Vranić, V.: Cost-effective real-time parking space occupancy detection system. In: 2022 12th International Conference on Advanced Computer Information Technologies (ACIT), pp. 506–510. IEEE (2022)
14. Høye, T.T., et al.: Deep learning and computer vision will transform entomology. Proc. Natl. Acad. Sci. **118**(2), e2002545117 (2021)
15. Bura, H., Lin, N., Kumar, N., Malekar, S., Nagaraj, S., Liu, K.: An edge based smart parking solution using camera networks and deep learning. In: 2018 IEEE international conference on cognitive computing (ICCC), pp. 17–24. IEEE (2018)
16. Chen, L.-C., Sheu, R.-K., Peng, W.-Y., Jyh-Horng, W., Tseng, C.-H.: Video-based parking occupancy detection for smart control system. Appl. Sci. **10**(3), 1079 (2020)
17. Amarasooriya, P.M.D.S., Peiris, M.P.P.L., Herath, H.M.D.S.: Implementation of smart parking system using image processing (2023)
18. Ahad, A., Kidwai, F.A.: YOLO based approach for real-time parking detection and dynamic allocation: integrating behavioral data for urban congested cities. Innov. Infrastruct. Solut. **10**(6), 1–15 (2025)
19. Li, W., et al.: Where should i park? Real-time detection of parking space occupancy. Camb. Explor. Arts Sci. **1**(2) (2023)

Position Recognition in the Taikyoku Kata of Karate with Computer Vision

Andres Felipe Murcia and Alexander Cerón-Correa(✉)

Universidad Militar Nueva Granada, Bogotá, Colombia
alexander.ceron@unimilitar.edu.co

Abstract. This study presents the development of a computer vision-based system to detect and recognize the fundamental postures and blocks of the Taikyoku kata, a basic sequence of movements in Karate. This system employs geometrical considerations, classifiers and specialized computer vision libraries to identify and analyze key positions from previously recorded videos. The solution focuses on providing a detailed and reliable analysis of postures, contributing to the teaching and practice of karate without the need for specialized equipment. In the proposed method, geometrical features are considered in a machine learning approach. This approach increases accessibility and reduces costs by avoiding the use of additional sensors. The proposed method is intended for beginner practitioners, where form requirements are more important than speed.

Keywords: Computer vision · Karate · pose detection · martial arts

1 Introduction

Karate is a martial arts discipline that demands precision in the execution of technical movements and specific postures, fundamental elements in both teaching and the technical progress of its practitioners. In particular, the kata Taikyoku, one of the most basic forms of this practice, incorporates postures such as Zenkutsu-dachi and Kokutsu-dachi, as well as basic blocks such as Gedan-barai and Uchi-uke. The correct execution of these movements is essential for developing a solid technical foundation. However, their evaluation often depends on subjective observation by instructors, which can lead to inconsistencies in technical feedback [1,2].

In recent years, advances in computer vision and machine learning have been applied to motion analysis in sports [3–5]; however, these applications have focused primarily on disciplines such as taekwondo or judo, or in general action recognition contexts. Karate, and in particular the structured analysis of its traditional katas, remains underexplored area [1,6]. Although there is preliminary research addressing similar tasks, a significant gap remains in the detailed identification of specific postures and blocks within the kata practice *Taikyoku*.

D. Chaves et al. (Eds.): CIARP 2025, LNCS 16528, pp. 102–111, 2026.
https://doi.org/10.1007/978-3-032-23161-1_8

On the other hand, some technical analysis systems in martial arts employ specialized hardware, such as inertial sensors [7], or motion capture equipment with multiple high-speed cameras. While these systems offer high accuracy, their cost and complexity hinder their widespread adoption in everyday training environments.

In this context, the development of an accessible tool based exclusively on computer vision is proposed. This tool is capable of automatically detecting and classifying fundamental stances and blocks in the Taikyoku kata. This approach seeks to contribute to technical standardization and support both students and instructors through objective and automated feedback on technical performance.

Karate, as a martial art, demands rigorous learning of fundamental stances and blocks that are essential for the technical progress of practitioners. In the case of Taikyoku kata, one of the first sequences taught, the proper execution of stances such as Zenkutsu-dachi and blocks such as Gedan-barai is crucial to laying a solid foundation for the student. However, currently, the evaluation of these postures relies primarily on subjective observation by instructors, which can lead to inconsistencies in teaching and learning.

The drawback lies in the lack of accessible and specific technological tools for the objective evaluation of karate postures and movements.

The primary objective is to generate new knowledge in the application of computer vision to the analysis of karate katas. This knowledge has a social impact, by facilitating technical learning for practitioners, and an economic impact, by offering an accessible tool that does not require specialized hardware.

This project significantly impacts the karate community by promoting more uniform and objective teaching, reducing the reliance on subjective evaluations by instructors, and increasing access to technological tools in academies.

The system's implementation is highly convenient, as it leverages open-source tools and widely available hardware. The main beneficiaries will be:

- Karate students, who will be able to perfect their techniques through an objective analysis of their movements.
- Instructors, who will have additional support to evaluate their students.
- Researchers, by having a new application that expands the use of computer vision in sports.

2 Related Work

The use of computer vision for human motion recognition has been widely explored in recent years, although applications to traditional martial arts remain limited. Below we present a summary of relevant contributions in karate, other sports, and related computer vision applications.

2.1 Karate and Martial Arts Recognition

Several studies have specifically addressed karate performance analysis. Pang et al. [8] presented a review of computer vision applications in martial arts. Bianco et al. [6] proposed a method based on skeletal motion using a Kinect sensor. More recently, Yu et al. [9] employed Kinect together with classifiers such as SVM and CNN for martial arts action recognition.

Mostafa et al. [1] proposed an automated evaluation of practitioners' style during the execution of the first kata, using deep learning and pose estimation. Their work demonstrates the feasibility of applying modern machine learning to karate performance evaluation. Earlier, Hachaj et al. [2] investigated Oyama karate tournament recordings with annotated videos, applying data mining and PCA techniques to recognize karate movements. These contributions provide the initial groundwork for karate recognition, although datasets and applications remain scarce compared to other sports. An application of computer vision and LSTM networks for pose detection in multiple-view Karate videos using OpenPose library is shown in [10]. Additionally, in [11] a Multi-Stream Karate Action Recognition Dataset (MS-KARD) that uses multiple vision perspectives, and other sensors accelerometer and gyroscope is presented.

2.2 Sports Training and Posture Estimation

Beyond karate, many works have explored pose detection in sports and fitness. Nguyen et al. [4] developed an automatic tool for yoga pose grading using skeleton representations, while Dhakate et al. [5] focused on real-time yoga pose analysis and personalized feedback using PoseNet and MediaPipe. Similarly, Bhamidipati et al. [3] presented an AI gym trainer for intelligent posture estimation, relying on Mediapipe and OpenCV to deliver real-time feedback. These studies highlight the applicability of skeleton-based pose estimation for technical training in sports.

The reviewed literature shows that while posture recognition using computer vision has advanced significantly in yoga, fitness, and other sports, research specifically targeting karate is still limited. Prior works in karate either focus on small-scale experiments or lack real-time implementations. The present work contributes to this gap by developing an accessible, low-cost system for recognizing fundamental stances and blocks of the Taikyoku kata, relying exclusively on computer vision techniques without the need for specialized sensors.

3 Methodology

3.1 Dataset Construction

The dataset was created from videos recorded in HD resolution with a fixed camera and clean background. Two intermediate-level karate practitioners performed the *Taikyoku* kata from different angles. From these videos, frames were extracted and processed using the *MediaPipe Pose* library, which detects 33 anatomical landmarks of the human body in 3D space (x, y, z).

The final dataset contained approximately 400 labeled frames, balanced across four classes:

- Two stances: *Zenkutsu-dachi*, *Kokutsu-dachi.*
- Two blocks: *Gedan-barai*, *Uchi-uke.*

Labels were manually assigned by the authors based on the correct execution of the techniques. Frames with incorrect or ambiguous detections were discarded. This ensured that only clean, representative samples were used in training and evaluation.

3.2 MLP Model Architecture

Two independent *Multilayer Perceptron* (MLP) classifiers were trained: one for stances and one for blocks. Both networks received as input a feature vector of dimension 99 (33 landmarks × 3 coordinates), normalized with *StandardScaler* to ensure uniform scale across all features.

Each MLP shared the same architecture:

- **Input layer:** 99 normalized coordinates corresponding to the 3D landmarks detected by MediaPipe.
- **Hidden layer 1:** 128 neurons with `ReLU` activation, followed by a `Dropout` layer with a rate of 30% to reduce overfitting.
- **Hidden layer 2:** 64 neurons with `ReLU` activation, refining learned representations.
- **Output layer:** 2 neurons with `softmax` activation, producing probability distributions over the two possible classes:
  - For stances: *Zenkutsu-dachi* vs. *Kokutsu-dachi.*
  - For blocks: *Gedan-barai* vs. *Uchi-uke.*

The models were trained using the `Adam` optimizer with a learning rate of 0.001 (default value), and the `sparse categorical crossentropy` loss function. Training was performed over 30 epochs with a batch size of 32. To improve reproducibility, experiments were conducted with a fixed random seed.

The dataset was divided in a stratified manner, with 70% of the samples used for training and 30% reserved for testing. This ensured preservation of class balance across subsets.

All experiments were run on a workstation with an Intel Core i7 CPU, 16 GB of RAM, and an NVIDIA GPU, using TensorFlow/Keras as the deep learning framework.

3.3 Rule-Based Geometric Baseline

As a baseline, a rule-based method was implemented. This approach relied on geometric relationships between landmarks, such as joint angles and relative distances, to classify techniques. For example:

- For **Zenkutsu-dachi**: the front knee angle (ankle–knee–hip) between 85° and 100°.
- For **Kokutsu-dachi**: extended front leg (angle close to 160°) with torso vertical.
- For **Gedan-barai**: forearm angled downward 30°–60° relative to the trunk.
- For **Uchi-uke**: elbow–shoulder–wrist angle greater than 90° in front of the torso.

This strategy does not require training and serves as an interpretable baseline to compare with the MLP models.

3.4 System Evaluation

Model performance was evaluated using standard metrics derived from the confusion matrix: accuracy, precision, recall, and F1-score. Additionally, confusion matrices were reported per class to analyze common misclassifications (e.g., between *Uchi-uke* and *Gedan-barai*).

Finally, the trained models were tested on complete kata videos, demonstrating real-time prediction capabilities.

4 Results

During the development of the system, two classification models based on multilayer perceptron (MLP) neural networks were trained: one for stance recognition (*Zenkutsu-dachi*, *Kokutsu-dachi*) and one for block recognition (*Gedan-barai*, *Uchi-uke*). Both models used as input the 99 normalized coordinates corresponding to the 33 3D keypoints detected with MediaPipe (see Fig. 1).

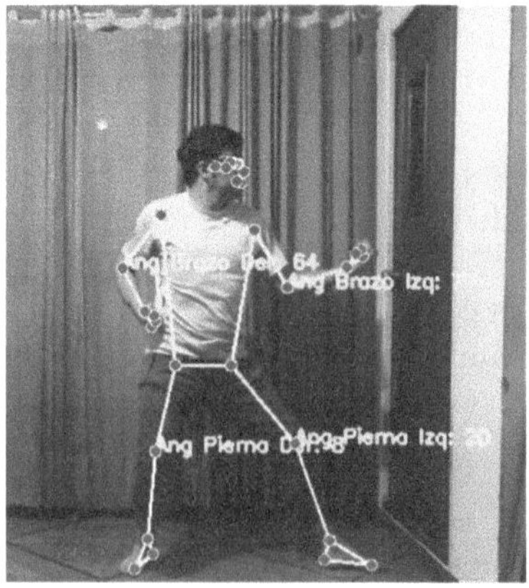

Fig. 1. Example of keypoints detected using MediaPipe.

4.1 Evaluation of the MLP Models

To assess performance, the dataset was split into 70% for training and 30% for testing, maintaining class balance. Evaluation was carried out using standard metrics: accuracy, precision, recall, and F1-score, calculated from the confusion matrices.

Stance Classification Model. The stance model achieved high accuracy, particularly for *Zenkutsu-dachi*. The confusion matrix (Fig. 2) shows that misclassifications were minimal and symmetrically distributed between the two stances. This indicates that the network effectively captured the geometric differences between the two postures.

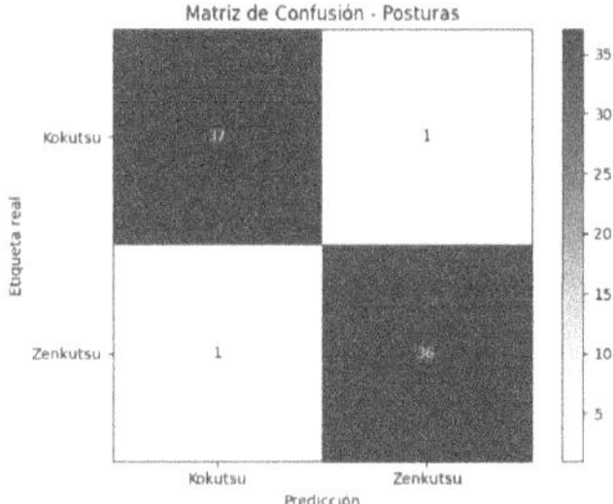

Fig. 2. Confusion matrix for the stance classification model.

Block Classification Model. The block model achieved lower overall accuracy compared to stances, with most errors arising from confusion between *Uchi-uke* and *Gedan-barai*. This reflects the visual similarity between both techniques when executed, making them harder to distinguish based solely on skeletal keypoints (Fig. 3).

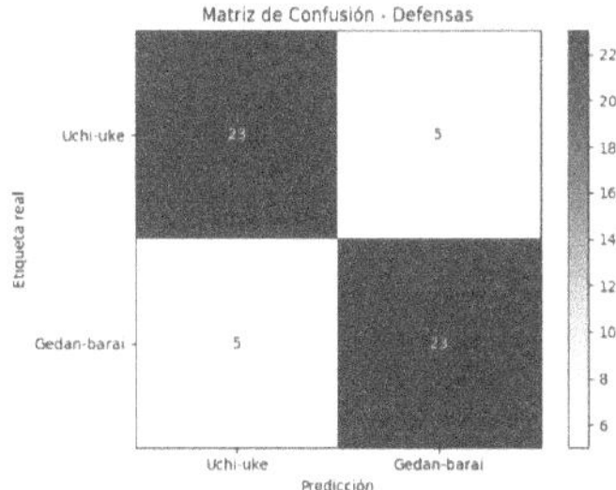

Fig. 3. Confusion matrix for the block classification model.

Quantitative Results. In numerical terms, the MLP models obtained the following metrics:

- **Stances:** accuracy of 97%, with balanced precision, recall, and F1-score across both classes.
- **Blocks:** accuracy of 82%, with symmetric confusion between classes but still consistent performance above baseline.

These results confirm that the neural network approach significantly outperforms the geometric rule-based method, which only achieved between 47% and 66% accuracy depending on the technique.

4.2 Real-Time Predictions

The trained models were tested on complete kata videos to validate their real-time applicability. Predictions were generated frame by frame and overlaid on the video sequences. The system displayed minimal latency and stable recognition of both stances and blocks, as illustrated in Figs. 4 and 5.

Fig. 4. Real-time prediction of *Zenkutsu-dachi* and *Gedan-barai* techniques.

4.3 Identified Limitations

Despite promising results, several limitations were observed:

- **Dependency on MediaPipe:** errors in landmark detection (deviations of 2–5 pixels) directly affect classification accuracy, especially for angle-based features.
- **Recording conditions:** changes in lighting, attire, or camera perspective reduce system robustness.

Fig. 5. Real-time prediction of *Kokutsu-dachi* and *Uchi-uke* techniques.

- **Class similarity:** blocks such as *Uchi-uke* and *Gedan-barai* often exhibit visual overlap, leading to misclassifications.
- **Dataset size:** only two practitioners and 10 videos were used, limiting generalization to real dojo conditions with diverse practitioners.

Overall, the MLP models provided reliable recognition for stances and acceptable performance for blocks, validating the feasibility of deep learning combined with skeleton-based features in karate movement recognition.

5 Conclusions and Future Work

The results of this study demonstrate that deep learning models can be trained to recognize basic stances and blocks from the *Taikyoku* kata with promising accuracy, even with a small dataset. In particular, the stance classifier achieved an accuracy of 97%, while the block classifier reached 82%. These results indicate that skeleton-based features extracted with MediaPipe, combined with a lightweight *MLP* model, can effectively capture the biomechanical patterns that characterize karate techniques.

The comparison against the geometric rule-based method confirmed the advantages of data-driven learning: while the geometric approach was useful for interpretability and for highlighting key angular variations, its performance was limited (47–66%). In contrast, the MLP models proved more robust to small execution differences and noise in keypoint detection.

Nevertheless, the study has important limitations. The dataset is small (two performers, 10 videos, 400 frames) and collected under controlled conditions.

This restricts the generalization of results to real dojo scenarios, where lighting, clothing, and body types vary considerably. Moreover, some techniques (particularly *Gedan-barai* and *Uchi-uke*) remain visually similar, which increases the chance of misclassification.

Future Work Includes

- Expanding the dataset with more practitioners, diverse recording conditions, and longer sequences to better capture temporal dynamics.
- Incorporating multi-camera setups to improve 3D motion estimation and reduce occlusion errors.
- Exploring temporal models (e.g., LSTM, TCN, or Transformers) to move beyond frame-by-frame recognition and improve consistency across full kata sequences.
- Developing real-time feedback interfaces that provide correction suggestions based on standardized technical references.
- Extending the approach to other karate katas or even to other sports domains to assess broader applicability in human motion analysis.
- Designing hybrid systems that combine MLP-based classification with geometric rules to provide both robustness and interpretable feedback.
- Comparing different pose estimation libraries such as OpenPose and BlazePose to evaluate precision and robustness of keypoint detection.

In summary, this work establishes a proof of concept for the automatic recognition of karate movements, showing that even simple neural models trained on skeletal keypoints can provide reliable classification of stances and blocks. While the scope is limited, the results point toward the feasibility of combining computer vision and machine learning for sports analysis and educational applications in martial arts.

Acknowledgments. The authors would like to thank the Universidad Militar Nueva Granada (UMNG) for providing academic and research training spaces, such as the SIVICO research seedbed, which belongs to the ACCEDER research group.

References

1. Mostafa, K., Hany, M., Atia, A.: Automated evaluation of karate practitioners' style during the first kata performance using deep learning and pose estimation. In: 2023 5th Novel Intelligent and Leading Emerging Sciences Conference (NILES), pp. 238–241, 2023. https://doi.org/10.1109/NILES59815.2023.10296663
2. Hachaj, T., Ogiela, M.R., Koptyra, K.: Learning from annotated video: an initial study based on oyama karate tournament recordings. In: 2015 10th International Conference on P2P, Parallel, Grid, Cloud and Internet Computing (3PGCIC), pp. 358–362, 2015. https://doi.org/10.1109/3PGCIC.2015.16
3. Bhamidipati, V.S.P., et al.: Robust intelligent posture estimation for an AI gym trainer using mediapipe and OpenCV. In: 2023 International Conference on Networking and Communications (ICNWC), pp. 1–7, 2023. https://doi.org/10.1109/ICNWC57852.2023.10127264

4. Nguyen, T.N., Tran, T.H., Vu, H.: An automatic tool for yoga pose grading using skeleton representation. In: 2022 9th NAFOSTED Conference on Information and Computer Science (NICS), pp. 187–192, 2022. https://doi.org/10.1109/NICS56915.2022.10013455
5. Dhakate, H., et al.: Enhancing yoga practice: real-time pose analysis and personalized feedback. In: 2024 International Conference on Emerging Systems and Intelligent Computing (ESIC), pp. 35–40, 2024. https://doi.org/10.1109/ESIC60604.2024.10481659
6. Bianco, S., Tisato, F.: Karate moves recognition from skeletal motion. In: Three-Dimensional Image Processing (3DIP) and Applications 2013. Ed. by Atilla M. Baskurt and Robert Sitnik, vol. 8650. International Society for Optics and Photonics. SPIE, 86500K, 2013. https://doi.org/10.1117/12.2006229
7. Sharmila, P., et al.: IoT-oriented gesture automation with mesh detection through OpenCV and pyfirmata protocol using ResNet-Mediapipe. In: 2023 Innovations in Power and Advanced Computing Technologies (i-PACT), pp. 1–5, 2023. https://doi.org/10.1109/i-PACT58649.2023.10434872
8. Pang, Y., et al.: Analysis of computer vision applied in martial arts. In: 2022 2nd International Conference on Consumer Electronics and Computer Engineering (ICCECE), pp. 191–196, 2022. https://doi.org/10.1109/ICCECE54139.2022.9712803
9. Yu, K.: Research on martial arts action recognition and matching based on Kinect data. In: 2025 Asia-Europe Conference on Cybersecurity, Internet of Things and Soft Computing (CITSC), pp. 191–197, 2025. https://doi.org/10.1109/CITSC64390.2025.00042
10. Ait-Bennacer, F.-E., et al.: Applying deep learning and computer vision techniques for an e-sport and smart coaching system using a multiview dataset: case of Shotokan Karate. In: International Journal of Online and Biomedical Engineering (iJOE) 18.12 (Sept. 2022), pp. 35–53 (2022). https://doi.org/10.3991/ijoe.v18i12.30893. https://online-journals.org/index.php/i-joe/article/view/30893
11. Kumar Yadav, S., et al.: MS-KARD: a benchmark for multimodal karate action recognition. In: 2022 International Joint Conference on Neural Networks (IJCNN), pp. 1–8, 2022. https://doi.org/10.1109/IJCNN55064.2022.9892646

Uncovering Artistic Aspects in Artworks Recognition with Deep Neural Networks

Alex Torquato Souza Carneiro(✉) and Carolina Farnetani de Almeida

Impacta Faculty of Technology, São Paulo, SP, Brazil
alex.carneiro@faculdadeimpacta.com.br,
carolina.almeida@aluno.faculdadeimpacta.com.br
https://www.impacta.edu.br/

Abstract. The study of visual arts reveals interesting aspects of events, culture and moral values experienced by people in each period. From the prehistoric arts to the contemporary and digital arts, the artists aim to express themselves, their beliefs, social values, and how they see reality in paintings and handcrafted artworks. To do so, artists explore distinct aspects of painting, such as lighting effects, color palettes, textures, geometric patterns, preferred scenes and represented emotions. According to the sociocultural context, artworks present some similarities and these similarity patterns define an art style. Art experts may recognize an art style by analyzing a single painting as deep learning models do, but the question "what do deep learning models learn to recognize a paintings?" remains unanswered. In this work, we investigate what modern deep learning models learn to perform painting recognition in order to uncover artistic aspects in the learned features. Our findings suggest these models may learn features that correlate to artistic aspects and have potential to open a range of possibilities for applications in arts fields and new research works in explainable models.

Keywords: Deep Neural Networks · Artworks Recognition · Computer Vision · Features Learning · Explainable AI

1 Introduction

People around the world have expressed themselves through distinct art forms like literature, music, and visual arts across history. Artworks, shaped by specific sociocultural contexts, define styles like Baroque, Fauvism, and Cubism [1]. Studying these styles is a key research area in the arts, where skilled artists identify aspects like textures, lighting, emotions, and scenes in a painting.

Recently, deep learning (DL) models have advanced rapidly in computer vision (CV), excelling in visual pattern recognition. However, their ability to interpret the complex traits of artworks remains underexplored. The visual and emotional layers of paintings challenge models to grasp physical features and intangible elements like intent and culture [2,3]. This exploration could enhance explainable models with a concise, human-related feature set.

D. Chaves et al. (Eds.): CIARP 2025, LNCS 16528, pp. 112–126, 2026.
https://doi.org/10.1007/978-3-032-23161-1_9

Applying DL models to analyze and recognize art paintings and styles raises questions about their ability to decode these aspects during training. An art style is a distinctive manner or approach to creating visual works, characterized by specific techniques, visual elements, themes, and medium, reflecting social expressions and cultural, historical, or personal influences [4]. Defined by recurring traits, like color palettes, lines, textures, perspectives, and subjects, art styles evolve with contexts, e.g., Impressionism, Cubism, Abstract Expressionism, aiding art classification. Studies show DL can distinguish styles like Fauvism or Baroque and classify emotions via color and composition [5,6], yet a holistic approach to intrinsic elements remains underexplored. Recognizing lighting, texture, emotion, and style demands models grasp both visual patterns and contextual cues, aligning with human perception [7,8].

In this work, we address this gap by investigating the capacity of modern deep learning models to recognize intrinsic aspects of paintings for finding similarities among art styles. We built a set of relational graphs that focus on elements such as color, lighting, texture and perspective among art styles, and also a balanced labeled dataset with images of paints from distinct styles. Our goal is to assess the current capabilities and limitations of deep learning models in finding similarities across the addressed styles. This study contributes to the field of computational art analysis by highlighting potential areas where CV may yet fall short of human-like perception and identifying ways to improve the interpretative power of deep learning models in cultural and artistic contexts.

Several works have been conducted to propose solutions for art recognition, where deep learning models are employed to identify authors or art styles of paintings [9–16]. Our main goal is not to solve a classification task, but rather to find evidence that deep learning models are capable of finding similarities between distinct art styles and correlating these similarities with artistic aspects present in related literature [17,18]. This is a first study that addresses the explainable models under the art's point of view.

The next sections are organized as follows: Sect. 2 describes the aspects present in painting arts, Sect. 3 compiles a set of works that represent the state of the art for applications of DL and CV in art, Sect. 4 describes the dataset employed for the development of this study and how the DL models have been trained for the conducted experiment, Sect. 5 describes our results and how they correlate with the theoretical relationship models, and Sect. 6 has the study conclusion discussing the results and suggesting future works.

The main contributions of this study are:

- the theoretical relationship diagrams presented in Sect. 4;
- the balanced dataset with the collected images from WikiArt dataset;
- the outcomes obtained from the trained models and their relations to the theoretical relationship diagrams.

2 Artistic Aspects

There are a number of artistic aspects, from subjective ones, such as preferred scenes and represented emotions, to others like color, lighting, texture, and perspective. These aspects are key to visual representation, aiding perception, interpretation, and emotional response. They serve as tools for expression and engage human visual-cognitive systems. Throughout art history, artists have manipulated these to convey meaning, depth, material simulation, and psychology. Understanding their scientific basis offers insight into perception, linking art, neuroscience, and design, as detailed below.

Color is perhaps the most impactful visual element with symbolic, physical, and psychological significance. In neuroscience, it stems from receptors in the retina, processed by the brain's visual cortex [19]. Colors are used to depict reality, evoke emotions, e.g., blue for tranquility and red for passion, and convey meaning. Color theory, including complementary, analogous, and triadic relations, has long guided artists in achieving balance or tension in compositions.

Lighting is key in shaping artistic expression and perception. Artists use it to evoke emotions, direct attention, and add depth. Scientifically, it affects how people detect contrast and spatial orientation [20]. The direction, intensity, and color temperature of light interact with surfaces to set mood, from natural daylight to stylized artificial light.

Texture in art refers to the perceived or actual surface quality of a work, e.g., whether it is the rough impasto of a Van Gogh painting or the smooth polish of marble in classical sculpture. Studies in neuroscience show that texture perception involves both the visual and somatosensory systems, even when touch is not involved [21]. In visual art, texture can be simulated using tools like brushwork and layering, evoking sensations of softness, hardness, or fluidity. Artists often exploit textural contrast to enhance realism or create visual tension.

Perspective is a visual element that defines spatial relationships and depth in artworks. It leverages the human eye's ability to interpret converging lines and diminishing sizes, processed by the brain to create a three-dimensional effect. Artists use perspective, such as linear, atmospheric, or multiple viewpoints, to guide viewers through scenes, enhance realism, or express emotion. Techniques like one-point perspective for stability or distorted angles for drama have shaped artistic practices, enabling compositions that range from natural landscapes to abstract interpretations [22].

Manipulating color, lighting, texture and perspective enables creating complex structures in artworks, such as people, animals, nature, scenes and so on. In this way the analysis of these highly complex structures incorporates the mentioned artistic aspects and the interactions between them and other ones, e.g., shapes, emotions and symbolism [18,23].

In next the sections, we address the DL models as extractors of features and investigate associations between the patterns learned by the models and the aspects described in this section.

3 Applications of Deep Learning in Arts

Recent DL and CV advances have opened new ways to explore art styles, emotions, and techniques. An example is the style-transfer app SelfieArt introduced by Alegre and Oliveira (2020) [24], which applies distinct styles (e.g., Cubism or Baroque) to images, like the Mona Lisa (usually classified as High Renaissance). Though playful, these results highlight style differences effectively.

One relevant work is authored by Cetinic et al. (2018) [25] in which authors investigated the accuracy achieved by some CNN architectures, such as CaffeNet, Hybrid-CNN and MemNet, for art painting classification according to distinct sets of categories: artist, style, genre and nationality. The best accuracy score was 79.10%, observed for artist recognition with Hybrid-CNN. Similarly, Yu and Shi (2024) [26] applied an improved CNN-based approach, which introduces multi-scale analysis, to classify paintings into various artistic styles. Their study used WikiArt dataset, which contains more than 80,000 artworks distributed into multiple periods and genres. The authors demonstrated that DL models could achieve high classification accuracy while also identifying influential features associated with specific styles.

Recognition of subjective aspects of art paintings, more specifically emotions and visual patterns in faces, has also been investigated. For example, in Zhao et al. (2014) [27], the authors proposed a set of hand-crafted features to identify emotions represented in art paintings. In this work the authors proposed an outperforming classifier that recognizes specific emotions, such as anger, fear and disgust, that has been evaluated over three datasets: International Affective Picture System (IAPS), ArtPhoto and Abstract. They also hypothesized some interesting relations between image elements and emotions, e.g., snakes are usually associated to fear, these kinds of associations may be learned by DL models. Another work is Dalmoro et al. (2022) [28], where the authors investigated the relations between visual features in faces and the artists' styles.

Another work that investigated intrinsic features of artworks was authored by Saleh and Elgammal (2016) [5], in this work the authors compile a list of features that are fused for classification of art paintings using SVM models. The classification targets were style, genre and artist. The features explored in this work were GIST, Classeme, Picodes and CNN-based (AlexNet). The achieved results present accuracy up to 63.06% for artist classification, which meant an improvement of at least 9.17% compared to previous approaches.

Saleh and Elgammal (2016) [5] and Cetinic et al. (2018) [25] reported best results for artist recognition, suggesting unique artist features. We hypothesize it happens due to specific features introduced by each artist in their works. These features are considered discriminative for an art genre or style by human experts but these results show they are hard to be learned by DL models. Unlike their classification focus, we explore feature-artistic aspect correlations for explainability, motivated by their findings on DL limitations. By this way, the findings of these authors reinforce our motivation for uncovering artistic aspects that are recognized by experts in paintings but are not easily learned by DL models.

Similar to these works, Imran et al. (2023) [16] combined deep and shallow neural networks for painting style classification, achieving up to 93.6% accuracy on WikiArt, highlighting similarities in using CNN-based architectures for high accuracy. However, their approach differs from ours by emphasizing ensemble methods for discrimination, while we prioritize explainability through aspect correlations. This work proposes a more complex approach, in which ensembles reduce variance but obscure feature interpretability. Our findings extend these by showing how simpler models (e.g., VGG16) exhibit stronger aspect affinities, possibly because complex ensembles overfit to non-artistic features like noise in large datasets.

4 Methodology

In this work we explored artistic aspects in artworks and their relations to features learned by DL models. Thus, we needed a vector-like representation of the styles for comparing them to each other. The first step to obtain these vectors was to train a classifier to discriminate images among a set of art styles.

4.1 Models Training and Usage

We trained five DL models on the full WikiArt dataset (81,432 images) to classify 27 art styles. By using the entire dataset, the models could learn a large variety of features. Models were loaded with ImageNet weights and had top layers replaced with a single output, later they were trained over 20 epochs using batches of 128 RGB images (224 × 224 pixels). Only the last five layers were updated using Adam optimizer (learning rate 0.001) and Cross Entropy loss to prevent overfitting. The training results are reported in Table 1 and Fig. 1. The training and evaluations were done using Tensorflow framework over Google Colab environment with the following hardware configuration: Intel® Xeon® CPU, 12.7 GB of RAM and Nvidia GPU Tesla T4 with 15 GB of dedicated memory.

Table 1. Training results of the image classifiers.

Architecture	Loss	Accuracy
Xception	0.109	0.971
VGG16	0.316	0.901
InceptionV3	1.511	0.794
ResNet50	1.523	0.510
EfficientNetB4	2.857	0.184

These classifiers have been trained to be used as extractors of features, to do so we removed the output layer to access the vector of features learned by the models. We calculated the mean vectors (centroids) associated to the feature vector of each art style. These centroids correspond to the vector-like representation of the art styles. ResNet50 and EfficientNetB4 were excluded from the study due to lower accuracy.

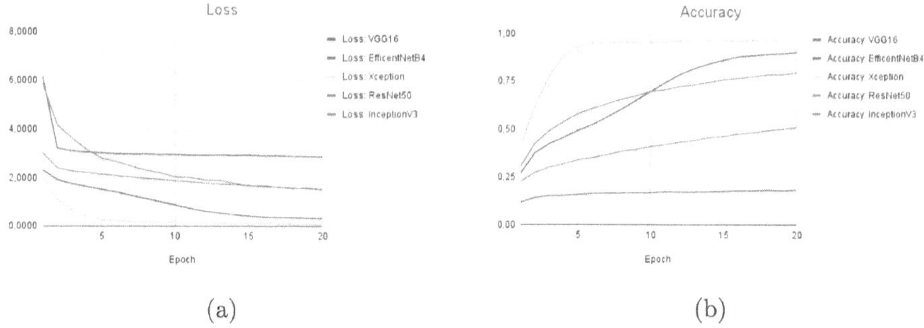

Fig. 1. Evolution of loss and accuracy for the models training.

4.2 Evaluation Dataset

We also collected 2026 images of eight art styles from WikiArt, labeled by style, forming the balanced **Art Styles dataset**[1]: this dataset contains between 239 and 296 images per style (mean 253.25, median 250). The chosen art styles were: Baroque, Realism, Rococo, Romanticism, Cubism, Expressionism, Fauvism and Impressionism, because they span from classical (e.g., Baroque's dramatic chiaroscuro and Rococo's ornate pastels) to modern (e.g., Cubism's geometric fragmentation and Fauvism's bold colors) patterns, testing model robustness and interpretability [29]. Figure 2 presents some samples of images collected from WikiArt to compose the Art Styles dataset.

Post-Renaissance Art and Modernism have been chosen because both are important landmarks in the history of art and reflect transformative periods: the former evolves classical ideals with expressive diversity, the latter breaks tradition with innovation and abstraction [18,30]. Together, these periods' styles offer a rich comparative framework that challenges classification models due to artists' multi-style works and the large variety of features present in the artworks. Their strategic choice highlights potential biases and diverse artistic aspects.

4.3 Theoretical Relationships between Art Styles

We built four relationship diagrams, presented in Figs. 3a–3d, according to the similarities between the art styles considering the addressed artistic aspects: color, lighting, texture, and perspective. These diagrams represent the theoretical relationships between the art styles and were built according to our qualitative observations over the paintings, related references (such as **The History of Art** by Ernst Gombrich (2006) [18]), and some support from www.grok.com platform.

The best performing models, i.e., Xception, VGG16 and InceptionV3, which likely learned the most discriminative features, were chosen to produce our results. The next step was to calculate the distances between the centroids and

[1] Art Styles dataset is available for download at https://2cm.es/1gkNX.

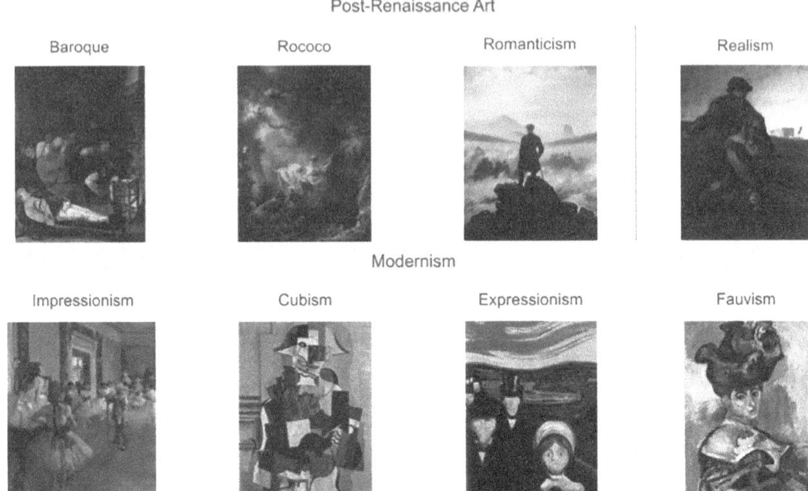

Fig. 2. Samples of images from the Art Styles dataset.

compile the closest styles to each other for verifying the correspondences according to the diagrams in the Fig. 3 as presented in next section.

5 Results

The Xception, VGG16 and InceptionV3 models provide, respectively, vectors with 100,352, 25,088 and 51,200 components from an RGB image of 224×224 pixels. We used the Euclidean and Cosine distances to determine the top-3 closest art styles to each other.

For each combination of model and distance, we built a table that contains all art styles with the respective top-3 closest styles. For example, suppose the style Baroque and its top-3 closest art styles for a given model and distance are Rococo, Expressionism and Cubism. Table 2 illustrates how Baroque appears in a row in the table followed by the top-3 closest art styles. The closest styles also indicate the correspondent relationships according to the diagrams of the Fig. 3 using superscripts (e.g., Rococo$^{2,3,\underline{4}}$).

Table 2. Illustrative example of top-3 closest art styles.

Art style	Closest art styles		
Baroque	Rococo$^{2,3,\underline{4}}$	Expressionism^{1,2}	Cubism$^{\emptyset}$

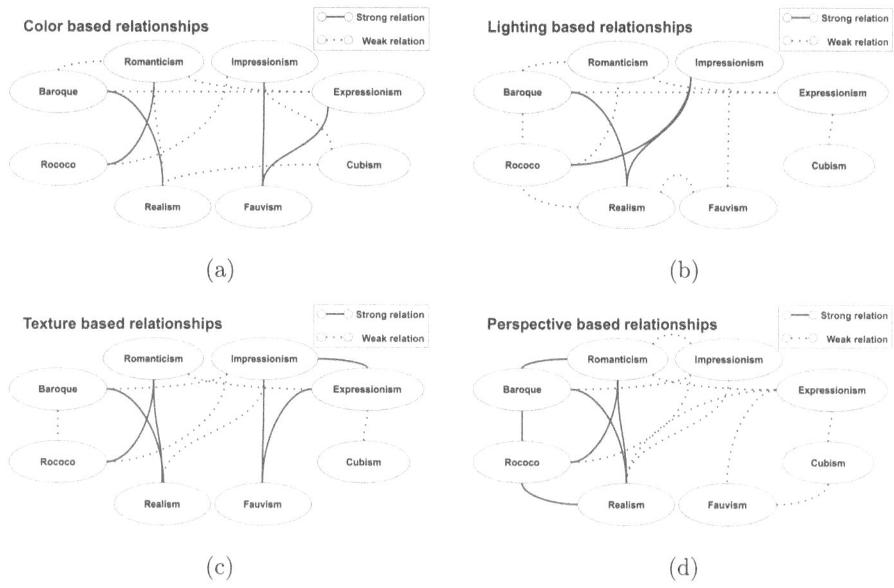

Fig. 3. Relationships between art styles based on artistic aspects: (a) Colors, (b) Lighting, (c) Texture and (d) Perspective.

The superscripts in the table use the codes: 1 $\rightarrow$ Color, 2 $\rightarrow$ Lighting, 3 $\rightarrow$ Texture and 4 $\rightarrow$ Perspective. Thus we see in Table 2, at right to Baroque, the art styles Rococo$^{2,3,\underline{4}}$, Expressionism^{1,2} and Cubism$^{\emptyset}$, it means:

- Baroque and Rococo are related (according to Fig. 3) in terms of Lighting, Texture and Perspective, represented by the values 2, 3 and $\underline{4}$ respectively, in this case the value 4 is underlined because there is a strong relationship between Baroque and Rococo in terms of Perspective;
- Baroque and Expressionism are related in terms of Color and Lighting, represented by the values 1 and 2 respectively, both values appear without underline because these relationships are weak;
- Baroque and Cubism are not related according to the addressed artist aspects, it is represented by the empty set symbol $\emptyset$.

The results of each pair of models and distances are presented following.

5.1 Xception Model

We might infer from the results reported in Table 3 how the centroids that represent the art styles are positioned in relation to the others and plausible biases acquired by the Xception model in the training. For example, the top-3 closest art styles to Baroque using Euclidean distance (upper half of the Table 3) are Realism, Rococo and Cubism, thus verifying these styles with correspondent relationships in the diagrams of Fig. 3 we observe that Realism presents strong relationships to Baroque in terms of all artistic aspects, Rococo presents weak relationships in terms of Lighting and Texture but strong in terms of Perspective, and Cubism does not present any relationship to Baroque.

Table 3. Results for the proximity between centroids using features learned by Xception model.

Art style	Closest art styles using Euclidean distance		
Baroque	Realism$^{\underline{1},\underline{2},\underline{3},\underline{4}}$	Rococo$^{2,3,\underline{4}}$	Cubism$^{\emptyset}$
Realism	Rococo$^{2,\underline{4}}$	Expressionism^{4}	Cubism^{1}
Rococo	Realism$^{2,\underline{4}}$	Impressionism$^{1,\underline{2},3,4}$	Fauvism$^{\emptyset}$
Romanticism	Impressionism^{4}	Rococo$^{\underline{1},2,\underline{3},\underline{4}}$	Fauvism$^{\emptyset}$
Cubism	Expressionism^{2,3,4}	Fauvism^{4}	Realism^{1}
Expressionism	Cubism^{2,3,4}	Fauvism$^{1,\underline{3},\underline{4}}$	Realism^{4}
Fauvism	Expressionism$^{\underline{1},\underline{3},4}$	Impressionism$^{\underline{1},2,\underline{3}}$	Rococo$^{\emptyset}$
Impressionism	Fauvism$^{\underline{1},2,\underline{3}}$	Rococo$^{1,\underline{2},3,4}$	Expressionism$^{\underline{3}}$
Art style	**Closest art styles using Cosine distance**		
Baroque	Cubism$^{\emptyset}$	Realism$^{\underline{1},\underline{2},\underline{3},\underline{4}}$	Expressionism^{1,2}
Realism	Rococo$^{2,\underline{4}}$	Impressionism$^{\underline{2},3,4}$	Fauvism^{2}
Rococo	Realism$^{2,\underline{4}}$	Impressionism$^{1,\underline{2},3,4}$	Fauvism$^{\emptyset}$
Romanticism	Rococo$^{\underline{1},2,\underline{3},\underline{4}}$	Impressionism^{4}	Realism$^{1,\underline{3},\underline{4}}$
Cubism	Expressionism^{2,3,4}	Baroque$^{\emptyset}$	Fauvism^{4}
Expressionism	Fauvism$^{\underline{1},\underline{3},4}$	Cubism^{2,3,4}	Realism^{4}
Fauvism	Impressionism$^{\underline{1},2,\underline{3}}$	Expressionism$^{\underline{1},\underline{3},4}$	Realism^{2}
Impressionism	Fauvism$^{\underline{1},2,\underline{3}}$	Realism$^{\underline{2},3,4}$	Rococo$^{1,\underline{2},3,4}$

Comparing the reports of the distinct distance methods, we observe some interesting clues about possible biases acquired by Xception model, for example: the Lighting aspect comprehends 11 relations presented with Euclidean distance and 15 relations with Cosine distance, it suggests using Cosine distance increases the affinity of Xception model to Lighting aspect. Otherwise, the other aspects have been less impacted by changing the distance method, although Cosine distance increased the number of correspondences to the artistic aspects in general. A deeper analysis about these results is presented in the ending of this section.

5.2 VGG16 Model

The results reported in Table 4 provide insights relative to the VGG16 model after trained to classify the addressed art styles.

The perspective-based relationships, presented in Fig. 3d, mostly include the top-3 closest art styles according to VGG16 model, especially using Euclidean distance with 21 occurrences, this suggests a strong affinity of VGG16 to features that evoke perspective patterns when the model is employed with Euclidean

Table 4. Results for the proximity between centroids using features learned by VGG16 model.

Art style	Closest art styles using Euclidean distance		
Baroque	Realism$^{\underline{1},\underline{2},\underline{3},\underline{4}}$	Romanticism$^{1,2,\underline{4}}$	Rococo$^{2,3,\underline{4}}$
Realism	Romanticism$^{1,\underline{3},\underline{4}}$	Baroque$^{\underline{1},\underline{2},\underline{3},\underline{4}}$	Rococo$^{2,\underline{4}}$
Rococo	Romanticism$^{\underline{1},2,\underline{3},\underline{4}}$	Impressionism$^{1,\underline{2},3,4}$	Realism$^{2,\underline{4}}$
Romanticism	Realism$^{1,\underline{3},\underline{4}}$	Rococo$^{\underline{1},2,\underline{3},\underline{4}}$	Baroque$^{1,2,\underline{4}}$
Cubism	Fauvism^{4}	Expressionism^{2,3,4}	Impressionism^{1}
Expressionism	Fauvism$^{\underline{1},\underline{3},4}$	Impressionism$^{\underline{3}}$	Cubism^{2,3,4}
Fauvism	Expressionism$^{\underline{1},\underline{3},4}$	Impressionism$^{\underline{1},2,\underline{3}}$	Cubism^{4}
Impressionism	Realism$^{\underline{2},3,4}$	Romanticism^{4}	Baroque^{3,4}
Art style	**Closest art styles using Cosine distance**		
Baroque	Realism$^{\underline{1},\underline{2},\underline{3},\underline{4}}$	Romanticism$^{1,2,\underline{4}}$	Rococo$^{2,3,\underline{4}}$
Realism	Romanticism$^{1,\underline{3},\underline{4}}$	Impressionism$^{1,\underline{2},3,4}$	Baroque$^{\underline{1},\underline{2},\underline{3},\underline{4}}$
Rococo	Romanticism$^{\underline{1},2,\underline{3},\underline{4}}$	Baroque$^{2,3,\underline{4}}$	Realism$^{2,\underline{4}}$
Romanticism	Realism$^{1,\underline{3},\underline{4}}$	Impressionism^{4}	Rococo$^{\underline{1},2,\underline{3},\underline{4}}$
Cubism	Fauvism^{4}	Impressionism^{1}	Expressionism^{2,3,4}
Expressionism	Fauvism$^{\underline{1},\underline{3},4}$	Cubism^{2,3,4}	Impressionism$^{\underline{3}}$
Fauvism	Expressionism$^{\underline{1},\underline{3},4}$	Cubism^{4}	Impressionism$^{\underline{1},2,\underline{3}}$
Impressionism	Realism$^{\underline{2},3,4}$	Fauvism$^{\underline{1},2,\underline{3}}$	Expressionism$^{\underline{3}}$

distance. Also several correspondences are observed between the texture-based relationships and the closest pairs reported in Table 4, where we highlight the results reported with Cosine distance, with 18 correspondences, that comprehend all strong theoretical relationships within 12 correspondences for them.

5.3 InceptionV3 Model

Finally the results reported in Table 5 provide insights relative to the InceptionV3 model after trained to classify the addressed art styles.

As observed with VGG16, InceptionV3 model has also reported results that suggest strong affinity to the Perspective aspect, but not as strong as VGG16, with 19 and 18 occurrences of perspective-based relationships with Euclidean and Cosine distances, respectively. We also highlight InceptionV3's affinity for texture-based relationships, with 16 occurrences identified using both distances.

5.4 Results Analysis

Compiling the results reported in Tables 3, 4 and 5, we found evidences that suggest the addressed artistic aspects represent information with relevant level of affinity to the features learned by the DL models as following described.

The pair **VGG16/Euclidean** distance reported results with 21 occurrences relative to Perspective aspect being the highest observed value. All of strong

Table 5. Results for the proximity between centroids using features learned by InceptionV3 model.

Art style	Closest art styles using Euclidean distance		
Baroque	Realism$^{\underline{1},\underline{2},\underline{3},\underline{4}}$	Rococo$^{2,3,\underline{4}}$	Expressionism^{1,2}
Realism	Rococo$^{2,\underline{4}}$	Baroque$^{\underline{1},\underline{2},\underline{3},\underline{4}}$	Impressionism$^{1,\underline{2},3,4}$
Rococo	Realism$^{2,3,\underline{4}}$	Impressionism$^{1,\underline{2},3,4}$	Romanticism$^{\underline{1},2,\underline{3},\underline{4}}$
Romanticism	Rococo$^{\underline{1},2,\underline{3},\underline{4}}$	Impressionism^{4}	Realism$^{1,\underline{3},\underline{4}}$
Cubism	Expressionism^{2,3,4}	Impressionism^{1}	Fauvism^{4}
Expressionism	Fauvism$^{\underline{1},\underline{3},4}$	Cubism^{2,3,4}	Impressionism$^{\underline{3}}$
Fauvism	Expressionism$^{\underline{1},\underline{3},4}$	Cubism^{2,3,4}	Impressionism$^{\underline{1},2,\underline{3}}$
Impressionism	Rococo$^{1,\underline{2},3,4}$	Cubism^{1}	Realism$^{\underline{2},3,4}$
Art style	**Closest art styles using Cosine distance**		
Baroque	Realism$^{\underline{1},\underline{2},\underline{3},\underline{4}}$	Rococo$^{2,3,\underline{4}}$	Cubism$^{\emptyset}$
Realism	Cubism^{1}	Rococo$^{2,\underline{4}}$	Impressionism$^{1,\underline{2},3,4}$
Rococo	Realism$^{2,3,\underline{4}}$	Impressionism$^{1,\underline{2},3,4}$	Romanticism$^{\underline{1},2,\underline{3},\underline{4}}$
Romanticism	Realism$^{1,\underline{3},\underline{4}}$	Rococo$^{\underline{1},2,\underline{3},\underline{4}}$	Impressionism^{4}
Cubism	Expressionism^{2,3,4}	Fauvism^{4}	Impressionism^{1}
Expressionism	Cubism^{2,3,4}	Fauvism$^{\underline{1},\underline{3},4}$	Impressionism$^{\underline{3}}$
Fauvism	Expressionism$^{\underline{1},\underline{3},4}$	Cubism^{4}	Impressionism$^{\underline{1},2,\underline{3}}$
Impressionism	Expressionism$^{\underline{1},\underline{3},4}$	Rococo$^{1,\underline{2},3,4}$	Realism$^{1,\underline{2},3,4}$

perspective-based relationships are in the top-3 closest art styles reported with VGG16 model and both distances, while the results of the other models do not include all strong perspective-based relationships. Despite their importance, the results obtained do not suggest strong affinity with the other aspects as observed in Table 6, where we reported the rate of occurrences of the relationships per art style in the top-3 closest sets of each model. Because Tables 3, 4 and 5 contain 8 art styles and their respective top-3 closest ones, there are 24 occurrences at maximum.

Table 6. Rate of occurrences of the relationships in the models' results.

Model	Artistic aspect			
	Color	Lighting	Texture	Perspective
Xception/Euclidean	41.7% (10)	45.8% (11)	45.8% (11)	62.5% (15)
Xception/Cosine	41.7% (10)	62.5% (15)	54.2% (13)	66.7% (16)
VGG16/Euclidean	54.2% (13)	58.3% (14)	66.7% (16)	**87.5% (21)**
VGG16/Cosine	50.0% (12)	58.3% (14)	75.0% (18)	**79.2% (19)**
InceptionV3/Euclidean	54.2% (13)	62.5% (15)	66.7% (16)	**79.2% (19)**
InceptionV3/Cosine	45.8% (11)	54.2% (13)	66.7% (16)	75.0% (18)

The highest rate in Table 6 is observed for the pair **VGG16/Euclidean** that contains 21 occurrences of perspective-based relationships into the closest art styles. It means 87.5% of the occurrences in top-3 closest styles correspond to perspective-based relationships. We also observed in Table 6 the second highest rate is presented by the pairs **VGG16/Cosine** and **InceptionV3/Euclidean**, with same number of 19 occurrences that means 79.2% of top-3 closest styles. Although these are the most relevant results, their statistical significance should be evaluated to consolidate our findings and hypothesis.

In addition, the strength of the relationships is an important analysis factor. Figure 4 presents the numbers of occurrences divided into weak and strong relationships. Following we compare the outcomes of each pair **model/distance** for all aspects and provide a vision for occurrences of the relationships.

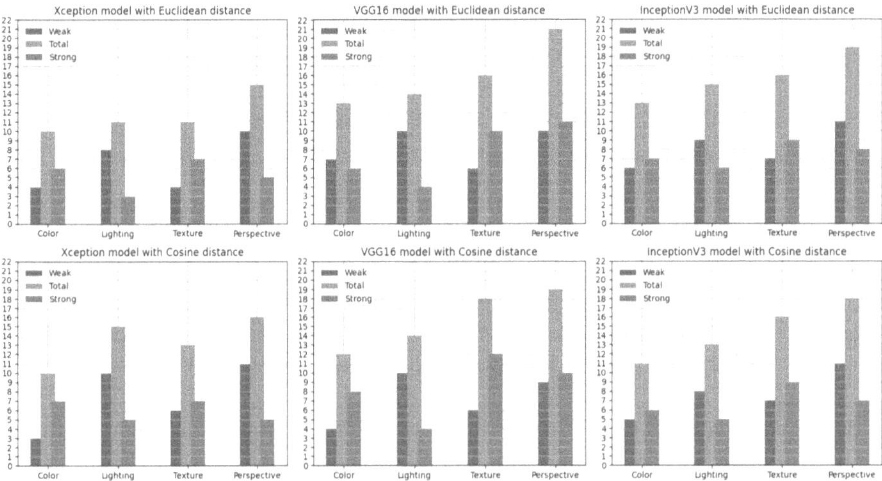

Fig. 4. Counting of occurrences of the weak and strong relationships in the models' results.

To assess differences in affinity strengths for the artistic aspects depicted in Fig. 4 across the six model-distance pairs, a Wilcoxon Signed-Rank Test was performed. The test analyzed the number of occurrences of the artistic aspects across the model-distance pairs (e.g., Xception/Euclidean and VGG16/Cosine). The null hypothesis (H_0) assumed no difference in median ranks of occurrences among these pairs. The largest difference reported by the test was for **Perspective** aspect with the **VGG16/Euclidean** pair, that resulted in a W-statistic of 1.420 and a p-value of 0.217, indicating no statistically significant difference. Although VGG16/Euclidean exhibited the highest occurrence rate (87.5%), this finding was not significant, suggesting a potential influence of its simpler feature extraction architecture that warrants further exploration with a larger dataset or scope of aspects.

Because this is an initial study, our findings suggest affinities between learned features and artistic aspects, though not definitive. Table 6 and Fig. 4 highlight perspective's prevalence, indicating stronger model affinity for it. Even though Xception's training performance is superior, it shows lower aspect affinity, with more $\emptyset$ top-3 styles. This suggests Xception learns diverse features beyond aspects, fitting its complexity and larger vector. Complex models thus favor diversity, reversing the trend to few focused aspects.

To verify the reverse affinity trend, a Pearson correlation test was conducted on the sum of weak and strong occurrences from each artistic aspect and feature vector sizes. We consider models that produce larger feature vector sizes as more complex to enable the Pearson correlation test. The test yielded correlation scores of -0.999 and $p < 0.01$ for perspective with both euclidean and cosine distances, indicating a statistically significant negative correlation in these cases. Other correlation scores were closer to 0 (zero) and $p > 0.1$ for all cases. Considering all artistic aspects together, the test yielded correlation scores of -0.956 and -0.971, and $p = 0.188$ and $p = 0.152$ for euclidean and cosine distances respectively. We highlight the complexity of the models also follow the reverse order of the affinity strength, due to the model's depth, parameter count, architectural innovation, and computational demand.

6 Conclusions

In this work, we investigated how deep learning (DL) models learn features for painting recognition and also looked for uncovering artistic aspects using trained Xception, VGG16, and InceptionV3 models. There are the top performers from five models. Also Euclidean and Cosine distances were evaluated with these models. We created relationship graphs for art styles based on color, lighting, texture, and perspective, comparing them to the top-3 closest styles from DL models. The highest affinity was between VGG16/Euclidean and perspective, indicating a bias toward learning perspective features. Perspective's frequent top-3 occurrence highlights its prominence in learned features.

We also highlight the results of Xception, that is the most complex model and provides the largest vector of features. Some of the top-3 closest art styles for Xception do not present any correspondence to the addressed artistic aspects and this model presented the lowest number of occurrences into the relationships when compared to the other models, independently of the distance method. This finding suggests a reverse affinity trend as the models provide larger feature vectors. Model complexity also follows this reverse trend, driven by architecture depth, number of parameters, and processing demand. However, no significant evidence suggesting a predominance of any other aspect than Perspective.

Despite these insights, our approach faces limitations. The WikiArt dataset has human-made biases, like curator choices and digitization, favoring Western styles and boosting color affinity while likely underrepresenting texture, due to non-digital artifact gaps. Model complexity affects results: simpler VGG16 shows higher affinity (87.5% for perspective) via focused extraction, while complex

Xception overgeneralizes, increasing $\emptyset$ matches. The relatively small test dataset size (2,026 images) might limit our observations.

Our findings have implications for computational art and explainable models. In computational art, the uncovered affinities enable applications like automated museum curation, where models recommend exhibits based on aspect similarities (e.g., linking Baroque lighting to Romanticism's emotions), or educational tools that visualize style evolutions. For explainable models, the reverse affinity trend highlights how complexity affects interpretability, aiding bias detection in AI art generation. Future investigations might consider using larger datasets, employing advanced optimization techniques for models training, exploring other approaches to obtain extractors of features, and analyzing the models' weights to find correlations between the artistic aspects and the learned features.

Acknowledgments. The authors would like to thank Impacta Faculty of Technology, the institution directors and Prof. Dr. Renato Santos for the support provided during the development of this work.

Disclosure of Interests. The authors have no competing interests to declare that are relevant to the content of this article.

References

1. Wölfflin, H., Hottinger, M.D.: Principles of art history: the problem of the development of style in later art. Dover New York, 1950
2. Tan, M., Le, Q.: Efficientnet: rethinking model scaling for convolutional neural networks. In: International Conference on Machine Learning, pp. 6105–6114. PMLR, 2019
3. Cetinic, E., Lipic, T., Grgic, S.: A deep learning perspective on beauty, sentiment, and remembrance of art. IEEE Access **7**, 73694–73710 (2019)
4. Riggle, N.: Personal style and artistic style. Philos. Q. **65**(261), 711–731 (2015)
5. Saleh, B., Elgammal, A.: Large-scale classification of fine-art paintings: learning the right metric on the right feature. Int. J. Digit. Art History (2016)
6. Santos, I., Castro, L., Rodriguez-Fernandez, N., Torrente-Patino, A., Carballal, A.: Artificial neural networks and deep learning in the visual arts: a review. Neural Comput. Appl. **33**, 121–157 (2021)
7. Manovich, L.: Computer vision, human senses, and language of art. AI Soc. **36**(4), 1145–1152 (2021)
8. Spee, B.T., Mikuni, J., Leder, H., Scharnowski, F., Pelowski, M., Steyrl, D.: Machine learning revealed symbolism, emotionality, and imaginativeness as primary predictors of creativity evaluations of western art paintings. Sci. Rep. **13**(1), 12966 (2023)
9. Lombardi, T.E.: The classification of style in fine-art painting. Pace University, 2005
10. Blessing, A., Wen, K.: Using machine learning for identification of art paintings. Stanford University Technical Report, 2010
11. Čuljak, M., Mikuš, B., Jež, K., Hadjić, S.: Classification of art paintings by genre. In: 2011 Proceedings of the 34th International Convention MIPRO, pp. 1634–1639. IEEE, 2011

12. Smirnov, S., Eguizabal, A.: Deep learning for object detection in fine-art paintings. In: 2018 Metrology for Archaeology and Cultural Heritage (MetroArchaeo), pp. 45–49. IEEE, 2018
13. Sandoval, C., Pirogova, E., Lech, M.: Two-stage deep learning approach to the classification of fine-art paintings. IEEE Access **7**, 41770–41781 (2019)
14. Liu, S., Yang, J., Agaian, S.S., Yuan, C.: Novel features for art movement classification of portrait paintings. Image Vis. Comput. **108**, 104121 (2021)
15. Cascone, L., Nappi, M., Narducci, F., Russo, S.L.: Classification of fragments: recognition of artistic style. J. Ambient Intell. Humaniz. Comput. **14**(4), 4087–4097 (2023)
16. Imran, S., et al.: Artistic style recognition: combining deep and shallow neural networks for painting classification. Mathematics **11**(22), 4564 (2023)
17. Gompertz, W.: What are you Looking at? 150 Years of Modern Art in the Blink of an Eye/Will gompertz. Penguin Books Ltd., Great Britain (2012)
18. Gombrich, E.: The History of Art, 16th ed. Phaidon Press, London (2006)
19. Wandell, B.A., Dumoulin, S.O., Brewer, A.A.: Visual field maps in human cortex. Neuron **56**(2), 366–383 (2007)
20. Boyce, P.R.: Human Factors in Lighting. CRC press, Boca Raton (2003)
21. Klatzky, R., Lederman, S.: Object recognition by touch. In: Blindness and Brain Plasticity in Navigation and Object Perception, pp. 197–220. Psychology Press, 2007
22. Kemp, M., et al.: The science of art: optical themes in western art from Brunelleschi to Seurat. Yale University Press New Haven, 1990
23. Mather, G.: The Psychology of Visual Art: Eye, Brain and Art. Cambridge University Press, Cambridge (2013)
24. Alegre, L.N., Oliveira, M.M.: Interactive multi-style transfer for selfies and videos with soft transitions. In: 2020 33rd SIBGRAPI Conference on Graphics, Patterns and Images (SIBGRAPI), pp. 17–22. IEEE, 2020
25. Cetinic, E., Lipic, T., Grgic, S.: Fine-tuning convolutional neural networks for fine art classification. Expert Syst. Appl. **114**, 107–118 (2018)
26. Qing, Y., Shi, C.: An image classification approach for painting using improved convolutional neural algorithm. Soft Comput. **28**(1), 847–873 (2024)
27. Zhao, S., Gao, Y., Jiang, X., Yao, H., Chua, T.S., Sun, X.: Exploring principles-of-art features for image emotion recognition. In: Proceedings of the 22nd ACM International Conference on Multimedia, pp. 47–56, 2014
28. Dalmoro, B., Monteiro, C., Musse, S.R.: Measuring the influence of painters through artwork facial features. In: 2022 35th SIBGRAPI Conference on Graphics, Patterns and Images (SIBGRAPI), vol. 1, pp. 37–42. IEEE, 2022
29. Alpaieva, Y.: A comparative study of selected machine learning methods for art images classification. Master's thesis, Univerzita Karlova, Matematicko-fyzikální fakulta, 2025
30. Allen, J.: Art since 1900: modernism, antimodernism, postmodernism. Ref. Rev. **20**(2), 43 (2006)

The Role of Cyclopean-Eye in Stereo Vision

Sherlon Almeida da Silva[1,2(✉)], Davi Geiger[2], Luiz Velho[2,3], and Moacir Antonelli Ponti[1]

[1] Instituto de Ciências Matemáticas e de Computação (ICMC), Universidade de São Paulo (USP), São Carlos, SP, Brazil
sherlon@usp.br, moacir@icmc.usp.br

[2] Courant Institute of Mathematical Sciences, New York University (NYU), New York, NY, USA
dg1@nyu.edu

[3] Instituto de Matemática Pura e Aplicada (IMPA), Rio de Janeiro, RJ, Brazil
lvelho@impa.br

Abstract. This work investigates geometric foundations of modern stereo vision systems, focusing on how 3D structure and human-inspired perception contribute to depth reconstruction. We revisit state-of-the-art stereo vision systems, analyzing their feature matching in the Cyclopean Eye coordinate system and adopting a Bayesian formulation to describe the surface reconstruction problem. We provide a theoretical understanding of occlusions and discontinuities in the Cyclopean domain, introducing novel geometric constraints. Through both theoretical insights and empirical studies, we demonstrate that combining geometric priors with learned features yields internal abstractions for understanding stereo vision systems, going beyond data matching to reveal underlying structures in occlusions, textureless/homogeneous regions, and repetitive patterns. Our analysis also includes an evaluation of stereo features from deep learning models and an examination of the role of visual attention mechanisms in recovering 3D surfaces.

Keywords: Explainable Stereo Vision · 3D Scene Understanding · Depth Estimation · Visual Human Perception

1 Introduction

This paper addresses the problem of stereo vision, recovering a 3D scene from the left (L) and right (R) images of the scene. With the advent of deep learning (DL) and the availability of datasets, a high level of accuracy has been achieved in stereo by RAFT-Stereo [15], CREStereo [13], DLNR [26], Selective-IGEV [22], LoS [14], and more recently by FoundationStereo [23], DEFOM-Stereo [11], MonoStereo [4], StereoAnywhere [1], and S^2M^2 [17]. In these approaches, during training the input is a stereo image pair, and the loss function is based on the discrepancy between the model predicted disparity map and the ground truth

D. Chaves et al. (Eds.): CIARP 2025, LNCS 16528, pp. 127–141, 2026.
https://doi.org/10.1007/978-3-032-23161-1_10

(GT) disparity. Once trained, a DL-based stereo model can infer the disparity from a new pair of images. In close examination, these methods first extract features from the left and right images and then perform disparity estimation at each pixel. Despite DL-based stereo achieving state-of-the-art (SOTA) performance in benchmarks, there is a lack of **understanding** on how to go from features to disparity output.

Understanding this problem means uncovering meaningful internal representations that capture its structure, as is the case of image stereo features. Not only enables better generalization to novel situations, it also facilitates broader application of the underlying principles. Our goal is to deepen our understanding of stereo vision in addition to feature extraction. By developing abstract 3D representations of reality, we aim in the future to improve the performance and efficiency of stereo algorithms and also to lay the groundwork for using such 3D models in other areas of computer vision.

1.1 Geometrical Ideas in Stereo Vision

In our pursuit of a deeper understanding of stereo vision, we adopt a Bayesian framework in which the problem is formulated as estimating the most probable 3D surface, S^{3D}, given a stereo pair of input images, I^L and I^R. Specifically, stereo vision aims to solve:

$$\arg\max_{S^{3D}} P(S^{\mathrm{3D}}|I^L, I^R) = \arg\max_{S^{3D}} \left[P(I^L, I^R|S^{3D})\, P(S^{3D})\right]$$

where $P(I^L, I^R|S^{\mathrm{3D}})$ models the likelihood of observing the image pair given a hypothesized 3D surface, and $P(S^{\mathrm{3D}})$ is a prior distribution that encapsulates geometric regularities of natural surfaces.

1.2 Brief History of Geometric Ideas in Stereo

The study of such geometric principles long predates Bayesian formalism. As early as 1499, Leonardo da Vinci, in Treatise on Painting, observed that different parts of the background become occluded depending on whether one views the scene with the left or right eye [5]. Later, Hermann von Helmholtz [9] laid foundational work in binocular vision, emphasizing its role in depth perception. He introduced the concept of the Cyclopean Eye – a notional single eye positioned centrally on the head – representing the brain's fused interpretation of binocular input. This idea also serves as a useful geometric abstraction: a coordinate system centered between the eyes for representing 3D surfaces.

Further contributions came from Julesz [12], who championed the Cyclopean framework and a bottom-up approach to stereo vision, later advanced by Marr and Poggio [16] through a computational model of stereopsis. These studies, and much of the computational modeling up to the early 2000's, laid a conceptual foundation that preceded the current dominance of purely data-driven DL methods.

1.3 Our Contribution

In this work, we revisit and extend geometric ideas to model $P(S^{3D})$, present in Eq. (5), and analyze these ideas with data from the Middlebury dataset [19] and RAFT-Stereo [15] features that yield FM_*.

Our main contribution is on **Foundational Vision Understanding** by deepens our understanding of 3D scene geometry, paving the way for more interpretable, generalizable and reliable vision systems.

More specifically, the contributions are as follows:

- **Features:** an analysis of the quality of the features when learned and extracted by DL algorithms from the L/R images.
- **Geometric Constraints:** a proposal of novel geometric constraints for stereo vision.
- **Cyclopean Eye Model:** its role in better understanding the geometry of stereo vision.
- **Attention:** we argue for the need of an attention mechanism to recover 3D surfaces.

2 The Cyclopean 3D View

We begin by briefly reviewing the Cyclopean coordinate system (XD), which provides a natural framework to describe the geometry of the 3D world. In particular, the relation between the match of the left and right features and the depth assignment is elucidated, so that in the next section we propose novel geometrical constraints.

2.1 Cyclopean Coordinate System (XD)

Consider a left and right images I^L, I^R both with height and width $M \times N$. A pixel coordinate system (CS) for the images are $\{(e, l), (e, r)\}$ where $e \in (0, 1, \dots M-1)$ index their respective epipolar lines and $l, r \in (0, 1, \dots N-1)$ (see Fig. 1). For many datasets, such as Middlebury [19], the images have been rectified by the epipolar lines, which are then simply the horizontal lines of the images. The space $(e, x, d) \in M \times 2N \times D$ allows us to describe the matching of $(e, l) \leftrightarrow (e, r)$ as an assignment of a disparity d to (e, x) (see Figs. 1, 2, and 3). Such matching and associated assignment is described by the invertible coordinate transformation:

$$\begin{pmatrix} d \\ x \end{pmatrix} = \begin{pmatrix} 1 & -1 \\ \frac{1}{2} & \frac{1}{2} \end{pmatrix} \begin{pmatrix} l \\ r \end{pmatrix} . \tag{1}$$

One consequence of this transformation is that the discrete Cyclopean width coordinate $x \in (0, \frac{1}{2}, 1, \frac{3}{2}, \dots, N-1, N-\frac{1}{2})$ has subpixel resolution, twice as much as the image pixel width resolution of $l, r \in (0, 1, \dots, N-1)$ (see Fig. 1).

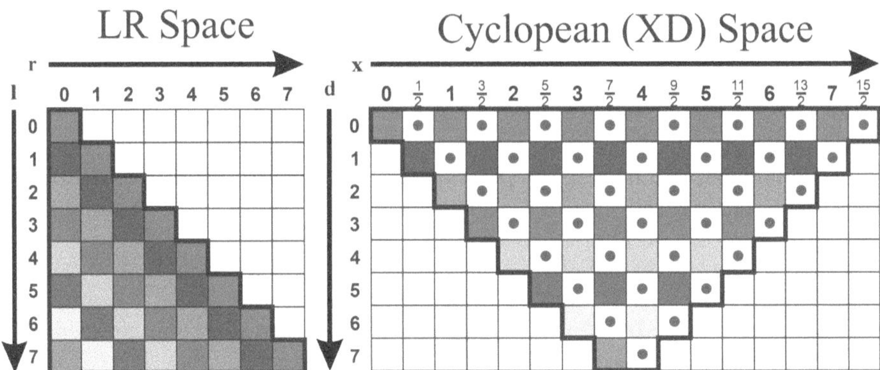

Fig. 1. Space Transformation from L/R CS (left) to the XD (right). The colors represent a disparity value. Empty positions in XD space are disallowed, while the 'red dot' data are obtained via a bilinear interpolation from LR data. The XD space has twice the resolution of the LR space, for each epipolar line. (Color figure online)

The XD provides a depth value $\mathcal{D}(e, x)$, for each match $(e, l) \leftrightarrow (e, r)$, as follows

$$\mathcal{D}(e, x) = f \frac{B}{d(e, x)}, \tag{2}$$

where f is the focal length, B is the baseline (distance between the left and right camera centers), see Fig. 2.

2.2 Depth from L, R, and XD

From Fig. 2 we infer the following triangle relations

$$\mathcal{D}^L(e, l) = \sqrt{\mathcal{D}^2(e, x) + \left(\frac{B}{2} - x\right)^2} \quad \text{and}$$
$$\mathcal{D}^R(e, r) = \sqrt{\mathcal{D}^2(e, x) + \left(\frac{B}{2} + x\right)^2} \tag{3}$$

These relationships introduce a bias in the depth estimation from the L and R cameras relative to the XD depth values. To the best of our knowledge, these bias equations have not been derived previously. The impact becomes evident when comparing depth estimates with man-made GT scenarios. A consequence of mismatches in L/R depth estimation yield, for example, motion sickness in virtual reality [24].

Proposition 1 (Datasets Depth in XD). *The depth provided by the Middlebury dataset, and most man-made datasets, is obtained by the match generated by a light projector (or laser) through Formula* (2)*. Thus, it is not the depth from the L or R coordinate system.*

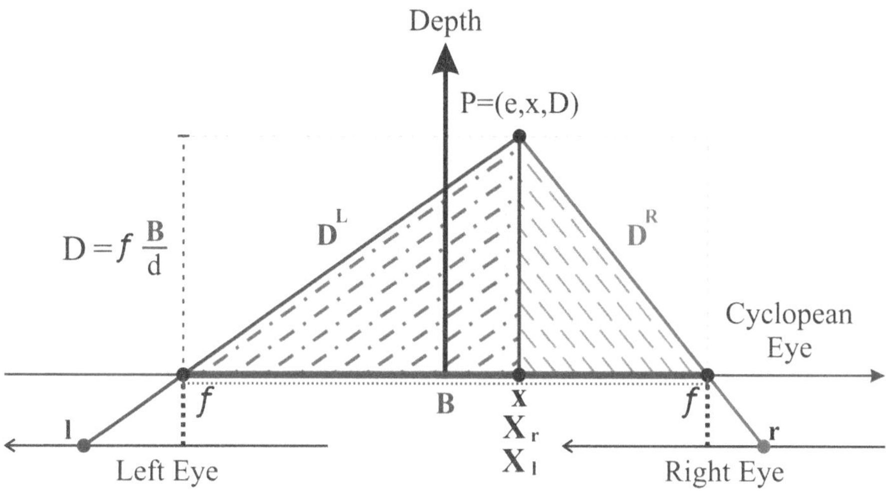

Fig. 2. $D^{L/R}(e, l/r)$ is the depth from L/R CS, respectively. A point P in 3D is described by the XD as $P = P^C(e, x, Z = \mathcal{D}(e, x))$. The same point can be described by the L/R CS as $P = P^{L,R}(e, X_{l,r}, \mathcal{D}(e, x))$, where $X_{l,r} \neq l, r$, since l, r are the projective projection of P into the L/R CS, while $X_{l,r}$ is the simpler orthogonal projection of P into the L/R CS. Note that $B = |X_l - X_r|$. The distance to P measured by the L, R, and Cyclopean Eye are $\mathcal{D}^L(e, l), \mathcal{D}^R(e, r), \mathcal{D}(e, x)$, respectively, and they are all different.

3 Image Features for Stereo Matching

Image features are the first stage of processing for DL algorithms and in particular RAFT-Stereo [15] architecture allows other DL algorithms to extract their features, making them one of the most used features in deep learning.

A Bayesian interpretation for the use of features F_*^L and F_*^R from left and right images, respectively, is as follows

$$
\begin{aligned}
P(I^L, I^R | S^{3\mathrm{D}}) &= \sum_{F^L, F^R} P(I^L, I^R, F^L, F^R | S^{3\mathrm{D}}) \\
&= \sum_{F^L, F^R} P(I^L, I^R | F^L, F^R) P(F^L, F^R | S^{3\mathrm{D}}) \\
&\approx P(I^L, I^R | F_*^L, F_*^R) P(F_*^L, F_*^R | S^{3\mathrm{D}}) \\
&= P(I^L, I^R | F_*^L, F_*^R) P(FM_* | S^{3\mathrm{D}}),
\end{aligned}
$$

where F_*^L, F_*^R dominate the sum and RAFT-Stereo [15] also does the following computations

$$
\begin{aligned}
FM(e, x, d(e, x)) &= 1 - \frac{FMS(e, x, d(e, x))}{max(FMS(e, x, d(e, x))}, \quad \text{where} \\
FMS(e, x, d(e, x)) &= F^L(e, x - \frac{1}{2} d(e, x)) \cdot F^R(e, x + \frac{1}{2} d(e, x)),
\end{aligned}
$$

and $d(e,x)$ is the Cyclopean disparity that describes the 3D surface. Thus, the feature matching similarity (FMS) must be large when there is a good feature match and $FM(e,x,d(e,x)) \in [0,1]$ is used so that good matches render these values small and bounded to the $[0,1]$ range.

In the Bayesian view, SOTA DL algorithms, e.g., RAFT-Stereo [15], break the stereo problem

$$\arg\max_{S^{3D}} P(S^{3\mathrm{D}}|I^L, I^R) \tag{4}$$

in two stages, first extracting features F_*^L, F_*^R from left and right images and then performing in the second stage

$$\arg\max_{S^{3D}} \left[P(S^{3\mathrm{D}}|FM_*)\right] = \arg\max_{S^{3D}} \left[P(FM_*|S^{3\mathrm{D}})\, P(S^{3D})\right] \tag{5}$$

Figures 4, 5, 7, and 8 show the quality of their extracted $FM_*(e,x,d(e,x))$ yielding a fairly sparse description in the matching space. In various regions, where data matching is available, it allows a visualization of the $S^{3\mathrm{D}}$ solution that maximizes the likelihood $P(FM_*|S^{3\mathrm{D}})$.

Our data investigations led us to Fig. 8 where multiple matches occur associated with a left image feature or with a right image feature. However, a final solution must choose only one of them to be correct. In Fig. 5 multiple matches occur for the same Cyclopean Eye coordinate x and in this case, human perception chooses one match. Also, there are no good feature matches in regions of occlusions such as in Fig. 4 and neither on homogeneous regions such as in Fig. 7. Thus, all these scenarios lead us to conclude that for a S^{3D} solution, given the features, there is a need for a prior $P(S^{3D})$ to help solve for the stereo problem, which we study next.

4 Geometrical Constraints (GCs)

We will now address three geometrical constraints that provide the basis for our understanding and analysis of geometrical properties in stereo vision. When creating a stereo algorithm these constraints could be expressed by a prior $P(S^{3D})$ present in Eq. (5). However, in this paper we do not attempt to create a stereo algorithm. The format we present this section is by introducing the definitions needed for us to articulate a geometrical constraint proposition followed by an analysis with real datasets. See more details on the project page[1]

Definition 1 (Opaque Surfaces and Stereo). *3D opaque surfaces do not let light pass through them. Consider a 3D point $P = (X, Y, Z)$ that belongs to an opaque surface. If it is visible by L or R, we can describe it as $(e, l, \mathcal{D}^L(e, l))$ or as $(e, r, \mathcal{D}^R(e, r))$, respectively.*

[1] Github: https://github.com/SherlonAlmeida/cyclopean-stereo.git.

Note that if P is visible by both eyes, as illustrated in Fig. 2, we have $(e, l) \leftrightarrow (e, r = l - d^L(e, l))$ and so (i) the disparities satisfy $d^L(e, l) = d^R(e, r = l - d^L(e, l))$, but in general $\mathcal{D}^L(e, l) \neq \mathcal{D}^R(e, r)$; and (ii) the assignment $d(e, x)$ derived from (1) yields $\mathcal{D}(e, x)$ from (2), but $\mathcal{D}(e, x) \neq \mathcal{D}^L(e, l), \mathcal{D}^R(e, r)$ as seen from (3).

Definition 2 (Transparent Surfaces and Stereo). *Transparent surfaces allow some light to pass through. Two distinct points $P_{1,2} = (X_{1,2}, Y_{1,2}, Z_{1,2})$ are in a transparent pair state if both can be seen by the XD and share the same (e, x) coordinates. Transparent stereo surfaces are made of a set of contiguous transparent-pair states.*

With these two definitions, we state a geometrical constraint

Proposition 2 (Opaque-GC). *For opaque surfaces each Cyclopean coordinate (e, x) has one and only one disparity, i.e., d is a function $d : (e, x) \to \mathbb{R}$.*

Note that this constraint is to be interpreted as prior geometrical constraint. Figure 3 illustrates this constraint, while Fig. 4 shows data that validate this constraint.

However, a scenario with opaque surfaces may not satisfy this rule. More precisely, it is possible in some special scenarios for two 3D points P_1, P_2, visible by the left eye as $(e, l_1), (e, l_2)$ and by the right eye as $(e, r_1), (e, r_2)$, to satisfy that both have the same $x = \frac{l_1 + r_1}{2} = \frac{l_2 + r_2}{2}$ but different disparities, for example, for the case of thin objects in front of backgrounds, as shown in Fig. 5. Thus, it is suggested that this rule is applied to recover surfaces. In these scenarios the above constraint, used as prior knowledge, will then force the solution to choose either one pair of matches or the other, but not both.

Human perception faces this "dilemma" when, for example, placing a finger in front of a scene (see Fig. 6). In this case, humans can either focus on the finger or in the background and reconstruct the 3D structure of the finger (ignoring background) or background (ignoring the finger). It is the attention mechanism that makes the choice of how to apply the GC-constraint and ignore the alternative solution. Previous studies of visual attention in human vision include [3,18]. However, today we do not see any attention mechanism being used by SOTA stereo.

Definition 3 (Occlusions and Discontinuities). *R- (L-) occlusions are regions that are seen by the R (L) eye but not by L (R). R- (L-) discontinuities are places where jumps of disparity (or of depth) occur in the R (L). L- and R-occlusions are termed half-occlusions [2,21].*

As first observed by Da Vinci 3D opaque surfaces follow GCs that link discontinuities to occlusions, see also [2,7,8,10]. We next describe our new proposed GCs.

Proposition 3 (Da Vinci-GC). *In the Cyclopean Eye the jumps along an epipolar line must occur along the 45 degree angles. Equivalently, the size of the jump of a R- (L-) discontinuity is equal to the size of the L- (R-) occlusion.*

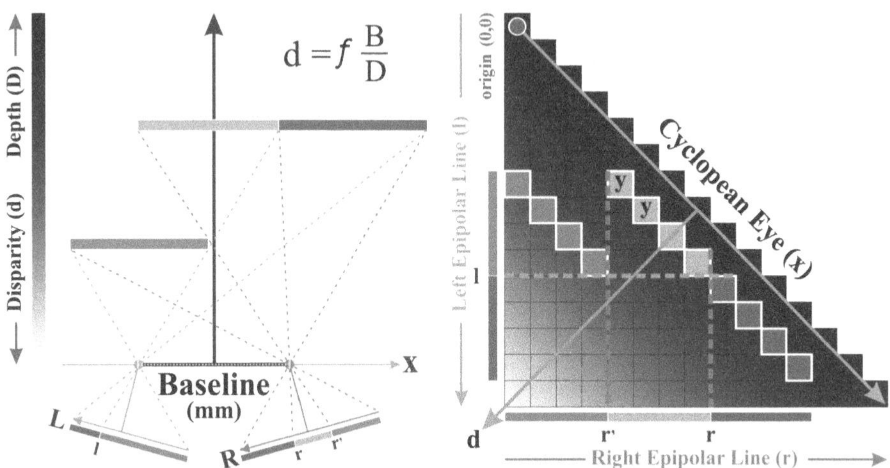

Fig. 3. An epipolar slice of a surface with left occlusion region and its description by the XD. Left. A top view of the epipolar slice of the surface and the two eyes projections. The baseline **B** connects L to R focal centers. The depth axis describe the inverse of disparity (Eq. (2) depicted). Right. A discrete XD, a rotation of the L/R CS described by Eq. (1). Note that the L CS is pointing down. The two red vertical dashed lines delimit the R occlusion area which are associated with a L discontinuity along the horizontal blue dashed line with a jump of the same size as the right occlusion, as described by Da Vinci-GC in Proposition 3. Note that the only two (2) light green squares (the ones without an "y" in them) are seen by the XD associated with the R occlusion, satisfying one disparity per coordinate x (as postulated by Opaque-GC), which is **half** of the size of the R occlusion area. (Color figure online)

Figure 3 illustrates this constraint. Figure 4 and Fig. 5 show and provide an analysis of the disparity jumps of 45 degrees. Da Vinci-GC was proposed in the works [7,10]. To the best of our knowledge, Da Vinci-GC in the Cyclopean Eye has not been proposed before. Figure 5 illustrates the consistency between Da Vinci-GC and Opaque-GC, where a solution that satisfies one constraint also satisfies the other constraint, both constraints forcing a solution to choose which data to focus on and which data to ignore.

Note that if a prior with this constraint is to be applied at some stage of a stereo algorithm, then the final disparity values assigned at such occlusion locations must be processed at a later stage. Feature matching also fails at homogeneous regions as we discuss next.

Definition 4 (Homogeneous Regions). *Homogeneous regions are areas in an image that lack texture. All stereo feature responses in these regions provide equally poor matches.*

Figure 7 illustrates homogeneous regions in real data. Feature matching at homogeneous regions does not discriminate which disparity to choose. Thus, feature matching does not provide data do discriminate the disparity for homogeneous regions nor for occlusion regions.

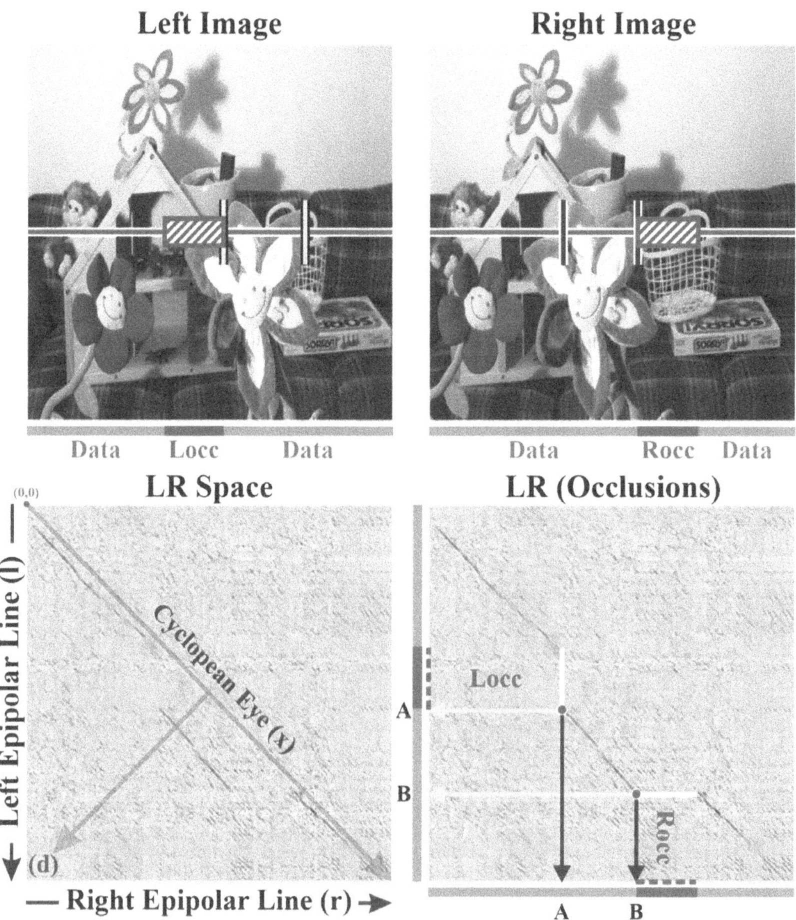

Fig. 4. Occlusions: The LR space displays FM distances associated with the epipolar lines $e = 128$ (red), where $FM_{e,x}(d) \in [0,1]$. Dark regions (low values of FM) represent good matches. Da Vinci-GC is verified with L/R-occlusions in dashed-red, associated with the L/R-discontinuities in white (jumps). In the Cyclopean Eye these are 45 degrees jumps. Moreover, the Opaque-GC (unique disparity per Cyclopean Eye coordinate) is also verified. This figure also presents an association of the LR Space data/occlusions with the RGB LR images, where the points A and B in the LR Space denotes the start/end of the blue flower. (Color figure online)

Definition 5 (Repetitive Patterns). *Objects with similar textures may create repeated patterns (multiple matching possibilities) in the LR and XD space. Although stereo feature responses in these regions may provide equally good matches, only the correspondence that correctly aligns the beginning and end of the object represents the true solution.*

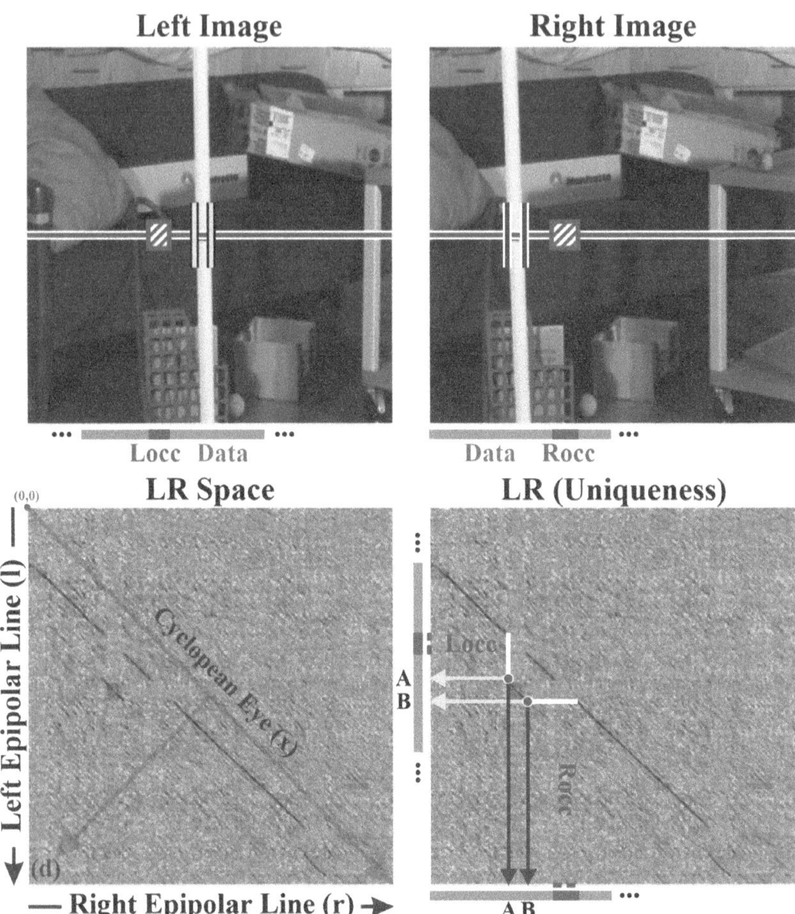

Fig. 5. Uniqueness: The LR Space displays FM distances associated with the epipolar lines $e = 128$ (red), where $FM_{e,x}(d) \in [0, 1]$. The Opaque-GC requires that either the solution recovers the thin pole in front or ignore it. On the right we show the solution that recover the thin pole in front with Da Vinci-GC, i.e., with the L/R-discontinuities in white (jumps). From the Cyclopean Eye these are 45 degrees changes. The Opaque-GC is consistently applied. Note that L/R-occlusions in dashed-red end up being expanded to regions where other possible good feature matches exist (low FM values). This figure also presents an association of the LR Space data/occlusions with the RGB LR images, where the points A and B denotes the start/end of the pole. (Color figure online)

Figure 8 illustrates repetitive patterns caused by an object composed of similar visual structures and textures. This example demonstrates a scenario where feature matching alone is insufficient to determine the correct disparity. Unlike homogeneous regions, which lack strong matching candidates, repetitive pattern regions yield multiple false-positive matches.

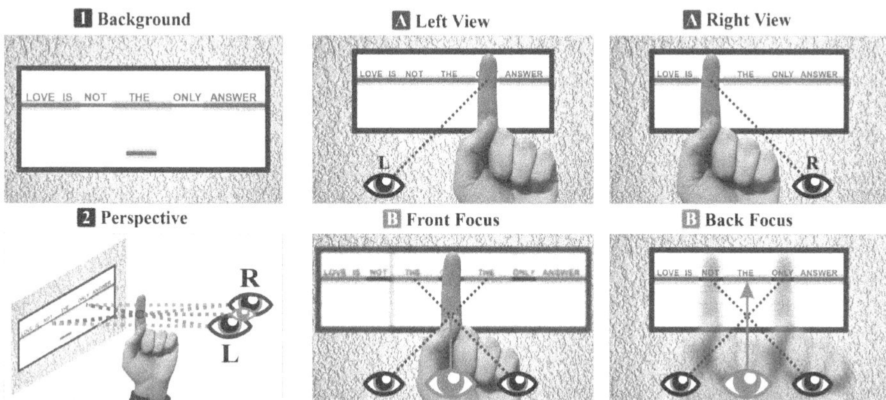

Fig. 6. Human perception experiment: This figure illustrates a stereo vision experiment with two objects in a scene: a left hand and finger (front) and a text frame on a wall (back). 1) The top-left subfigure shows the wall, and the text frame containing the phrase "Love is not the only answer". The green regions (e.g. the sequence of words 'Love', 'is', 'the', and 'answer') represent the text seen by both L/R views, while the red regions (e.g. the words 'not', and 'only') represent half occlusions. 2) The bottom-left subfigure shows the experiment in another perspective, where the red intersection point represents the position of the front object (finger) exactly in the middle of L/R eyes, that is, facing the Cyclopean Eye (orange). The finger cause an occlusion into the right and left eyes in the words 'not' and 'only', respectively. When analyzing the left and right views individually (subfigures Blue-A) the left eye sees the phrase 'Love is not the answer', the right eye sees the phrase 'Love is the only answer'. The human Cyclopean Eye (subfigures Orange-B) decides which object to pay attention, either directing the focus into the finger (front) or in the text frame (back). If the attention (Front Focus) is at the finger (front), the background seems to be blurred and duplicated. However, if the attention (Back Focus) is at the Background (wall + text frame), the finger (front) seems to be transparent, blurred and duplicated. Interestingly, by the Cyclopean view it is possible to combine the data from the L/R view, then it is possible to see the entire phrase 'Love is not the only answer'. Observe that this experiment is in accordance with the propositions 2 and 3. Also, the choice of focusing on the finger data offers a solution similar to the scenario in Fig. 5. (Color figure online)

Proposition 4 (Filling-GC). *For regions where the data matching does not offer a clear solution, a prior geometrical constraint satisfying Da Vinci-GC should be applied to recover the disparity function* $d(e,x)$ *everywhere in the Cyclopean Eye.*

We conclude this section by pointing out the importance of Gaussian Curvature as a 3D prior for Filling-GC. While Dong et al. [6] proposes CAD models reconstruction by enforcing zero Gaussian Curvature in developable surface patches, da Silva et al. [20] demonstrated that the best Stereo-Vision techniques nowadays on the main benchmarks tend to reconstruct surfaces with low Gaussian Curvature as a prior. The new generation of stereo approaches leverage

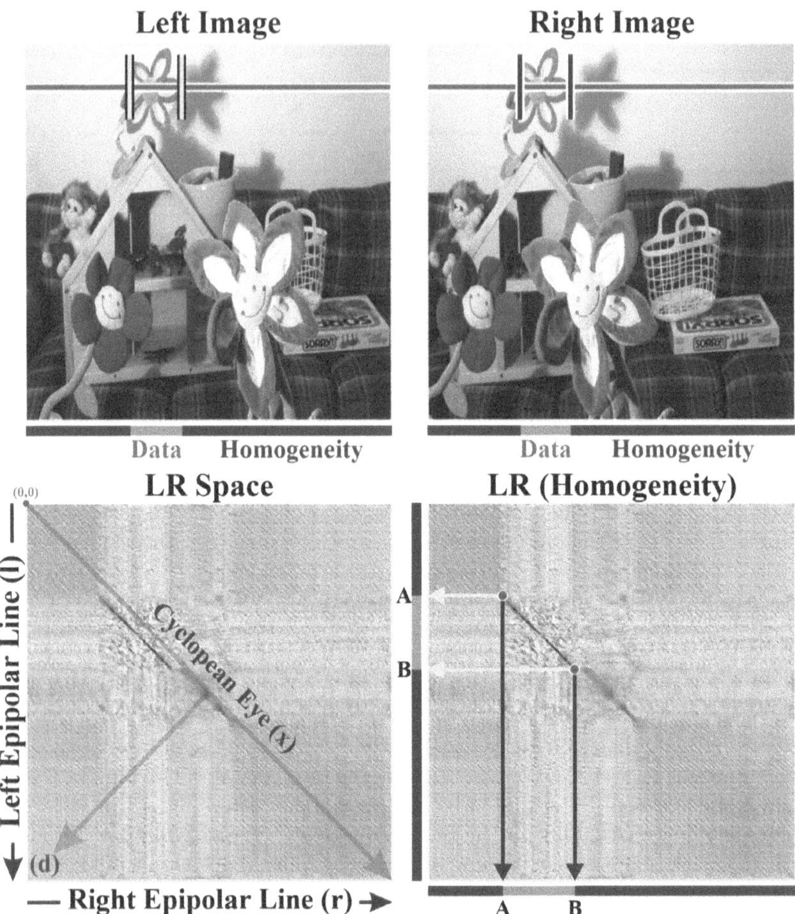

Fig. 7. Homogeneity: The LR space displays FM distances associated with the epipolar lines $e = 30$ (red), where $FM_{e,x}(d) \in [0, 1]$. Dark regions represent low FM and good matches. Note that shadows do yield good matches. However, in order to fill in homogeneous regions, surface priors become necessary. (Color figure online)

stereo matching, 3D geometry priors, and also provide surface consistency by combining monocular features (e.g. from DepthAnythingV2 [25]).

This work is primarily analytical, focusing on theoretical insights and empirical observations. As a limitation, the practical integration of our Cyclopean Eye model and the proposed geometric constraints into modern stereo pipelines remains unexplored. For future work, we aim to incorporate these ideas into end-to-end stereo architectures, enabling direct evaluation of their impact on depth reconstruction.

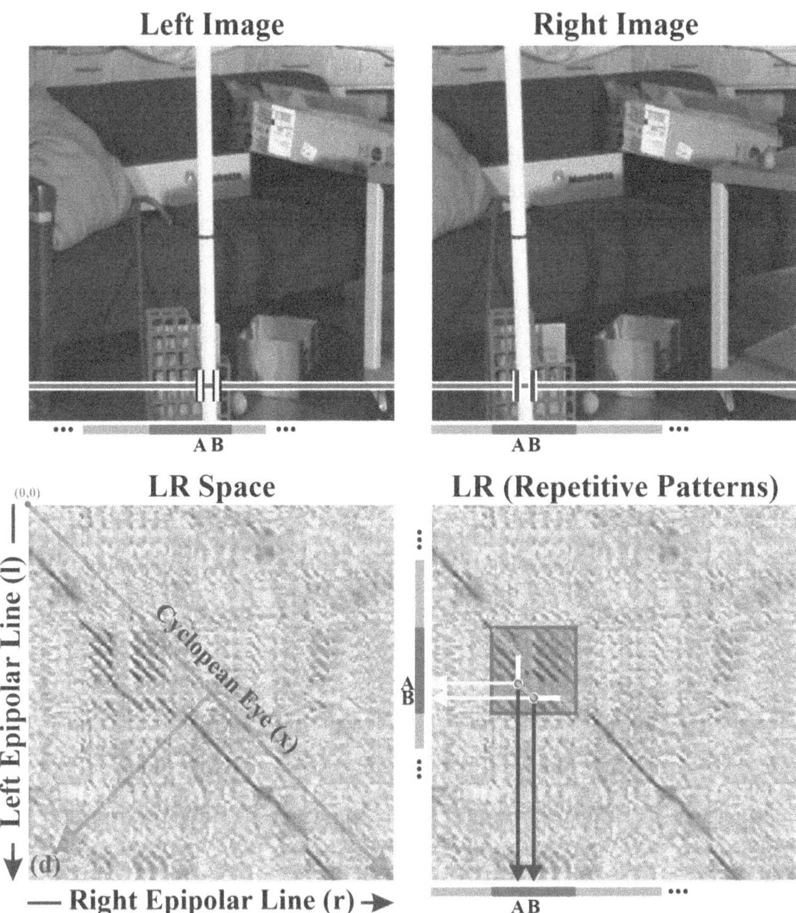

Fig. 8. Repetitive Patterns: The LR space displays FM distances associated with the epipolar lines $e = 464$ (red), where $FM_{e,x}(d) \in [0, 1]$. Dark regions represent low FM and good matches. Note the red object behind the pole with a repetitive pattern in the LR space, causing multiple good matches. However, an optimal solution which align with the beginning (left-most part) of the object and satisfying the geometrical constraints is the ground truth solution represented by light-green inside the pink box in the right-most subfigure. (Color figure online)

5 Conclusion

We presented an analysis of the geometric foundations underlying 3D stereo vision. Starting with the Cyclopean Eye model, we incorporated geometric constraints that account for depth discontinuities and occlusions, showing how this model combines information from both left and right eyes into one unified perspective. The Cyclopean Eye, in turn, supports the attention-based mechanisms employed by human stereo vision.

Our investigation emphasized the role of geometric structure in understanding and interpreting 3D scene reconstruction. We analyzed the quality of stereo feature matching derived from deep learning models, proposed novel geometric constraints tailored to stereo vision, and highlighted how these constraints help interpret the performance of data driven methods.

Together, these contributions reinforce the idea that combining strong geometric priors with learned features leads to more interpretable, generalizable, and reliable stereo vision systems. Ultimately, our goal is to underscore the importance of 3D geometric modeling in capturing critical visual information and guiding the development of next-generation vision approaches.

Acknowledgments. This study was financed in part by the Coordenação de Aperfeiçoamento de Pessoal de Nível Superior - Brasil (CAPES) - Finance Code 001.

References

1. Bartolomei, L., Tosi, F., Poggi, M., Mattoccia, S.: Stereo anywhere: robust zero-shot deep stereo matching even where either stereo or mono fail. In: Proceedings of the Computer Vision and Pattern Recognition Conference, pp. 1013–1027 (2025)
2. Belhumeur, P.N.: A bayesian approach to binocularsteropsis. Int. J. Comput. Vis. **19**(3), 237–260 (1996)
3. Bruce, N.D., Tsotsos, J.K.: An attentional framework for stereo vision. In: The 2nd Canadian Conference on Computer and Robot Vision (CRV'05), pp. 88–95. IEEE (2005)
4. Cheng, J., et al.: Monster: marry monodepth to stereo unleashes power. In: Proceedings of the IEEE/CVF Conference on Computer Vision and Pattern Recognition (2025)
5. Da Vinci, L.: A Treatise on Painting. Read Books Ltd., Alcester (2014)
6. Dong, Q., et al.: Neurcadrecon: neural representation for reconstructing cad surfaces by enforcing zero gaussian curvature. ACM Trans. Graph. **43**(4) (2024). https://doi.org/10.1145/3658171
7. Geiger, D., Ladendorf, B., Yuille, A.L.: Occlusions and binocular stereo. In: Sandini, G. (ed.) Second European Conference on Computer Vision, pp. 425–433. Springer, Berlin, Heidelberg (1992)
8. Geiger, D., Ladendorf, B., Yuille, A.L.: Occlusions and binocular stereo. Int. J. Comput. Vis. **14**(3), 211–226 (1995)
9. von Helmholtz, H.: Treatise on Physiological Optics, Volume III. German edition and, Dover Publications, New York, [1962] (1962)
10. Ishikawa, H., Geiger, D.: Occlusions, discontinuities, and epipolar lines in stereo. In: Burkhardt, H., Neumann, B. (eds) Computer Vision — ECCV'98. ECCV 1998. LNCS, vol. 1406, pp. 232–248. Springer, Berlin, Heidelberg (1998). https://doi.org/10.1007/BFb0055670
11. Jiang, H., et al.: Defom-stereo: depth foundation model based stereo matching. In: Proceedings of the IEEE/CVF Conference on Computer Vision and Pattern Recognition (2025)
12. Julesz, B.: Foundations of cyclopean perception. University of Chicago Press (1971)

13. Li, J., et al.: Practical stereo matching via cascaded recurrent network with adaptive correlation. In: Proceedings of the IEEE/CVF Conference on Computer Vision and Pattern Recognition, pp. 16263–16272 (2022)
14. Li, K., Wang, L., Zhang, Y., Xue, K., Zhou, S., Guo, Y.: Los: local structure-guided stereo matching. In: Proceedings of the IEEE/CVF Conference on Computer Vision and Pattern Recognition, pp. 19746–19756 (2024)
15. Lipson, L., Teed, Z., Deng, J.: Raft-stereo: multilevel recurrent field transforms for stereo matching. In: 2021 International Conference on 3D Vision (3DV), pp. 218–227. IEEE (2021)
16. Marr, D., Poggio, T.: A computational theory of human stereo vision. Proc. R. Soc. Lond. **204**, 301–328 (1979)
17. Min, J., Jeon, Y., Kim, J., Choi, M.: S^2m^2: scalable stereo matching model for reliable depth estimation. In: Proceedings of the IEEE/CVF International Conference on Computer Vision (2025)
18. Ooi, T.L., He, Z.J.: Binocular rivalry and visual awareness: the role of attention. Perception **28**(5), 551–574 (1999)
19. Scharstein, D., et al.: High-resolution stereo datasets with subpixel-accurate ground truth. In: Jiang, X., Hornegger, J., Koch, R. (eds.) Pattern Recognition. GCPR 2014. LNCS, vol. 8753, pp. 31–42. Springer, Cham (2014). https://doi.org/10.1007/978-3-319-11752-2_3
20. da Silva, S.A., Geiger, D., Velho, L., Ponti, M.A.: Towards understanding 3D vision: the role of gaussian curvature. In: Proceedings of the 21st International Conference on Computer Vision Theory and Applications, vol. 3, pp. 548–559. Marbella, Spain (2025)
21. Wang, J., Zickler, T.: Local detection of stereo occlusion boundaries. In: Proceedings of the IEEE/CVF Conference on Computer Vision and Pattern Recognition, pp. 3818–3827 (2019)
22. Wang, X., Xu, G., Jia, H., Yang, X.: Selective-stereo: adaptive frequency information selection for stereo matching. In: Proceedings of the IEEE/CVF Conference on Computer Vision and Pattern Recognition, pp. 19701–19710 (2024)
23. Wen, B., Trepte, M., Aribido, J., Kautz, J., Gallo, O., Birchfield, S.: Foundation-stereo: zero-shot stereo matching. In: Proceedings of the IEEE/CVF Conference on Computer Vision and Pattern Recognition (2025)
24. Xia, Z., Liu, Y., Bai, Y., Zhang, Y., Cheng, C.: Mismatches between 3d content acquisition and perception cause more visually induced motion sickness. PRESENCE: Virtual Augmented Reality **33**, 481–492 (2024)
25. Yang, L., et al.: Depth anything v2. arXiv preprint arXiv:2406.09414 (2024)
26. Zhao, H., Zhou, H., Zhang, Y., Chen, J., Yang, Y., Zhao, Y.: High-frequency stereo matching network. In: Proceedings of the IEEE/CVF Conference on Computer Vision and Pattern Recognition, pp. 1327–1336 (2023)

1D-CNN Multikernel Architecture for DDoS Attack Classification in IoT Environments

Fray L. Becerra-Suarez[1(✉)] and Manuel G. Forero[2]

[1] Grupo de Investigación en Inteligencia Artificial (UMA-AI), Universidad Privada Norbert Wiener, Lima 51001, Peru
fray.becerra@uwiener.edu.pe

[2] Semillero Lún, Grupo D+Tec, Facultad de Ingeniería, Universidad de Ibagué, Ibagué 730001, Colombia
manuel.forero@unibague.edu.co

Abstract. The exponential growth of the Internet of Things (IoT) has become the main target for cybercriminals to orchestrate various attacks such as distributed denial of service (DDoS). Despite advances, there is still a need for accurate, low-resource solutions that are adaptable to the IoT environment. This paper presents a one-dimensional multikernel convolutional architecture (1D-CNN) and an LSTM optimized for the classification of traffic associated with different types of DDoS attacks. The models were trained and evaluated on CICIoT2023 using a stratified 1% per-class sample of the normal and DDoS classes (which together represent 75.15% of the dataset). Two classification scenarios were considered: multiclass and binary. The 1D-CNN achieved an F1 score of 0.9981 in multiclass and 0.9991 in binary, outperforming the LSTM, which obtained 0.9692 and 0.9989, respectively. The 1D-CNN contains 433,557 parameters, about 85% fewer than the LSTM, confirming its lightness and computational efficiency. These results support its suitability for real-time monitoring and detection deployments in IoT environments.

Keywords: cybersecurity · Internet of Things · IoT · DDoS attacks · intrusion detection · CNN · deep learning · CICIoT2023

1 Introduction

The Internet of Things (IoT) is not just about devices connected to a network; it also includes sensors, actuators, and other components that interact within a physical environment to collect, process, and transmit data in real time [1]. Its application has been widely adopted in various industrial sectors, such as healthcare, education, autonomous driving, agriculture, among others [2,3]. This has led to exponential growth in the usage of IoT, but it has also raised significant security concerns due to the limited computing capacity, decentralized architectures, and storage constraints of these devices, along with their constant exposure

D. Chaves et al. (Eds.): CIARP 2025, LNCS 16528, pp. 142–155, 2026.
https://doi.org/10.1007/978-3-032-23161-1_11

to public networks, making them vulnerable to various cyberattacks that occur daily across these networks [4,5].

Among the primary types of attacks, distributed denial-of-service (DDoS) attacks stand out. This type of cyberattack involves multiple compromised systems (often part of a botnet) sending large volumes of traffic or requests simultaneously to a targeted server, service, or network infrastructure with the intention of saturating its resources and causing disruption or degradation of legitimate services [6–8]. These attacks have increased significantly within IoT environments, where IoT devices can be exploited as attack nodes within botnets to launch large-scale DDoS attacks.

Early detection of these attacks in IoT traffic is critical for operational continuity and the security of smart networks. Traditional detection methods based on rules or signatures have limitations when faced with new attack vectors, whereas machine learning and deep learning approaches have shown high potential to identify malicious patterns in network traffic. For example, in [9], the authors proposed a model named BayesCNN, a Bayesian-based CNN, was developed and evaluated using the CICDDoS2019 dataset with dimensionality reduction via PCA. For multiclass classification, the model achieved an accuracy of 99.66% across 13 attack classes. Furthermore, BayesCNN, which integrates data fusion with 27 layers and Bayesian techniques for uncertainty management, achieved an even higher accuracy of 99.79%.

In the study by Abdel-Ghani et al. [10], the authors proposed an optimized CNN architecture built with two convolutional layers with 32 filters and a kernel size of 4, followed by a dense layer with 535 units and a learning rate of 0.0001279. This architecture was compared with SimpleCNN, MobileNetV2, EfficientNetV2B0, and IoTNet using the CICIoT2023 dataset. The overall accuracy results for multiclass and binary classification were 96% and 98%, respectively, representing the best performance among the evaluated models. In [11], CNNs were explored for detecting DDoS attacks in multicloud environments using the NSL-KDD dataset. The proposed CNN-based approach achieved an accuracy of 97.46%.

Another relevant study [12] presented a novel CNN-based framework for the detection and classification of DDoS attacks, addressing class imbalance and computational cost challenges while achieving an accuracy of 99.99% for binary classification and 98.44% for multiclass classification. The model employed techniques such as random sampling to address class imbalance and feature selection methods, including low information gain and principal component analysis, to optimize performance while maintaining low computational demands.

In [13], a DDoS detection method was proposed that combines a self-attention mechanism with a CNN-BiLSTM architecture to address the issues of high dimensionality, low precision, and high false-positive rates in network traffic. A random forest with Pearson correlation analysis was used to select relevant features, while the CNN and BiLSTM extracted spatial and temporal features, respectively. Experimental results demonstrated that when tested on the CIC-IDS2017 and CIC-DDoS2019 datasets, the proposed method achieved precision, accuracy, recall, and F1 scores of 95.670%, 95.824%, 95.904%, and 95.864%, respectively.

Additionally, in [14], a hybrid framework was proposed that combines a Tree-CNN to capture hierarchical dependencies, a deep autoencoder to reduce noise, and a spiral optimization algorithm to fine-tune system parameters. This framework was evaluated using three datasets: UNSW-NB15, CIC-IDS2017 and CIC-IDS2018. The model achieved binary classification accuracy of 96.02% on UNSW-NB15, 99.99% on CIC-IDS2017 and 99.96% on CIC-IDS2018.

Although the studies mentioned above have achieved very promising results in detecting DDoS attacks using traditional and hybrid architectures, challenges remain in training stability, adaptability to IoT environments with limited resources, and accuracy in real-time network traffic classification. In this context, this study proposes two optimized architectures based on 1D-CNN and LSTM for the classification of DDoS traffic in IoT environments. A total of 13 variants of DDoS attacks are included in the CICIoT2023 dataset, which has been taken as a reference to validate the proposed architectures. The experimental results demonstrate that the proposed architectures are viable for implementation in production scenarios where IoT devices are continuously faced with DDoS attacks on public networks.

2 Methodology

2.1 Dataset and Preprocessing

In this study, the CICIoT2023 dataset was used, which captures IoT traffic from a testbed with 105 devices and 33 types of attacks grouped into seven main categories [15]. The dataset is distributed as PCAP (raw captures) and CSV (features aggregated in packet windows), allowing for different processes to extract malicious and normal traffic patterns. We focused on the DDoS family and the normal class, which together account for 35,082,755 records (75.15% of the entire dataset). Due to the large amount of data, a stratified sample of 1% per class was extracted, preserving the original imbalance between classes. Table 1 summarizes the distribution of CICIoT2023 and the percentage of data used.

Initially, the 47 descriptors in the data set were evaluated to identify the features that had a unique value across all records. As a result of this process, 6 features ('ece_flag_number', 'cwr_flag_number', 'Telnet', 'SMTP', 'SSH', 'IRC') were identified as meeting this criterion, so they were eliminated. This resulted in a total of 40 input features; the label was encoded separately. Subsequently, an analysis was performed to detect the presence of duplicate values, nulls and infinite positive and negative values, but no records were found that met these conditions. To process the label, it was encoded using the LabelEncoder() method of the Scikit-learn library, which is widely used in supervised classification tasks. Finally, in order to normalize the scale of the independent variables and improve the efficiency of the training process, the StandardScaler() method was applied to all descriptors, excluding the label.

Table 1. Description of Benign Traffic and DDoS Attacks in the CICIoT2023 Dataset

Traffic Type	Total Records	Percentage Used (1%)
BenignTraffic	1,098,195	10,981
DDoS-ICMP_Flood	7,200,504	72,005
DDoS-UDP_Flood	5,412,287	54,122
DDoS-TCP_Flood	4,497,667	44,976
DDoS-PSHACK_Flood	4,094,755	40,947
DDoS-SYN_Flood	4,059,190	40,591
DDoS-RSTFINFlood	4,045,285	40,452
DDoS-SynonymousIP_Flood	3,598,138	35,981
DDoS-ICMP_Fragmentation	452,489	4,524
DDoS-UDP_Fragmentation	286,925	2,869
DDoS-ACK_Fragmentation	285,104	2,851
DDoS-HTTP_Flood	28,790	287
DDoS-SlowLoris	23,426	234
Total	35,082,755	350,820

2.2 Deep Learning Architectures

Two architectures are proposed: one based on one-dimensional multikernel convolutional neural networks, whose effectiveness lies in the use of convolution operations and filters that detect local and hierarchical patterns in the data [16], and a Long Short-Term Memory (LSTM) that incorporates memory cells and input, forget, and output gates to model long-range temporal dependencies, mitigating gradient vanishing [17].

The first 1D-CNN architecture (Fig. 1) takes the 40 descriptors as input; the next section of the network consists of three parallel convolutional branches, each followed by a batch normalization layer and a ReLU activation function. Each branch uses 32 convolutional filters with different kernel sizes (3, 5, and 7), using 'same' padding and L2 regularization of 1×10^{-4}. Subsequently, an additional convolutional layer with 64 filters and a kernel size of 5 is incorporated, followed by a batch normalization layer and ReLU activation, to continue extracting relevant patterns from the combined representation of the previous branches. To prevent overfitting, a Dropout layer with a rate of 0.25 is added, which randomly turns off certain activations during training and improves the generalization of the model.

The output of these layers is flattened using a Flatten layer to convert the three-dimensional output into a one-dimensional vector, facilitating its use in the following dense layers. Next, a dense layer of 156 neurons with ReLU activation is incorporated, allowing the model to learn abstract representations and nonlinear combinations of the previously extracted features. To continue with regularization and reduce the risk of overfitting, a second dropout layer with a rate of 0.25 is added. Finally, an output layer with a number of units equal to the number of classes in the dataset is included, using the Softmax activation function for multiclass classification.

The 1D-CNN Multikernel model has 433,557 parameters (433,237 trainable and 320 untrainable), occupying 1.65 MB, which allows an efficient implementation in environments with limited resources, ensuring high accuracy in traffic classification and DDoS attack detection.

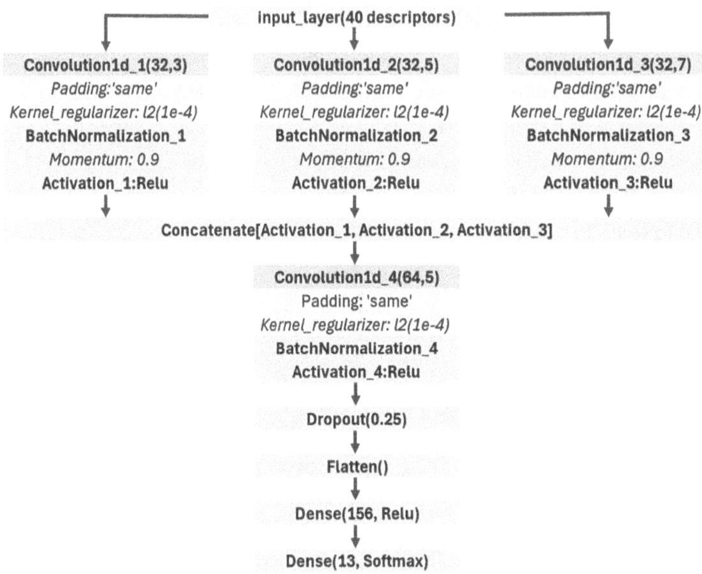

Fig. 1. Proposed 1D-CNN architecture.

The following LSTM architecture (Fig. 2) operates on the same input of 40 descriptors. The sequential trunk consists of two stacked LSTM layers of 64 units. The first is configured with return_sequences = True, retaining the output at each time step so that the second LSTM processes the entire dynamics before aggregating it into a global summary using its final state (return_sequences = False). After each LSTM, batch normalization (BatchNormalization) is incorporated to stabilize the distribution of activations during training, and Dropout (0.25) is used as a regularization mechanism that randomly turns off units and reduces the risk of overfitting. A dense intermediate layer of 156 neurons with ReLU activation is added to the aggregated representation; then, a second Dropout (0.25) is applied to reinforce generalization.

2.3 Training Setup

The models were compiled using the Adam optimizer with a learning rate of 1×10^{-4}, the sparse_categorical_crossentropy loss function which supports mul-

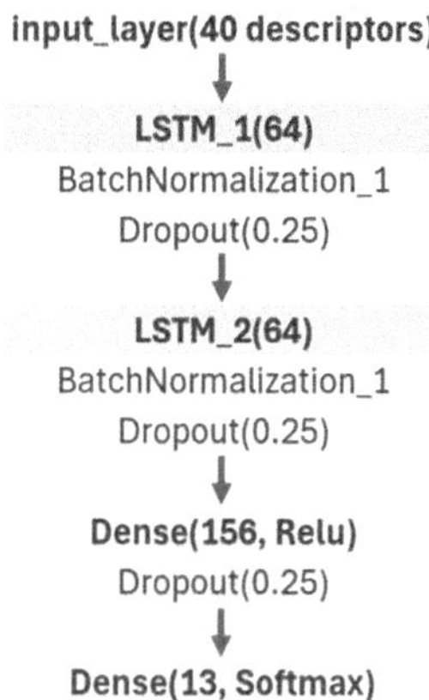

Fig. 2. Proposed LSTM architecture.

ticlass labels, and the accuracy metric was used for evaluation. Four essential callbacks were implemented. EarlyStopping, to stop training after 6 epochs without improvement in validation loss by restoring the best weights. CSVLogger was implemented to record metrics throughout this process. ReduceLROnPlateau was used to dynamically reduce the learning rate (factor 0.8) after 3 epochs without improvement, considering a minimum rate of 1×10^{-6}. ModelCheckpoint was used to save the optimal model in HDF5 format. The training was performed for 100 maximum epochs with a batch size of 128 samples, using 60% of the data for training, 20% for validation, and 20% for testing. To evaluate the model's performance, metrics such as accuracy, precision, recall, and F1 score were used, whose mathematical expressions are detailed in [18,19].

3 Results and Discussion

This section presents the results obtained for the proposed models for both multiclass and binary classification. The experiments were carried out on the Google Compute Engine backend with a T4 GPU and 12.7 GB of RAM.

3.1 1D-CNN Architecture

In Fig. 3(a–b), the multiclass learning curves show a steady upward trend in training and validation accuracy, accompanied by a monotonic decrease in loss. After a rapid initial phase, both metrics approach similar plateaus and maintain a small and stable generalization gap, indicating adequate optimization and effective regularization. The validation loss flattens slightly above the training loss, as expected in nontrivial multiclass settings, with no signs of divergence or late degradation. Small oscillations are observed after the first few epochs, which is typical of mini-batch training and class imbalance, but they remain limited and do not affect overall convergence.

Figure 3(c–d) shows the binary task with an even faster convergence profile. Training and validation accuracy increase significantly and stabilize within a few epochs, while both loss curves drop rapidly and remain close to zero with minimal variance. The near overlap of the training and validation trajectories, coupled with persistently low validation loss, suggests strong separability between normal traffic and DDoS and an absence of overfitting.

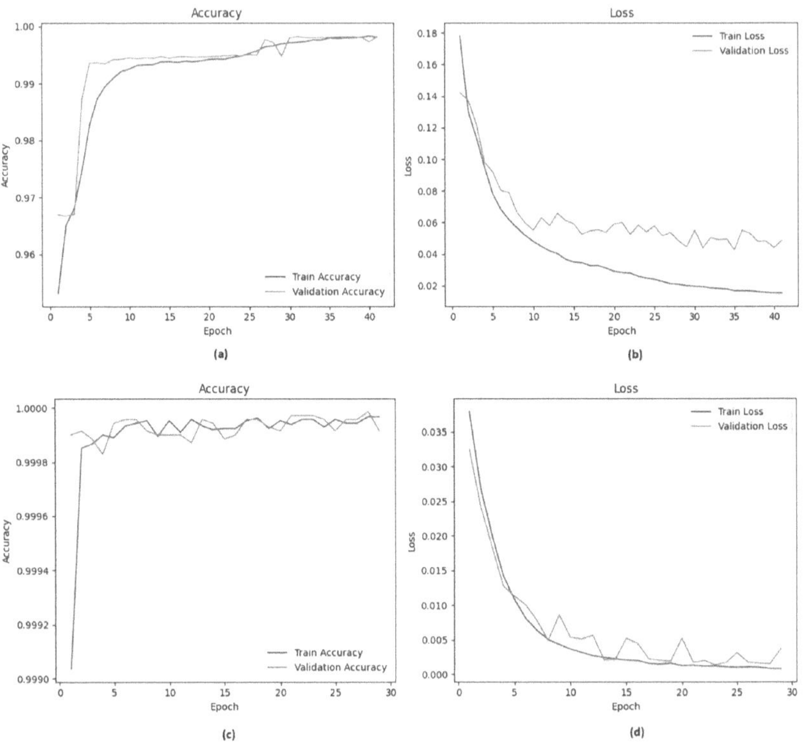

Fig. 3. Training and validation plots for multiclass and binary classification.

To evaluate the detailed performance of this architecture, confusion matrices were generated. Figure 4 corresponds to the confusion matrix for multiclass classification. This matrix shows that the vast majority of samples are correctly classified into their respective classes, with minimal presence of false positives and false negatives. This indicates that the proposed model is highly effective in differentiating multiple types of DDoS attacks. Regarding the results obtained from the confusion matrix for the binary classification (Fig. 5), a near-perfect performance is evidenced, with a very low number of false positives and no false negatives in the test set. This result confirms the robustness of the architecture to identify the presence of DDoS attacks in IoT environments.

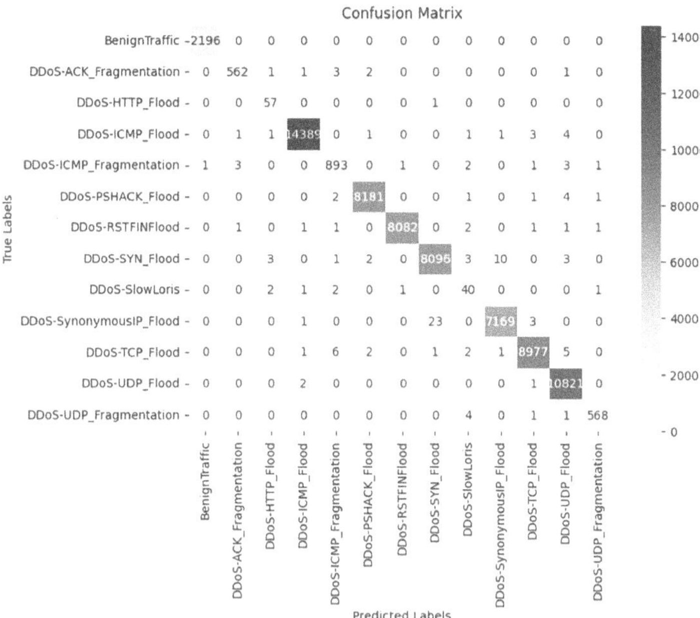

Fig. 4. Confusion matrix for multiclass classification - 1D-CNN.

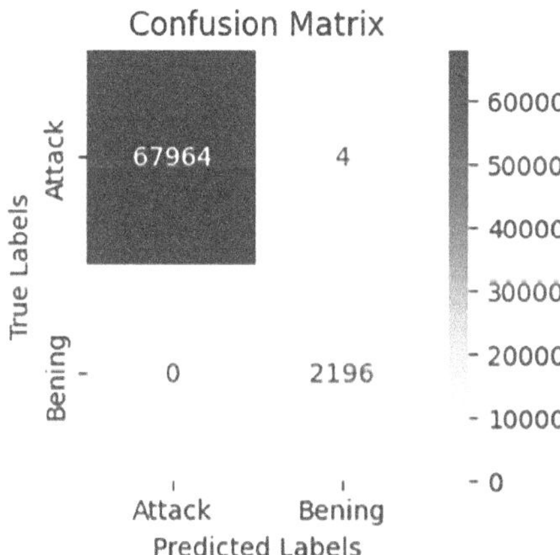

Fig. 5. Confusion matrix for binary classification - 1D-CNN.

3.2 LSTM Architecture

The learning curves of the LSTM model for multiclass classification (Fig. 6(a,b)) show a stable and monotonic evolution of Accuracy and Loss in training and validation, with early convergence and no signs of overfitting during the 100 epochs. The training and validation curves remain close throughout the epochs, except in epoch 20. However, these results suggest good generalization of the model on the training and validation set. Similarly, for binary classification (Fig. 6(c,d)), the behavior is similar to the multiclass scenario, with rapid and stable convergence of the metrics and a validation accuracy that reaches high values in just 18 epochs. This pattern indicates that LSTM is effective in separating normal traffic from malicious traffic (DDoS), maintaining stability during training.

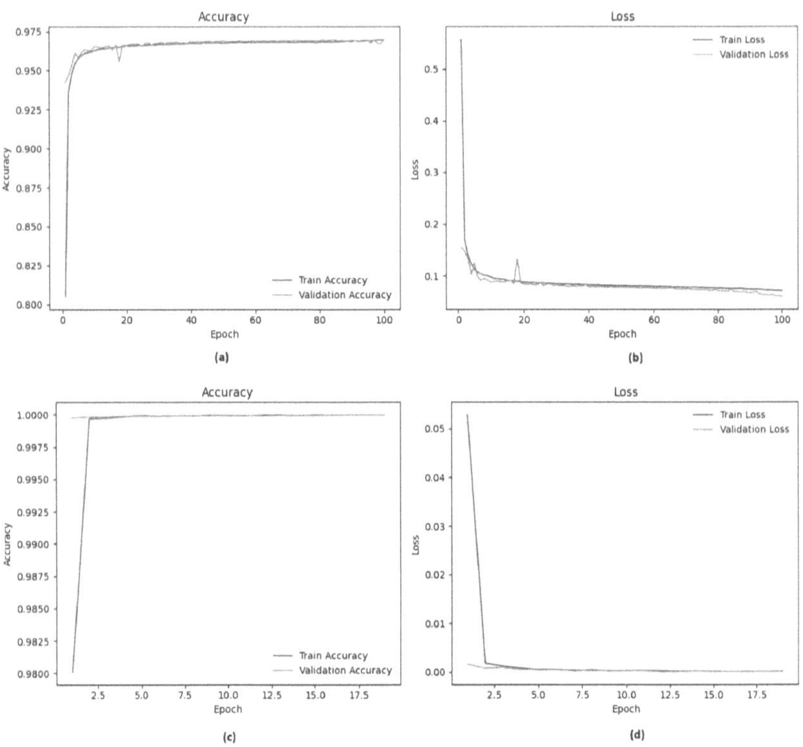

Fig. 6. Training and validation plots for multiclass and binary classification – LSTM.

The confusion matrices show that in the multiclass classification (Fig. 7), the main diagonal predominates and low rates of false positives and false negatives between classes are observed. Figure 8 (binary) shows behavior close to optimal, with minimal classification errors, which reaffirms the robustness of the model for DDoS detection.

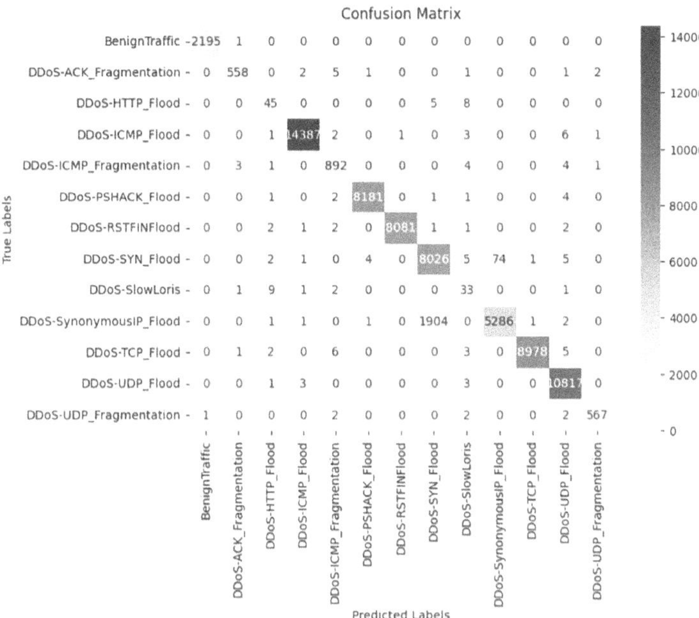

Fig. 7. Confusion matrix for multiclass classification - LSTM.

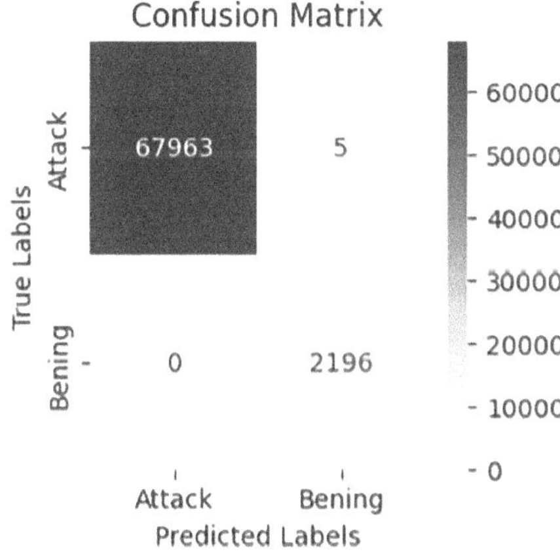

Fig. 8. Confusion matrix for binary classification - LSTM.

3.3 Comparative Analysis

Table 2 summarizes the results in the test set for both architectures. The 1D-CNN consistently achieved 0.9981 across all metrics for multiclass classification, suggesting very low false positive and false negative rates. In the binary scenario, its performance was even higher (accuracy = 0.9999, precision = 0.9982, recall = 1.0000, F1 score = 0.9991). For its part, LSTM showed solid but inferior performance in multiclass (accuracy = 0.9698, precision = 0.9746, recall = 0.9698, F1 score = 0.9692), while in binary it was practically equivalent to the 1D-CNN (accuracy = 0.9999, precision = 0.9977, recall = 1.0000, F1 score = 0.9989).

In terms of efficiency, the training-validation times for the 1D-CNN were 374 s (multiclass) and 346 s (binary), compared to 2000.5 s and 417.63 s for the LSTM. In terms of inference, the 1D-CNN required 5 s (multiclass) and 8 s (binary), while the LSTM took 7 s and 6 s, respectively. Overall, the 1D-CNN offers a better balance between accuracy and computational cost in the multiclass scenario, and both models are practically perfect in binary; the inference times (<10 s) support their use in IoT environments with fast response requirements.

Table 2. Results in the test set for multiclass and binary classification with 1D-CNN and LSTM.

	1D-CNN		LSTM	
Indicator	Multiclass	Binary	Multiclass	Binary
Accuracy	0.9981	0.9999	0.9698	0.9999
Precision	0.9981	0.9982	0.9746	0.9977
Recall	0.9981	1.0000	0.9698	1.0000
F1 score	0.9981	0.9991	0.9692	0.9989
Training & Validation Time	374 s	346 s	2000.5 s	417.63 s
Inference Time	5 s	8 s	7 s	6 s

The results of the proposed architectures, specifically 1D-CNN, are highly effective for detecting DDoS attacks in IoT environments, significantly outperforming the approaches reported in previous studies (Table 3). These values show a significant improvement over the results obtained by the Optimized CNN approach in [10], which reports an accuracy of 0.9800 and an F1 score of 0.9700, as well as compared to the CNN model proposed in [11], which achieves an accuracy of 0.9650 and an F1 score of 0.9780 in a binary scenario with the NSL-KDD dataset.

In the multiclass classification scenario, the 1D-CNN architecture also demonstrates superior performance, ranking above the values reported in [9], where an accuracy of 0.9979 is obtained, as well as in [10], where an accuracy of 0.96 and an F1 score of 0.96 are achieved, and those reported in [13] using a CNN-AttBiLSTM approach, where an accuracy of 0.9567 and an F1 score of 0.9586 are achieved using the CIC-IDS2017 and CIC-DDoS2019 datasets.

These results confirm that incorporating a multikernel scheme into the 1D-CNN architecture significantly contributes to improving accuracy and recall in identifying legitimate and malicious traffic, without neglecting the LSTM architecture. This performance supports the practical applicability of the proposed model in intrusion detection systems with high effectiveness and low resource consumption in IoT environments, considering that the 1D-CNN architecture that performed best uses a total of 433,557 parameters (1.65 MB), of which 433,237 correspond to trainable parameters and 320 to non-trainable parameters, reflecting a computationally lightweight structure suitable for implementation on devices with hardware constraints.

Table 3. Comparison of results with similar models reported in previous studies.

Ref.	Approach	Dataset	Classification	Acc.	Prec.	Recall	F1 score
[9]	BaysFusCNN	CICDDoS2019	Multiclass	0.9979	0.9857	0.9855	0.9856
[10]	Optimized CNN	CICIoT2023	Binary	0.9800	0.9700	0.9700	0.9700
[10]	Optimized CNN	CICIoT2023	Multiclass	0.9600	0.9600	0.9600	0.9600
[11]	CNN	NSL-KDD	Binary	0.9650	0.9720	0.9850	0.9780
[13]	CNN-AttBiLSTM	CICDDoS2019	Multiclass	0.9567	0.9582	0.9590	0.9586
Our study	1D-CNN	CICIoT2023	Binary	0.9999	0.9982	1.0000	0.9991
Our study	1D-CNN	CICIoT2023	Multiclass	0.9981	0.9981	0.9981	0.9981

4 Conclusions

This study proposes two optimized architectures based on 1D-CNN Multikernel and LSTM for detecting DDoS attacks in IoT environments. The 1D-CNN multikernel showed superior performance compared to the results reported in other studies. The stable learning curves and confusion matrices show a dominant diagonal and very low error rate. This architecture outperformed LSTM in the multiclass scenario and achieved comparable results in the binary scenario, while maintaining a contained computational footprint and inference times suitable for near real-time response applications and devices with limited resources.

As future work, we propose evaluating recent architectures, in particular Transformers with efficient attention mechanisms and lightweight variants such as MobileViT, while preserving the memory and latency constraints inherent to the IoT. Likewise, priority will be given to the use of cutting-edge, multidomain datasets with cross-site and cross-device validation protocols that allow generalization to be measured and overfitting to a single distribution to be avoided. In parallel, compression techniques (structural pruning, quantization-aware training, and knowledge distillation) and robustness strategies against zero-day attacks through continual learning will be explored in order to improve decision reliability. These lines will be validated in heterogeneous scenarios and edge deployment conditions, using performance and computational cost metrics (latency, memory and energy usage), so that the system maintains its performance in the face of traffic evolution and the variability inherent in IIoT environments.

References

1. Elgazzar, K., et al.: Revisiting the internet of things: new trends, opportunities and grand challenges. Front. Internet. Things **1**, 1073780 (2022). https://doi.org/10.3389/friot.2022.1073780
2. Amjad, M., Arulmozhi, E., Shin, Y.-H., Kang, M.-K., Cho, W.-J.: A review of IoT and machine learning for environmental optimization in aeroponics. Agronomy **15**(7), 1627 (2025). https://doi.org/10.3390/agronomy15071627
3. de Anda-Suárez, J., et al.: An integrated blockchain framework for secure autonomous vehicle communication system. Information **16**(7), 557 (2025). https://doi.org/10.3390/info16070557
4. Dargaoui, S., Azrour, M., Allaoui, A., Guezzaz, A., Alabdulatif, A., Alnajim, A.: Internet of things authentication protocols: comparative study. CMC **73**(1), 65—91 (2024). https://doi.org/10.32604/cmc.2024.047625
5. Seyedi, B., Postolache, O.: Securing IoT communications via anomaly traffic detection: synergy of genetic algorithm and ensemble method. Sensors **25**(13), 4098 (2025). https://doi.org/10.3390/s25134098
6. Hirsi, A., et al.: Comprehensive analysis of DDoS anomaly detection in software-defined networks. IEEE Access **13**, 23013–23071 (2025). https://doi.org/10.1109/ACCESS.2025.3535943
7. Raghunath, K.M.K., Kumar, V.V., Venkatesan, M., Singh, K.K., Mahesh, T.R., Singh, A.: XGBoost regression classifier (xrc) model for cyber attack detection and classification using inception V4. J. Web Eng. **21**(4), 1295—1322 (2022). https://doi.org/10.13052/jwe1540-9589.21413
8. Karthikeyani, R., Karthikeyan, E.: A review on distributed denial of service attack. AJRCoS **16**(4), 133–144 (2023). https://doi.org/10.9734/ajrcos/2023/v16i4378
9. AlSaleh, I., Al-Samawi, A., Nissirat, L.: Novel machine learning approach for DDoS cloud detection: bayesian-based CNN and data fusion enhancements. Sensors **24**(5), 1418 (2024). https://doi.org/10.3390/s24051418
10. Abdel-Ghani, M., Zakraoui, J., Belhi, A., Al-Ali, A., Rahme, S., Bouras, A.: Analysis of lightweight cnn-based intrusion detection models in IoT. In: 2024 IEEE/ACM International Conference on Big Data Computing, Applications and Technologies (BDCAT), Sharjah, United Arab Emirates: IEEE, pp. 314—323 (2024). https://doi.org/10.1109/BDCAT63179.2024.00056
11. Verma, R., Jailia, M., Kumar, M., Kaliraman,B.: CNN-based detection of DDoS attacks in multi-cloud environments. In: 2024 International Conference on Communication, Computer Sciences and Engineering (IC3SE), pp. 1011–1016 (2024). https://doi.org/10.1109/IC3SE62002.2024.10593351
12. Najar, A.A., Sugali, M.N., Lone, F.R., Nazir, A.: A novel CNN-based approach for detection and classification of DDoS attacks. Concurr. Comput.: Pract. Exp. **36**(19), e8157 (2024). https://doi.org/10.1002/cpe.8157
13. Zhao, J., Liu, Y., Zhang, Q., Zheng, X.: CNN-AttBiLSTM mechanism: a DDoS attack detection method based on attention mechanism and CNN-BiLSTM. IEEE Access **11**, 136308–136317 (2023). https://doi.org/10.1109/ACCESS.2023.3334916
14. Al-Dulaimi, R.T.A., Türkben, A.K.: A hybrid tree convolutional neural network with leader-guided spiral optimization for detecting symmetric patterns in network anomalies. Symmetry **17**(3), 421 (2025). https://doi.org/10.3390/sym17030421
15. Neto, E.C.P., Dadkhah, S., Ferreira, R., Zohourian, A., Lu, R., Ghorbani, A.A.: CICIoT2023: a real-time dataset and benchmark for large-scale attacks in IoT environment. Sensors **23**(13), 5941 (2023). https://doi.org/10.3390/s23135941

16. Akgun, D., Hizal, S., Cavusoglu, U.: A new DDoS attacks intrusion detection model based on deep learning for cybersecurity. Comput. Secur. **118**, 102748 (2022). https://doi.org/10.1016/j.cose.2022.102748
17. Krichen, M., Mihoub, A.: Long short-term memory networks: a comprehensive survey. AI **6**(9), 215 (2025). https://doi.org/10.3390/ai6090215
18. Becerra-Suarez, F.L., Jiménez-Fernández, L.J., Ticona-Tapia, J.R., Cárdenas-Gonzáles, J.R., Bustamante-Quintana, P.H.: SynKGen: a kernel PCA-Based oversampling method for enhanced credit card fraud detection. Revista Científica de Sistemas e Informática **5**(2), e952 (2025). https://doi.org/10.51252/rcsi.v5i2.952
19. Bolia, C., Joshi, S.: Optimized deep neural network for high-precision psoriasis classification from dermoscopic images. Revista Científica de Sistemas e Informática **5**(2), e966 (2025). https://doi.org/10.51252/rcsi.v5i2.996

Design of Variable PI Controllers for a DC MIMO Microgrid Using Machine Learning

Javier A. Romero R.[1(✉)], Charles E. Hunt[1], and Hector Vargas[2]

[1] Electrical and Computer Engineering Department, University of California, Davis, CA 95616, USA
jaromerori@ucdavis.edu

[2] Software Department, Universidad Manuela Beltran, Bogota, Colombia

Abstract. This document presents a new methodology to design an intelligent variable-gain PID controller for a MIMO DC microgrid. First, a set of optimal controller gains is found under specific grid conditions using Particle Swarm Optimization (PSO). Then, linear interpolation is performed over the optimal gains. Finally, a neural network is trained using the interpolated gains. The neural network is used to generate optimal gains for PID controllers, depending on the actual state of the microgrid. Three methods were compared for the controller gain selector: K-Nearest Neighbor (KNN), Interpolated function, and regression Artificial Neural Network (ANN). The ANN method showed the best performance among the three. The ANN learned patterns of behavior in the microgrid to create a general model of optimal gains for unseen conditions.

Keywords: Machine learning · Microgrid · Intelligent PID

1 Introduction

1.1 Problem Formulation

Microgrids are electrical networks formed by the union of Renewable Energy Sources (RES), Energy Storage Systems (ESS), the electrical grid, and the user or load [9]. The use of microgrids represents a change in the way energy is produced, transported, and distributed. If the microgrid is standalone, there is no need for a transport layer, and the generation is done at the same location as the user. Microgrids will require a control system to achieve goals such as maintaining stable voltage, reducing transient voltage changes, extending battery life, and obtaining the maximum power available from the RES. That is why several researchers have explored different control strategies, and no consensus has been reached on the best control strategy [6,9,11].

Even with modern control approaches, the most widely used device in the industry remains the PID (Proportional Integral Derivative) controller. Tuning the device parameters is a challenging task due to the system's non-linearity

D. Chaves et al. (Eds.): CIARP 2025, LNCS 16528, pp. 156–171, 2026.
https://doi.org/10.1007/978-3-032-23161-1_12

and multiple input and output characteristics. Iterative methods, such as Particle Swarm Optimization (PSO), have been employed to achieve the optimal set of gains for the controller. Since the gains are fixed, it cannot be optimal for all operating points, and a trade-off must be made [5]. Controllers designed with Artificial Neural Networks can overcome the control problems due to the non-linearity of the system [11]. However, replacing controllers with neural networks is challenging due to the industry's trust in traditional PID controllers. A solution to this problem is the use of an intermediate controller architecture that bridges the gap between traditional and intelligent controllers. In this paper, a methodology is proposed to design a PID controller with variable gains using an ANN.

1.2 Previous Work

Control strategies in Multiple Input Multiple Output (MIMO) systems can be classified as centralized, distributed, and decentralized [11]. In the centralized strategy, a central controller is used with measurement capabilities for all variables in the circuit. Some centralized control strategies for microgrids are Fuzzy Logic Controller (FLC) [15], Ruled Based Control (RBC) [2], Model Predictive Control (MPC) [7], Optimization Based Control (OBC) [3], and Artificial Neural Network (ANN) [4].

A classic architecture can be studied in [14], where a control strategy with PID controllers is presented for a MIMO microgrid. However, it is not explained how the system is tuned and evaluated at a single operating point. In [5], a similar network is analyzed, where several methods were explored to select the system gains. The analyzed methods are Pole Placement, Linear Matrix Inequality Approach, PSO, and Genetic Algorithm (GA). Among those, the PSO algorithm presented the best performance. However, with certain values of the load, the system was unstable. Iterative algorithms have been shown to be useful in finding optimal gains for the controller; however, those gains are optimal only for a specific state of the microgrid. This shows the need to create non-linear controllers for microgrids.

Controllers designed with Artificial Neural Networks are better for nonlinear systems [11]. In [16], a control strategy was created through the use of neural networks. The design process started with the analysis of the control system with PI controllers. Optimum gains were found using the Bat Search Algorithm. The signals generated by the controllers were then used to train a neural network. The ANN replaced the PI controllers, which obtained a better result than the classic controllers. In [13], a general system with DC-DC converters connected in parallel to a common bus is presented. The control strategy is implemented solely with Neural Networks, and training is conducted based on data generated by the PID controllers.

Many control strategies replace the PID controllers. However, it is more likely for the industry to accept new control systems if the PID is kept and mixed with a modern approach. In [10], a voltage regulation control system using a PID variable gain controller is presented. The selection of gains is performed through particle swarm optimization (PSO) and the coati optimization algorithm (COA). It was demonstrated that the ICOA algorithm converges in fewer steps. However, a disadvantage of this method is the need for multiple iterations with every change in the system. Real-time operation could not be achieved using this method. In [1], an AC voltage regulator control system is presented using PID+LP controllers and variable gains, where LP stands for low-pass filter. The gains of the controller are modified during operation using fuzzy logic. The strategy demonstrated better performance compared to fixed gains. However, the strategy was not evaluated in a MIMO system, and keeping the simple structure of the PID is more attractive to the industry. In [8], a variable-gain PID control strategy is presented for an inverter. The selection of gains is done using a Neural Network. Through several iterations, the best weights for the neural network are found using the ant lion iteration method. The control strategy demonstrated better performance compared to PID and FOPID (Fractional Order PID). This is a reason to evaluate neural networks for control in other systems.

In [12], an example of the use of neural networks to control microgrids is presented. The control strategy was designed for a MIMO microgrid using neural networks and FOPID controllers. The microgrid consisted of solar photovoltaic panels, wind generators, batteries, and a fuel cell. The control strategy employs reinforcement learning to determine the optimal weights of a neural network, thereby generating the gains of the controllers. The combination of neural network and PID was shown to be better than the classic approach; however, the models used to represent the components of the microgrid were linear, even when power converters are not.

1.3 Proposed Approach

In this paper, a new method is presented to design a variable-gain intelligent PID controller for a MIMO DC microgrid. A microgrid model was designed in Simulink. A neural network adjusts the optimal gains for three PID controllers in a DC microgrid. The design of the control system could be divided into three parts. First, particle swarm optimization was employed to generate a set of potential gains for the control system, contingent upon the actual state of the microgrid. Simulations of the circuit with non-linear models were used. In the second step, the optimal gains were interpolated for other states of the network, and finally, a neural network was trained with the interpolated data. In this work, three methods were tested to vary the gains of the PI controllers during operation. The methods are K-nearest neighbor, linear interpolation of gains, and a regression neural network. The proposed method offers an alternative to direct neural network training through iterative processes and complex circuit simulations. It was seen that the ANN learns patterns of behavior in the microgrid and creates a general model of optimal gains for unseen conditions.

2 Microgrid Design

The microgrid shown in Fig. 1 was modeled using MATLAB's Simulink software. Its components include an array of solar panels, a set of batteries, two DC-DC converters, and a load represented by a resistor. The solar panel model used is the "LG Electronics LG300N1W-G3", available in Simulink. Seven solar panels, connected in parallel, form the array, which has a peak power of 2119 W at an incident irradiance of 1 kW/m^2. A standard Lithium-Ion battery model available in Simulink was used, with a capacity of 2500 Ah. The initial state of charge for all simulations was set to 60 %.

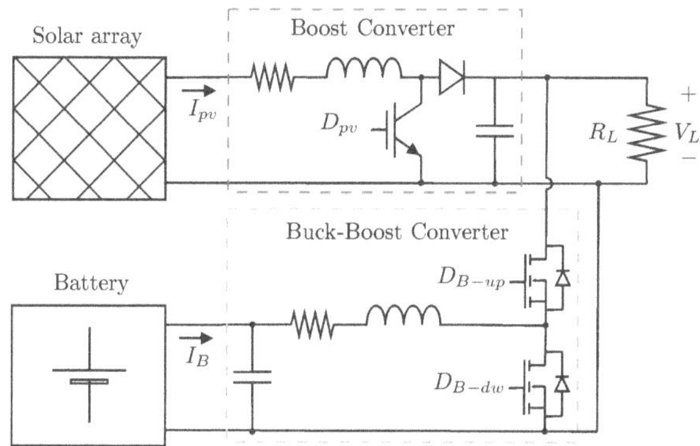

Fig. 1. Microgrid model.

In Fig. 1, the circuits for the DC-DC converters are presented. A boost converter is connected to the solar panel array to control the output current and achieve the maximum power extraction. This circuit is also known as a maximum power point tracker (MPPT). Even when the boost converter is used to increase the output voltage, its main purpose in the system is to regulate the current I_{pv}. To modify the output of the power converter, a pulse width modulation (PWM) signal must be applied to the port D_{pv}.

Since the voltage in the load must also be regulated, a set of batteries is required. Figure 1 shows a bi-directional Buck-boost converter connected between the batteries and the load. The converter has two functions. First, to regulate the charge and discharge current of the battery, and second, to regulate the voltage over the load. When the output power of the solar panels exceeds the required load power, the battery must be charged with the excess power. Otherwise, if the output power of the solar panels is not enough, the battery must supply the remaining power to the load.

The transistors in the circuits are controlled by Pulse Width Modulation (PWM) signals; the transistors are never activated simultaneously. The superior one is used to charge the battery, while the inferior one is used to discharge it.

The selected switching frequency for the transistors is 25 kHz. The switching frequency is selected by the circuit designer. The higher the frequency, the smaller the inductors and capacitors need to be. However, it means that the controllers must be faster. The signal D_{B-up} is used to extract current from the battery, and the signal D_{B-dw} is used to charge the battery. The simulation frequency of the microgrid must be selected to be higher than the switching frequency, capturing all the circuit's behavior. For the simulations, a frequency of 1 MHz was used. A microgrid model is available on Github: https://github.com/jaaromerori-col/microgrid_control

The initial control system was created with discrete PI controllers. The update rate of the control system was configured at 500 kHz. This rate was selected to account for the sample rate of the discrete microcontrollers used in the microgrids. Figure 2 presents the three control loops of the system, where the first one, located at the top of the figure, is used to extract current from the solar array and achieve the maximum extracted power. An optimal current, I_{pv-sp}, is defined according to the actual irradiance and is compared with the actual current from the solar array, I_{pv}.

The second loop, located at the bottom of the Fig. 2, presents two controllers. The first compares the load voltage V_L with the reference V_{L-sp} and defines an objective current for the battery. If the current is positive, the battery will be discharged. On the other hand, if it is negative, the battery will be charged. The third controller compares the desired battery current with the actual current I_B and generates the duty cycle for the two PWM signals of the buck-boost converter. Saturation blocks were used to limit the control signals to a range of 0 to 1. The control loops used for the battery can be reduced to a single PID block. However, it is useful for safety purposes to know the amount of current required for the battery.

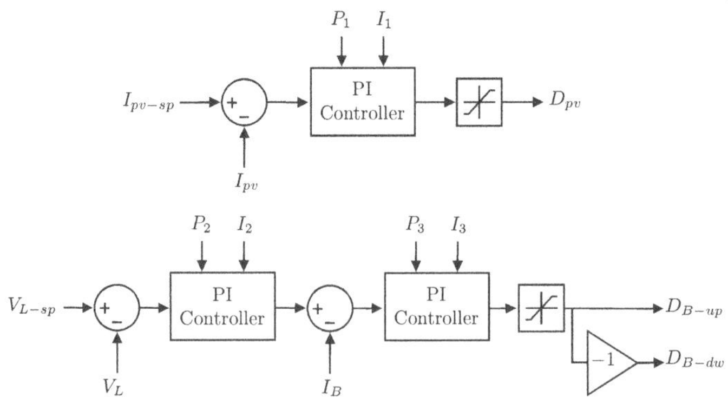

Fig. 2. Control system.

Selecting the proper gains for the controllers is a challenge due to the system's non-linearity. The behavior varies at different working points. Changes in load resistance R_L will result in a change in the system's transfer function. Additionally, changes in the current supplied by the solar array are perceived as disturbances to the voltage control loop. Fixed gains for the controller will work in a range of operations, but unexpected behaviors may occur at other operation points. That is why a better strategy is to adjust the system's gains according to the microgrid's state.

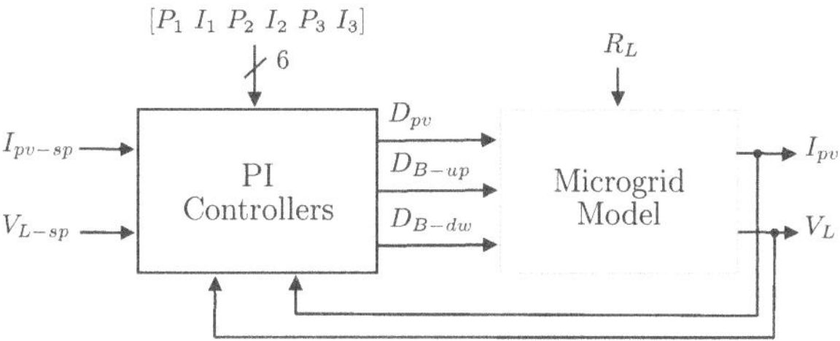

Fig. 3. Model of the system for the optimization with PSO.

3 Search for Ideal Gains Through PSO

To design a controller that overcomes the nonlinearity of the system, a search for ideal gains at different operating points was conducted. Figure 3 presents a model of the microgrid connected to the control system. The load resistance, the desired current for the solar array, and the load voltage are fixed inputs of the model. Then, using particle swarm optimization (PSO), ideal gains for PI controllers were found to minimize the cost function of Eq. 1.

The cost function J is the sum of the square tracking error in the load voltage and the current of the solar array in each run of the model simulation. The reference voltage was regulated to 48 V, and the reference current was selected based on the irradiance. The variable N is the total number of samples per simulation, and β is a weight factor used to adjust the units of terms in the cost equation. A unit value for β was used, as the current and voltage values were of similar magnitude.

$$J = \sum_{n=1}^{N} [(I_{pv}(n) - I_{pv-sp}(n))^2 + \beta(V_L(n) - V_{L-sp(n)})^2] \tag{1}$$

In each simulation, the transition between an initial current of zero and a specific current was made. Twenty milliseconds of the model were simulated, with the transition occurring at 10 ms. This time was sufficient to check the transient response of the system and achieve steady-state operation. Figure 4 illustrates the optimization process. A set of possible input variables for the system is identified. For each combination of input variables, an optimization is performed to minimize the cost function. The possible combinations of the input variables generate a total of 9 solutions. Each one is optimum for a different working point in the microgrid.

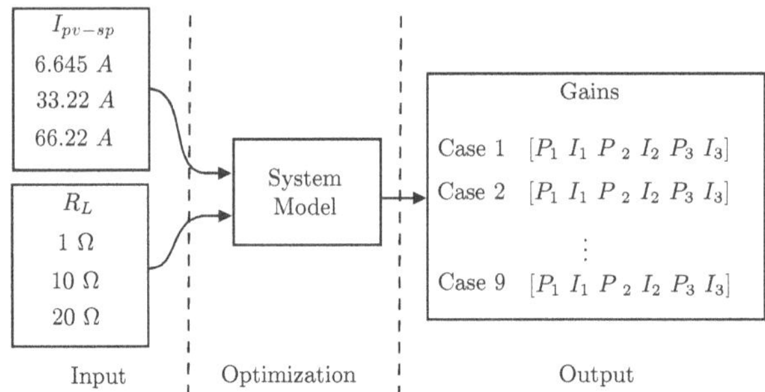

Fig. 4. Input and output variables for the optimization with PSO.

Table 1. Optimal gains for the system given the current and the load.

I_{pv-sp}	R_L	P_1	I_2	P_2	I_2	P_3	I_3
6.645A	1Ω	0.97	3000	0.01	2615	520.0	708.6
33.22A	1Ω	1.72	3000	43.21	0.01	583.9	949.9
66.22A	1Ω	486.1	1605	48.72	2994	770.2	2522
6.645A	10Ω	2.86	2103	24.58	2899	0.01	0.01
33.22A	10Ω	1.71	2516	85.26	3000	0.01	138.6
66.22A	10Ω	480.3	467.7	122.4	501.1	59.47	8.56
6.645A	20Ω	140.1	545.9	92.83	4.17	0.01	36.81
33.22A	20Ω	495.5	4974	32.64	3000	0.01	1073
66.22A	20Ω	163.8	596.6	103.9	3000	0.01	402.4

A total of 30 particles were used for each optimization process. The maximum number of iterations allowed was 60 for each combination of inputs in the micro-grid. Since the simulation of the model takes approximately 7 s, the maximum optimization time is 31 h. However, the number of iterations required was less. A method to reduce the time is to use the solutions of the previous optimization as initial particles for the next optimization. The optimization process required approximately 4 h. The set of optimal gains is presented in Table 1.

4 Gain Selector

Figure 5 illustrates the system model, where the gains of the controllers are adjusted according to the microgrid's state. The voltage and current of the load are measured. Then the load resistance is identified using Ohm's law, and the gains of the controllers are selected. To select the ideal gains, three methods were tested. The K-nearest neighbor (KNN), a linear interpolation method and a regression artificial neural network (ANN). Each method was tested using simulations with variable input values over an 8-ms period. Initially, the same input variables from the training were used; then, the system's response was observed when input conditions different from the training were applied.

4.1 KNN Gain Selector

The K-nearest neighbor is a supervised classification algorithm. To perform the classification task, the distance between the actual input and previous input data points is calculated. The category of the closest data points is selected as the algorithm output. In this case, the input variables are the solar array set point current and the load resistance. The output is the category that refers to a specific set of gains established in the optimization process. Figure 6 illustrates

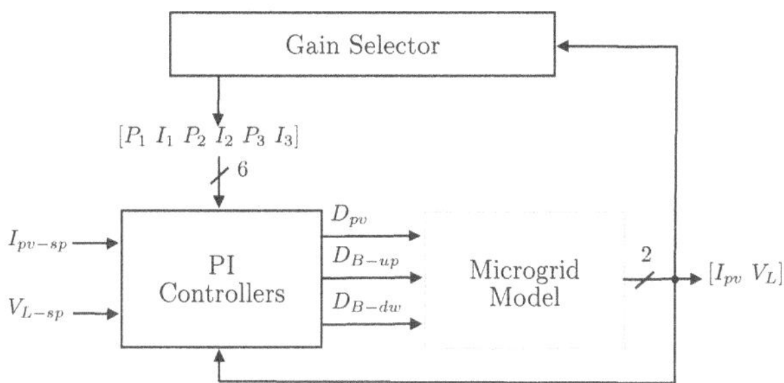

Fig. 5. Control scheme using a gain generator for PID controllers.

the system's response under the conditions used during the training process. The load resistance was fixed at 10Ω. The current of the solar array followed the reference, which was updated every 10 ms in response to changes in irradiance. A 5% interval around the voltage reference is graphed with blue dashed lines. The voltage of the load is regulated between the limits, and after the change at 50 ms, it is evident that there is a rejection of the perturbations generated by the changes in the current of the solar array.

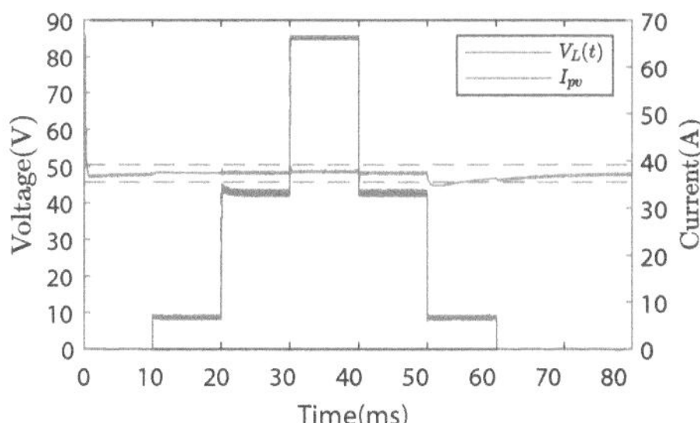

Fig. 6. Response of the system with KNN gain generator to the same conditions of the optimization. A constant resistance of 10Ω was used as load resistance.

Then, new combinations of inputs were used. The system's response was better than maintaining constant gains for all operation points. In Fig. 7, the system's response to a new load is illustrated. The resistance is far away from the points of Table 1, which is why there are more pronounced oscillations. Even with the voltage overshoot, the perturbations were rejected. However, there are operation points where voltage regulation was not achieved. In Fig. 8, a combination of inputs is presented where the control system is unable to regulate the load voltage.

Figure 7 illustrates the system's response when the load resistance is a value that was not encountered during the training process. The current reference for the solar array is still the same as the training. As a result, more oscillations and higher overshoots were observed in the load voltage; however, regulation was achieved. The voltage overshoots surpass the 5% limit, but the current is still following the reference without major concern. Finally, the model was simulated using new input variables for the current reference and load resistance. The result is presented in Fig. 8 where the load resistance changed every 10 ms to the values 18Ω, 10Ω, 19Ω, 1Ω, 14Ω, 16Ω, and 14Ω. There are operational points where voltage regulation was not achieved, and the system exhibits unstable behavior. The highest voltage overshoot was around 67%. Adjusting the controller gains using the KNN algorithm can lead to instability in the microgrid.

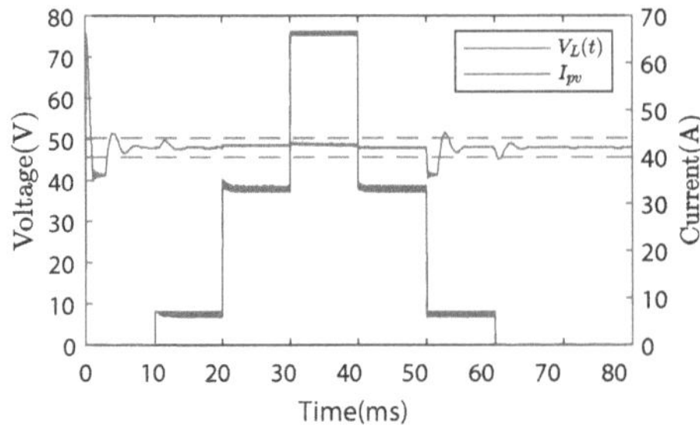

Fig. 7. Response of the system with KNN gain generator with a new load $R = 5\Omega$.

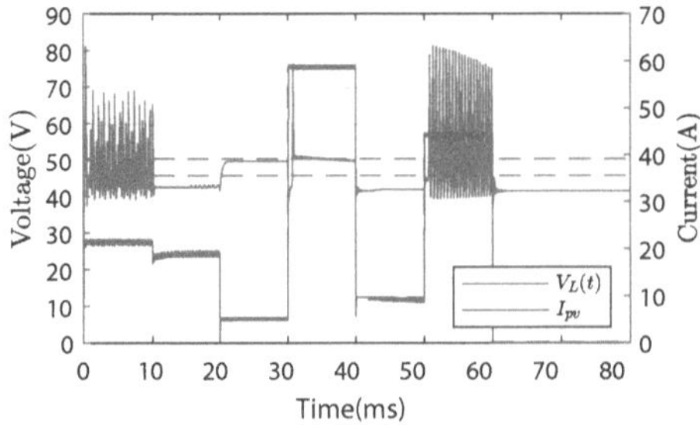

Fig. 8. Response of the system with KNN gain generator to different conditions unseen in the optimization.

4.2 Interpolator Gain Selector

In Fig. 9, the function for generating the integral gain for the voltage controller is shown when the KNN selector is used. When KNN is used, only the corners of the function are defined, and only those can be used during the control process. A second gain generator was created that interpolates the gains of Table 1. If the values of Table 1 are interpolated, a new function is created to define the integral gain with more possible values. The new interpolated function for choosing the gains is presented in Fig. 10.

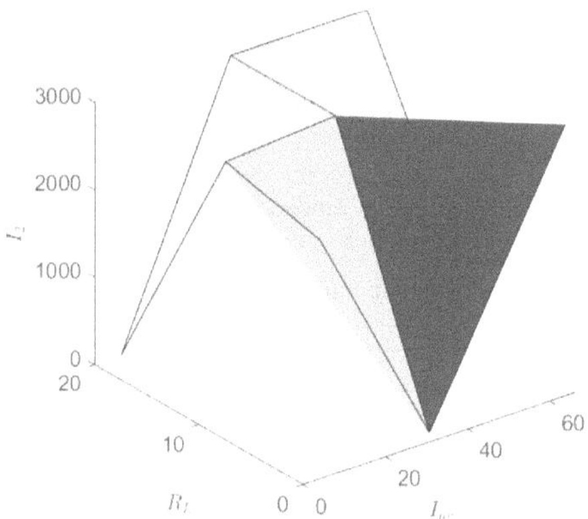

Fig. 9. Integral gain function for the KNN gain selector of the voltage controller.

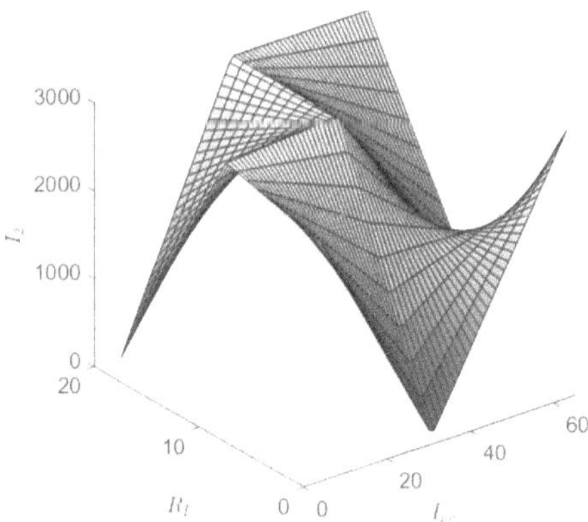

Fig. 10. Integral gain function for the interpolator gain selector of the voltage controller.

As a consequence of the use of an interpolated function for each gain, the voltage oscillations of the previous control system can be eliminated. Figure 11 shows the result of a microgrid simulation using interpolated gains, under the same conditions as the simulation result presented in Fig. 7. A reduction in voltage overshoots is observed. This means that interpolation can enhance the system's behavior for conditions not encountered during training. Also, the unwanted instability shown by the system in Fig. 8 was reduced but still present, as shown in Fig. 12.

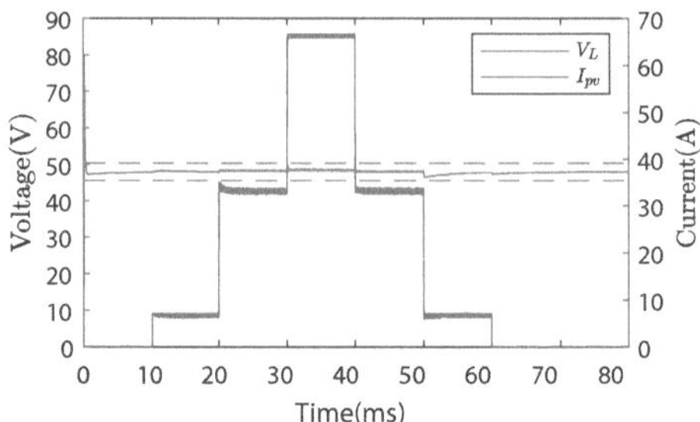

Fig. 11. Response of the system with the interpolated function gain generator with a load $R = 5\Omega$.

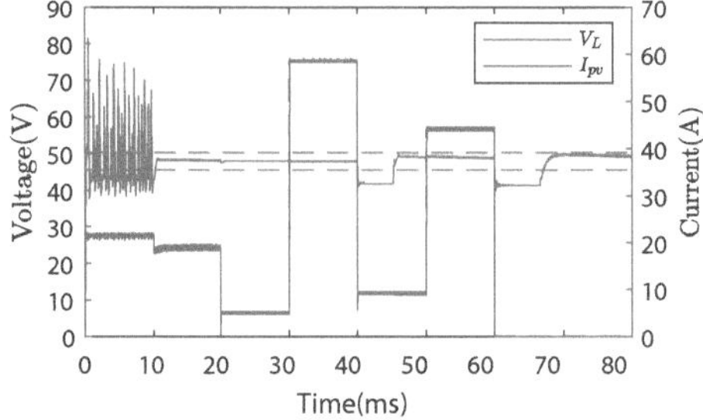

Fig. 12. Response of the system with the interpolated function gain generator with variable load and irradiance.

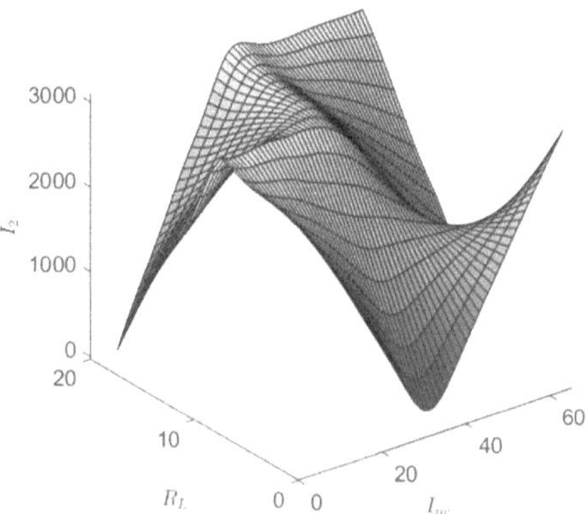

Fig. 13. Integral gain function for the ANN gain generator of the voltage controller.

4.3 ANN Gain Selector

With the data of the interpolated functions, an Artificial Neural Network (ANN) was trained to generate the gains for the controllers. The network has two inputs: the load resistance and the solar array current. 20 neurons were used in the hidden layer with hyperbolic tangent activation functions. Six neurons were used for the output, without the use of an activation function. The output of the network is the set of ideal six controller gains for each microgrid state. In Fig. 13, the possible values for the integral gain I_2 are generated by the artificial neural network. This allows for comparison of gain functions with the other methods; a clear difference between them is the reduction of sharp edges.

To evaluate the behavior of the ANN gain selector, several simulations were performed with the same conditions that created difficulties for the other control strategies. In contrast to previous methods, the ANN gain selector created the appropriate gains to maintain system stability and achieve current and voltage regulation. In Fig. 14, the response of the system under the same conditions as the simulations of Figs. 8 and 12. The system rejected all the perturbations and brought the variables to the desired state.

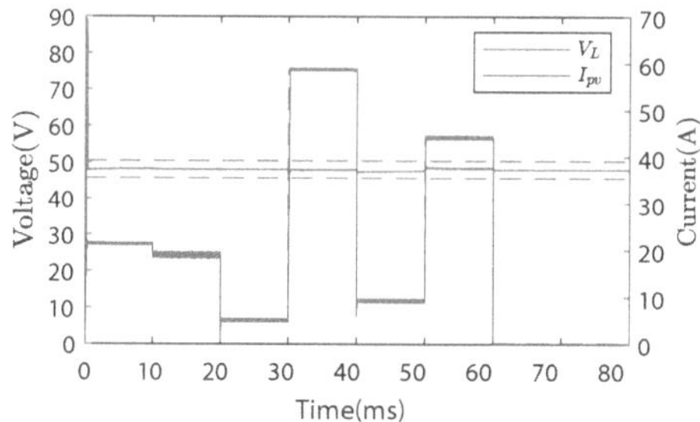

Fig. 14. Response of the system with the ANN gain generator with variable load and irradiance.

4.4 Comparison Between Gain Selectors

Three gain selectors were designed: the KNN gain selector, the interpolated gain selector, and the ANN gain selector. To compare the performance of the three control strategies, ten scenarios with random conditions were simulated. Each simulation replicates 8 ms of the microgrid, where the variables of irradiance, solar panel array reference current, and load resistance were changed randomly every 10 ms. The cost function for each control strategy was calculated based on the responses. Figure 15 presents the result of the evaluation. In nine out of ten scenarios, the ANN gain controller exhibited the best behavior. In simulation number four, the ANN was outperformed. However, we can see that in this scenario, all the control systems presented a low-cost function, and the three achieved the control objectives.

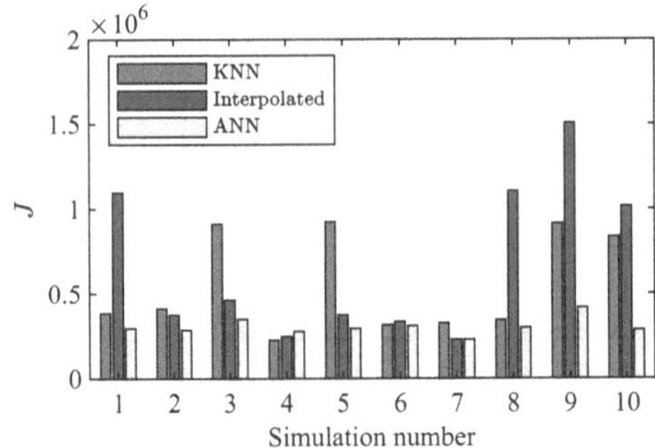

Fig. 15. Cost function of the response of the microgrid with three gain selectors.

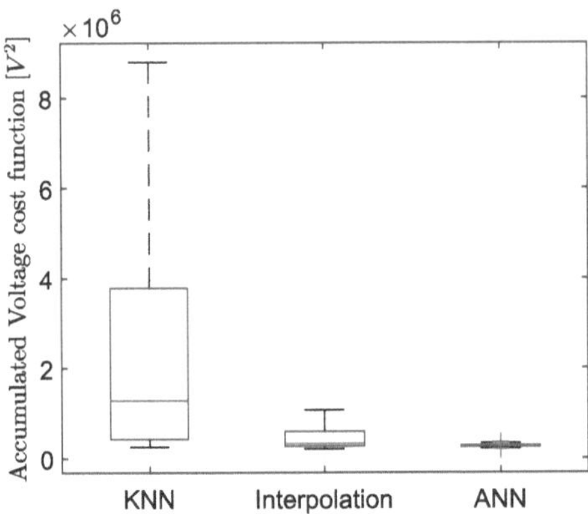

Fig. 16. Box plot of the accumulated square voltage error in each simulation.

Figure 16 presents a box plot of the voltage cost function in twenty simulations. Seven random changes of the voltage set point and seven random changes of the current set point were performed for each simulation. The microgrid control system was tested in 980 possible scenarios. Table 2 presents the mean and standard deviation of the cost function for each gain selector. The ANN gain selector demonstrated superior performance under various conditions of the test simulations. It presented the lowest mean and deviation in the cost function. A

Table 2. Optimal gains for the system given the current and the load.

Gain selector	Mean $[V^2]$	Standard deviation $[V^2]$
KNN	24.1×10^{-5}	2.60×10^{-5}
Interpolator	4.80×10^{-5}	3.07×10^{-5}
ANN	2.73×10^{-5}	4.08×10^{-5}

trade-off between the complexity of the algorithm and performance, since the KNN gain selector presented the lowest performance.

5 Conclusion

A new method was presented to design a variable-gain intelligent PID controller for a MIMO DC microgrid. Variable gains can show better performance than fixed gains due to the non-linearity of the system. The design of the control system could be divided into three parts. First, particle swarm optimization was employed to generate a set of potential gains for the control system, contingent upon the actual state of the microgrid. The PSO algorithm was used to determine the optimal set of parameters for the controllers at nine specific operation points. In the second step, the optimal gains were interpolated for other states of the network, and finally, a neural network was trained with the interpolated data.

Three methods were compared for the controller gain selector: K-Nearest Neighbor, an interpolation function, and a regression Artificial Neural Network. Each strategy changed the gains of the controllers depending on the measurements in the microgrid. The control strategies were compared using a cost function that considers the error in the voltage of the load resistance and the current of the solar array. The results were mixed between interpolation and the KNN method, as shown in Fig. 15. The interpolation performed better for system conditions that were not evaluated in the initial optimizations. Moreover, the KNN strategy's performance decreased as it moved further away from the training conditions. The mean and standard deviation for the cost function in all the evaluation simulations were around ten times smaller for the interpolated function method compared to the KNN method. We can conclude that it is better to interpolate gains instead of using KNN. However, instability in the microgrid was observed for both strategies.

The best strategy to select the gains of the controllers is the ANN. The most challenging task was regulating the voltage, but this task was successfully accomplished by the ANN gain selector. The mean and standard deviation for the cost function in all the evaluation simulations were around ten times smaller for the ANN function method compared to the interpolation method. The ANN gain selection strategy maintained the system's stability in those working points where the other strategies failed. The ANN performed better than the interpolated gain generator, which was an unexpected result, given that the ANN was trained using data generated by the interpolated gain generator. An explanation for this result is that the ANN behaves as a low-pass filter. Figure 10 presents the function to select the integral gain of the voltage controller using the interpolation method. The same map is presented in Fig. 13, but this time generated by the ANN. The sharp edges of the function were eliminated, and the ANN can be seen as a low-pass filter. Another explanation is that the data for the training process were divided into three data sets: training, validation, and testing. The idea of using the validation data set is to verify that the ANN is replicating a general model of the data and not overfitting to it. In this case, the ANN learned patterns of behavior in the microgrid and created a general model of optimal gains for unseen conditions, thereby enhancing the controllers' performance. In this model, outliers of the interpolated function were eliminated.

Acknowledgments. The author acknowledges the invaluable support from the Electrical and Computer Engineering Department and the Global Affairs Office at the University of California, Davis. Their efforts help create a welcoming environment for international students and enhance the university's research capabilities.

References

1. Ahmadnia, M., Hajipour, A., Tavakoli, H.: Robust variable-order fractional PID-LP fuzzy controller for automatic voltage regulator systems. Appl. Soft Comput. **167**, 112268 (2024)
2. Ahmed, M., Kuriry, S., Shafiullah, M., Abido, M.: Dc microgrid energy management with hybrid energy storage systems. In: 2019 23rd International Conference On Mechatronics Technology (ICMT), pp. 1–6. IEEE (2019)
3. Akram, U., Khalid, M., Shafiq, S.: An innovative hybrid wind-solar and battery-supercapacitor microgrid system–development and optimization. IEEE Access **5**, 25897–25912 (2017)
4. Benmouna, A., Becherif, M., Boulon, L., Dépature, C., Ramadan, H.S.: Efficient experimental energy management operating for fc/battery/sc vehicles via hybrid artificial neural networks-passivity based control. Renew. Energy **178**, 1291–1302 (2021)
5. Guentri, H., Allaoui, T., Mekki, M., Denai, M.: Power management and control of a photovoltaic system with hybrid battery-supercapacitor energy storage based on heuristics methods. J. Energy Storage **39**, 102578 (2021)
6. Hasan, M.A., et al.: A comprehensive review of control strategies and efficiency optimization for islanded AC microgrids. IFAC Journal of Systems and Control **33**, 100326 (2025). https://doi.org/10.1016/j.ifacsc.2025.100326
7. Hou, J., Song, Z., Hofmann, H., Sun, J.: Adaptive model predictive control for hybrid energy storage energy management in all-electric ship microgrids. Energy Convers. Manage. **198**, 111929 (2019)
8. Kashfi, R., Balochian, S., Alishahi, M.: Design of a optimal robust adaptive neural network-based fractional-order PID controller for H-bridge single-phase inverter. Appl. Soft Comput. **166**, 112142 (2024). https://doi.org/10.1016/j.asoc.2024.112142
9. Lin, X., Zamora, R.: Controls of hybrid energy storage systems in microgrids: critical review, case study and future trends. J. Energy Storage **47**, 103884 (2022)
10. Liu, S., Lin, Z., Feng, R., Huang, W., Yan, B.: Intelligent control method for automatic voltage regulator: an improved coati optimization algorithm-based strategy. Meas. J. Int. Meas. Confederation **252**(March), 117263 (2025). https://doi.org/10.1016/j.measurement.2025.117263
11. Marimuthu, R., et al.: Review on advanced control techniques for microgrids. Energy Rep. **10**, 3054–3072 (2023)
12. Nosrati, K., Skiparev, V., Tepljakov, A., Petlenkov, E., Levron, Y., Belikov, J.: Application of neural network based variable fractional order PID controllers for load frequency control in isolated microgrids. Elsevier Inc. (2023). https://doi.org/10.1016/B978-0-443-18426-0.00003-0

13. Saddriwala, M.A., Alam, M.: Artificial neural network based decentralized current-sharing control for parallel connected DC-DC converters in DC microgrid application. Comput. Electr. Eng. **120**(PA), 109731 (2024). https://doi.org/10.1016/j.compeleceng.2024.109731
14. Sathishkumar, R., Kollimalla, S.K., Mishra, M.K.: Dynamic energy management of micro grids using battery super capacitor combined storage. In: 2012 Annual IEEE India Conference (INDICON), pp. 1078–1083. IEEE (2012)
15. Shayeghi, H., Monfaredi, F., Dejamkhooy, A., Shafie-Khah, M., Catalão, J.: Assessing hybrid supercapacitor-battery energy storage for active power management in a wind-diesel system. Int. J. Electr. Power Energy Syst. **125**, 106391 (2021)
16. Singh, P., Lather, J.S.: Dynamic power management and control for low voltage dc microgrid with hybrid energy storage system using hybrid bat search algorithm and artificial neural network. J. Energy Storage **32**, 101974 (2020)

Ensemble Deep RVFL Neural Networks Integrating News and Protest Sentiment for Tourism Stocks

Nando Verelst and Pablo A. Henríquez(✉)

Departamento de Administración, Universidad Diego Portales, Santiago, Chile
nando.verelst@mail.udp.cl, pablo.henriquez@udp.cl

Abstract. In the context of financial forecasting, particularly within the tourism sector, the integration of sentiment analysis with traditional stock market indicators offers a promising avenue for enhancing prediction accuracy. Addressing this potential, we implement a sentiment-augmented methodology for stock trends prediction. Sentiment analysis from 6,464 CNN articles and 74,042 ACLED protest event data using FinBERT sentiment scores, alongside technical indicators, enhances prediction accuracy. The analysis was conducted on three major hotel chains: Hilton, Hyatt, Marriott, and three airlines JetBlue, United, and Delta from 2020–2024. Our evaluation indicates Deep ensemble RVFL networks outperform conventional methods with 76% accuracy (71% F1) for hotel chains and 61% (54% F1) for airlines, a 7% improvement over technical-only models. The RVFL variants maintained this edge in performance, requiring only 0.01-0.02 s training time, which is beneficial for real-time applications. These findings demonstrate the value of multi-source sentiment analysis with randomized neural networks for robust tourism stock prediction, revealing a greater predictability of hotel stocks over airline stocks.

Keywords: Sentiment analysis · Stock classification · Randomized neural networks · Tourism stocks · FinBERT

1 Introduction

The tourism industry has shown high vulnerability in reference to security risks. As an example, the 2015 Paris terror attack resulted in over €2 billion losses for European travel and hotel companies shares, one day before said incident [7]. This shows the strong effect negative incidents have on the tourism industry. It's high reactivity highlights the need for predictive models that are reliable, and take into consideration both qualitative information, for example, media sentiment, as well as quantitative indicators, for instance, technical indicators.

Moreover, previous research has shown that media framing effects influence consumer choices, including tourist behavior, particularly through risk-amplifying narratives [19]. These framing effects also have a direct impact on

D. Chaves et al. (Eds.): CIARP 2025, LNCS 16528, pp. 172–190, 2026.
https://doi.org/10.1007/978-3-032-23161-1_13

stock performance, providing critical information to traders that can influence stock values [6]. Given the interconnectedness of news media and market behavior, understanding how sentiment conveyed in news articles impacts stock prices becomes essential. The influence of sentiment extends beyond direct reporting on economic indicators; it also shapes broader market perceptions and investor psychology [5].

Traditional models of market prediction predominantly focus on technical indicators [18,24,26]. However, incorporating textual sentiment analysis has emerged as a significant advancement in understanding market dynamics [1]. Recent studies suggest that media sentiment, particularly from news outlets, serves as a leading indicator of market movements by capturing the sentiments of investors and consumers before they are reflected in financial metrics [8]. This demonstrates that negative events, along with their media portrayal, directly affect the stock values of companies, including those within the tourism industry. However, the effects of news sentiment on stock performance in the tourism industry remain underexplored.

The current study aims to integrate sentiment analysis from tourism-related news and protest sentiment to analyze their effects on stock classification for tourism-related companies. By utilizing sentiment scores derived from textual data, we seek to develop a predictive model that highlights the added value of sentiment analysis in forecasting stock movements. This approach leverages insights from both traditional financial indicators and the qualitative aspect of media sentiment, providing a comprehensive view of market behavior.

Our investigation builds upon existing literature by focusing specifically on the interplay between public sentiment and the valuation of tourism stocks. This focus is particularly relevant in a rapidly changing news landscape, where timely data and its interpretation can offer a competitive edge to investors. By meticulously classifying news articles and assessing their emotional connotations, we aim to validate how sentiment affects market predictions, thereby enhancing the predictive accuracy of stock movements in the tourism sector.

In summary, this study seeks to improve the accuracy of tourism stock predictions by integrating sentiment analysis from news and protest data with traditional technical indicators. We achieve this by developing a predictive model using sentiment scores derived from textual data and assess its impact on stock classification. By classifying news articles and evaluating their emotional implications, our research seeks to demonstrate the influence of sentiment on market predictions, ultimately improving the precision of forecasting stock movements in the tourism sector.

2 Literature Review

Previous studies have examined the prediction of financial markets using textual data, particularly from media sources [1]. Increased media coverage significantly impacts stock valuations, resulting in less synchronized stock prices and a more pronounced market reaction to individual news stories as opposed to overarching market trends [6]. This suggests that general market trends can be disrupted by the information disseminated by the media. In this context, both social and traditional media serve as valuable sources for predicting market trends [8]. The rapid dissemination of information generated by the media influences investor decisions, with the nature of those decisions varying depending on the content of the articles [20].

The tourism sector, and stocks from tourism related firms, are particularly sensitive to bad news [7]. For instance, news regarding COVID-19 had a substantial impact on travel and leisure (T&L) stocks, with media attention driving short-term price fluctuations and fear and uncertainty contributing to long-term volatility [30]. Also, feelings of investors have shown to cause irrational behavior and strong fluctuations of capital flows in the tourism sector [4,23], due to the high sensitivity of this industry. Moreover, as stated in [25], geopolitical risks (including social protests), are negatively related to T&L stock performance, although this may depend on the country. This is especially relevant considering that tourism is one of the most prominent casualties of geopolitical risk.

In that sense, neural networks (NN) are an useful approach for stock prediction, specially with text data mining techniques [1]. Moreover, the random vector functional link (RVFL) neural network has shown to be practical for stock prediction tasks, due to its combination of speed, accuracy, and robustness [2,3,28]. However, its use remains relatively novel in stock market prediction, without applications to the tourism industry.

3 Background

Feed-forward randomized neural networks (RNNs) are designed to overcome the challenges of slow training and convergence typically associated with backpropagated neural networks. By employing randomized configurations, these networks achieve efficient training dynamics [10,31]. This section elaborates on two architectures: RVFLwoDL and RVFL. It also explores their extensions, including deep and ensemble frameworks.

3.1 Random Vector Functional Link Without Direct Links (RVFLwoDL)

The RVFLwoDL model is a conventional RVFL network that excludes direct connections between the input and output layers [12]. Input weights and biases for the hidden neurons are initialized randomly, enabling the transformation of input data into a high-dimensional feature space. The output layer weights are computed analytically using the least squares method [9].

Shallow RVFLwoDL

RVFLwoDL with a single hidden layer, termed shallow RVFLwoDL, is depicted in Fig. 1. Let $\mathbf{X} \in \mathbb{R}^{k \times d}$ represent the input matrix, where k and d denote the number of samples and features, respectively. The enhanced feature matrix $\mathbf{H}$ is defined as:

$$\mathbf{H} = g(\mathbf{X}, \mathbf{W}), \tag{1}$$

where g is a nonlinear activation function, and $\mathbf{W}$ contains the hidden layer weights. Activation functions like sigmoid, hyperbolic tangent, or Gaussian are commonly utilized [17]. In this work, we adopt the sigmoid function:

$$g(\mathbf{X}, \mathbf{W}) = \frac{1}{1 + \exp(-\mathbf{X} \cdot \mathbf{W})}. \tag{2}$$

The network's output, $\hat{\mathbf{Y}}_{\text{RVFLwoDL}}$, is expressed as:

$$\hat{\mathbf{Y}}_{\text{RVFLwoDL}} = \mathbf{H} \cdot \boldsymbol{\beta}, \tag{3}$$

where $\boldsymbol{\beta} \in \mathbb{R}^{N \times 1}$ represents the output layer weights, and N is the number of hidden neurons. To derive $\boldsymbol{\beta}$, the least squares optimization problem is solved as follows:

$$\min_{\boldsymbol{\beta}} ||\mathbf{H}\boldsymbol{\beta} - \mathbf{Y}_{\text{RVFLwoDL}}||^2 + \lambda||\boldsymbol{\beta}||^2, \tag{4}$$

where λ is the regularization parameter, and $\mathbf{Y}_{\text{RVFLwoDL}}$ denotes the training targets. The optimal solution is [15]:

$$\hat{\boldsymbol{\beta}} = (\mathbf{H}^T\mathbf{H} + \lambda\mathbf{I})^{-1}\mathbf{H}^T\mathbf{Y}_{\text{RVFLwoDL}}, \tag{5}$$

where $\mathbf{I}$ is the identity matrix.

Deep RVFLwoDL

The deep variant of RVFLwoDL (DRVFLwoDL) includes L hidden layers ($L \geq 2$). illustrates its structure. The feature transformations at layer l are described as:

$$\mathbf{H}_l = \begin{cases} g(\mathbf{X}\mathbf{W}_1), & l = 1, \\ g(\mathbf{H}_{l-1}\mathbf{W}_l), & l \geq 2, \end{cases} \tag{6}$$

where $\mathbf{W}_1$ and $\mathbf{W}_l$ are the weight matrices connecting the input to the first hidden layer and subsequent hidden layers, respectively. The output after L layers is given by:

$$\hat{\mathbf{Y}}_{\text{DRVFLwoDL}} = \mathbf{H}_L \cdot \boldsymbol{\beta}. \tag{7}$$

Ensemble Deep RVFLwoDL

The ensemble deep RVFLwoDL (edRVFLwoDL) extends the deep RVFLwoDL (DRVFLwoDL) architecture by adding multiple output layers, whose predictions are combined to produce the final result. Figure 2 illustrates both the original DRVFLwoDL and the extended edRVFLwoDL models. For each layer l, the output is computed as:

$$\hat{\mathbf{Y}}_l^{\text{DRVFLwoDL}} = \mathbf{H}_l \cdot \boldsymbol{\beta}_l, \tag{8}$$

where $\boldsymbol{\beta}_l$ represents the output weights for the l-th layer. The final ensemble prediction is computed as:

$$\hat{\mathbf{Y}}_{\text{edRVFLwoDL}} = \frac{1}{L}\sum_{l=1}^{L}\hat{\mathbf{Y}}_l^{\text{DRVFLwoDL}}. \tag{9}$$

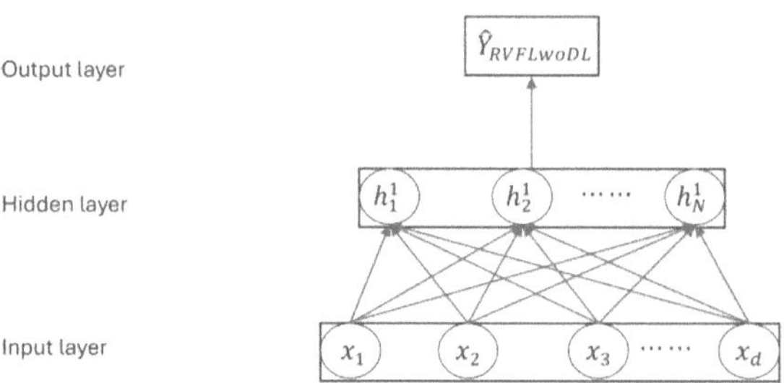

Fig. 1. The structure of RVFLwoDL.

3.2 Random Vector Functional Link (RVFL)

RVFL networks introduce direct connections between input and output layers, enhancing network performance by integrating original inputs with learned features [22]. Input-hidden weights are fixed and generated randomly, while output weights are optimized using least squares or pseudo-inverse techniques [13,21].

Shallow RVFL

In shallow RVFL, input features are enhanced and directly linked to the output layer (Fig. 3). The feature transformation is:

$$\mathbf{H} = g(\mathbf{X}, \mathbf{W}), \tag{10}$$

where $\mathbf{X} \in \mathbb{R}^{k \times d}$ is the input matrix, and $\mathbf{W}$ is the hidden layer weight matrix. By combining $\mathbf{X}$ and $\mathbf{H}$, we define $\mathbf{D} = [\mathbf{X}, \mathbf{H}]$. The output weights $\boldsymbol{\beta}$ are computed as:

$$\min_{\boldsymbol{\beta}} ||\mathbf{D}\boldsymbol{\beta} - \mathbf{Y}_{\text{RVFL}}||^2 + \lambda||\boldsymbol{\beta}||^2, \tag{11}$$

with the optimal solution:

$$\hat{\boldsymbol{\beta}} = (\mathbf{D}^T\mathbf{D} + \lambda\mathbf{I})^{-1}\mathbf{D}^T\mathbf{Y}_{\text{RVFL}}. \tag{12}$$

The final output is:

$$\hat{\mathbf{Y}}_{\text{RVFL}} = \mathbf{D} \cdot \boldsymbol{\beta}. \tag{13}$$

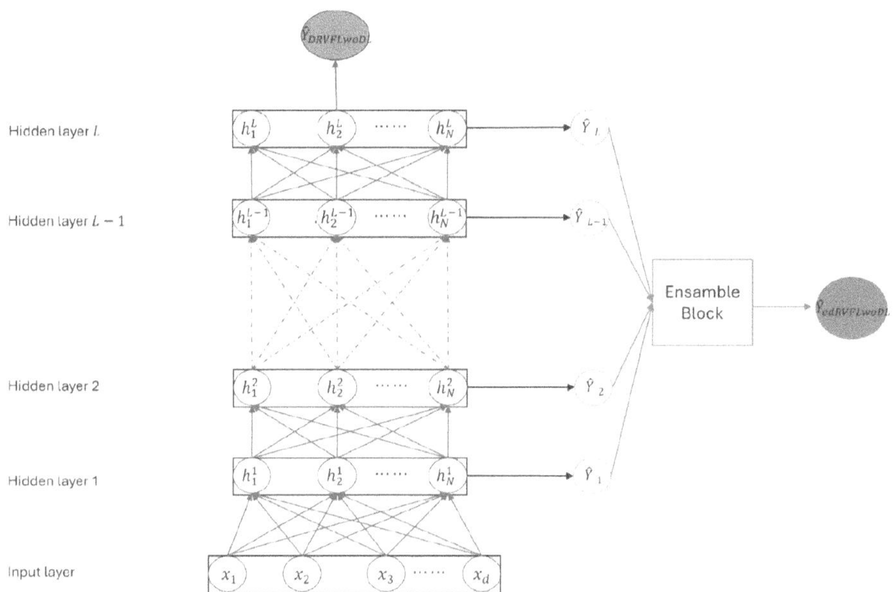

Fig. 2. Comparison of the original DRVFLwoDL (blue) and the ensemble edRVFLwoDL (red) architectures. Blue lines depict the deep RVFLwoDL structure, while red lines highlight the additional output layers introduced in the ensemble version. (Color figure online)

Deep RVFL

Deep RVFL (DRVFL) incorporates multiple hidden layers, with direct links connecting the input to each hidden layer [11]. Feature transformations are computed as:

$$\mathbf{H}_l = \begin{cases} g(\mathbf{X}\mathbf{W}_1), & l = 1, \\ g(\mathbf{D}_l\mathbf{W}_l), & l \geq 2, \end{cases} \tag{14}$$

where $\mathbf{D}_l = [\mathbf{X}, \mathbf{H}_{l-1}]$ combines input and previous layer features. The output weights $\boldsymbol{\beta}$ are determined as in Eq. (11). The network prediction is:

$$\hat{\mathbf{Y}}_{\text{DRVFL}} = \mathbf{D}_L \cdot \boldsymbol{\beta}. \tag{15}$$

Ensemble Deep RVFL

Ensemble deep RVFL (edRVFL) assigns an independent output layer to each hidden layer (Fig. 4). The l-th layer's output is [14,16,27]:

$$\hat{\mathbf{Y}}_l^{\text{DRVFL}} = \mathbf{H}_l \cdot \boldsymbol{\beta}_l, \tag{16}$$

and the overall prediction is the average of these outputs:

$$\hat{\mathbf{Y}}_{\text{edRVFL}} = \frac{1}{L} \sum_{l=1}^{L} \hat{\mathbf{Y}}_{l}^{\text{DRVFL}}. \tag{17}$$

4 Methodology

We employ a methodology that integrates stock market data, sentiment analysis from tourism-related news, and protest sentiment to predict trends in tourism-related stocks. Our approach combines technical indicators and sentiment scores to create a robust input dataset for machine learning models. The methodology is structured into three phases: data collection and preprocessing, feature engineering, and model development and evaluation.

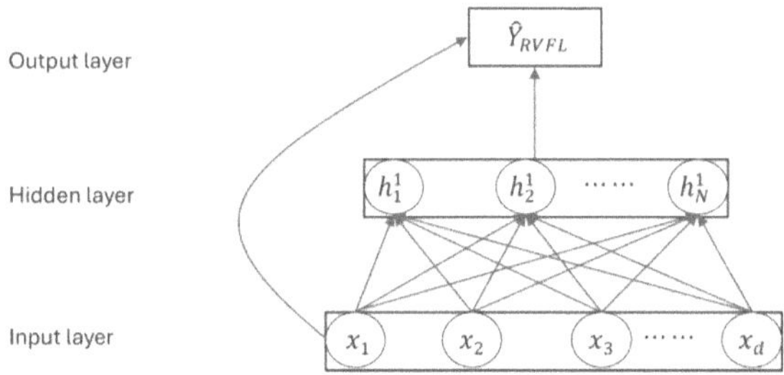

Fig. 3. The structure of RVFL.

4.1 Stock Market Data

Stock market data were retrieved for five prominent hotel chains (Hilton, Hyatt, Marriott) and three major airlines (JetBlue, United Airlines, Delta) from Yahoo Finance. The dataset spans from January 1, 2020, to September 9, 2024, and includes daily records of the Open, High, Low, Close, and Volume prices for each stock. We defined a multi-class classification target variable to represent the stock's daily trend: Buy or Rise (+1) if the variation in the Close price exceeded 1%, Hold (0) if the absolute variation was less than 1%, and Sell or Drop (−1) if the variation was less than −1%.

The preprocessing phase also involved calculating technical indicators using the `ta` library in Python. These indices capture the underlying trends and momentum of the stocks. They are mathematical calculations based on historical price and volume data, which help to filter out market "noise" and provide insights into potential future price movements. A Moving Average smooths out a stock's price data by creating a constantly updated average price over a specific period (e.g., 50 d). This helps identify the general direction of the trend. If the current stock price is above its moving average, it suggests an upward

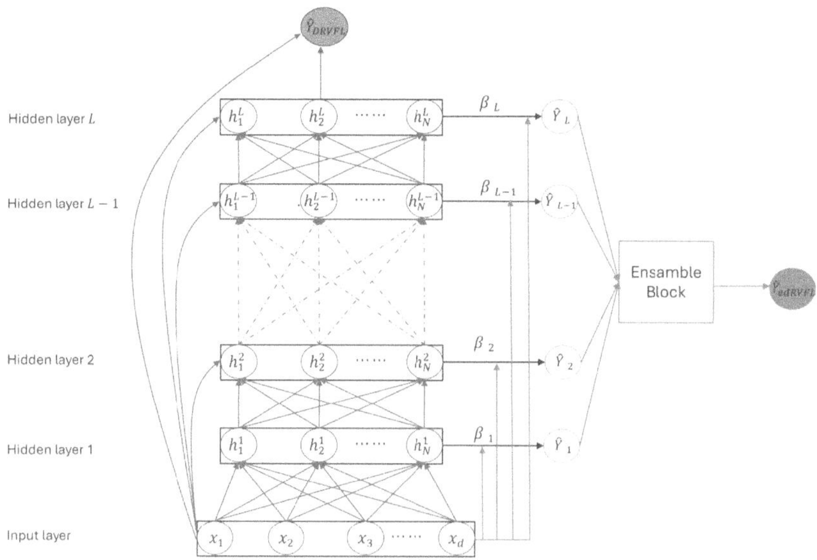

Fig. 4. Comparison of the Deep RVFL (DRVFL, blue) and the ensemble Deep RVFL (edRVFL, red). Blue lines indicate the direct connections between input and output layers present in both architectures. (Color figure online)

trend (bullishness). Conversely, if the price is below, it indicates a downward trend (bearishness). The RSI is a momentum oscillator that measures the speed and change of price movements. It operates on a scale from 0 to 100. The core idea is to identify when a stock has been bought too aggressively (becoming "overbought") or sold off too aggressively (becoming "oversold"). Bollinger Bands consist of three lines: a simple moving average (the middle band), and an upper and lower band that are standard deviations away from the average. Bollinger Bands create a dynamic "envelope" around the price. The bands expand during periods of high volatility (large price swings) and contract during periods of low volatility. When the price moves near the upper band, the stock is considered relatively high, and when it nears the lower band, it is considered relatively low. This helps assess volatility and identify potential reversal points as the price tends to stay within the bands.

By combining these indicators, we can move beyond simple price observation to analyze the strength of a trend (Moving Averages), the intensity of recent price changes (RSI), and the prevailing market volatility (Bollinger Bands).

Sentiment features from both news and protests were synchronized with the stock market data to ensure alignment of dates, creating a dataset that incorporated both quantitative and qualitative aspects.

4.2 Qualitative Data

Tourism-related news data were collected from CNN News using the Selenium package. The query term `tourism` was used to extract 6,464 articles published between January 1, 2020, and September 9, 2024. Each article was preprocessed by removing stop words, converting text to lowercase, and eliminating punctuation and special characters. We show examples of news and their sentiment score in Table 1.

Table 1. Examples of news and their respective sentiment index

News extract	Sentiment score
Protests in the streets. Graffiti warning tourists to go home. Local populations dwindling as short-term rentals mushroom and price residents out. It feels like this was the year that tourism turned nasty âĂŞ and local communities started pushing back. Venice has started charging daytrippers an entry fee, while one busy Swiss town has announced it wants to follow suit. Locals have staged protests in Mallorca and Barcelona. And while it has come to a head in Europe, this is a global phenomenon. A Japanese town overlooking Mount Fuji erected view-blocking barriers in May (then removed them in August). Bali introduced a tourist entry tax for foreign visitors in February. And US national parks are full to bursting – with 13 million more visits in 2023 than in 2022, according to NPS numbers. In peak season, visitors must book ahead to enter. Increased enthusiasm doesn't seem to correlate with increased respect for the landscape, however.	−0.614
As peak vacation season sails into view and the world shakes off the last shackles of the pandemic, it feels like the appetite for hitting the road has never been greater. International tourism reached 80% of pre-pandemic levels in the first quarter of 2023, according to the United Nations World Tourism Organization, with an estimated 235 million tourists traveling internationally in January, February and March. And experts are cautiously optimistic about a continued travel rebound. Demand is high, with many popular destinations booking out earlier in the year. Thankfully, there's so much out there still to see and do.	0.718

Protest data for the United States were obtained from the Armed Conflict Location and Event Data (ACLED) dataset, encompassing 74,042 recorded events from January 1, 2020, to September 9, 2024. We show examples of protests reports and their sentiment score in Table 2.

Table 2. Examples of protest events' descriptions and their respective sentiment index

Protest Description	Sentiment Score
On 27 November 2024, members of Teamsters gathered at Dulles International Airport (Virginia) to call for higher pay and better healthcare for United Airlines technicians.	0.0467
On 26 October 2024, an individual shot and wounded a Jewish man walking to a synagogue in West Rogers Park, Chicago (Illinois). The perpetrator targeted the victim based on his faith. The victim suffered non-life-threatening injuries. The perpetrator also shot at police and was wounded, subsequently being charged with a hate crime and terrorism.	−0.988

4.3 Sentiment Analysis

We construct a sentiment index for both news data and protests descriptions. For this, we applied sentiment analysis, using FinBERT, a Large Language Model created by further training the BERT language model in the finance domain. This model is adequate in the context of this study, due to its specialization in financial related texts. By using a model which was pretrained on financial data, the sentiment scores should be more accurate. FiBERT represents a state-of-the-art method, specially apt to capture contextual information from news articles, surpassing traditional dictionary approaches and other BERT related models. Both a categorical variable (positive, neutral, negative) and a sentiment score ranging from −1 to 1 are constructed trough the analysis of both texts. For this analysis, we use the sentiment score as an independant variable in the model.

4.4 Classification Task

The classification task can be mathematically expressed as follows:

$$\begin{aligned} y_{T+1} = F\Big(&x_T^{\text{TA}},\ x_T^{\text{SA-news}},\ x_T^{\text{SA-protests}},\\ &x_{T-1}^{\text{TA}},\ x_{T-1}^{\text{SA-news}},\ x_{T-1}^{\text{SA-protests}},\\ &\ldots,\ x_{T-K}^{\text{TA}},\ x_{T-K}^{\text{SA-news}},\ x_{T-K}^{\text{SA-protests}}\Big). \end{aligned} \tag{18}$$

where F represents the mapping function from inputs $x_T, x_{T-1}, \ldots, x_{T-K}$ (past K observations of technical indicators (TA), sentiment scores from news (SA-news), and sentiment scores from protests (SA-protests)) to the output y_{T+1} (the stock trend on day $T+1$).

4.5 Experiment Parameter and Evaluation Measure

To evaluate the proposed methodology, we used Random Forest (RF), XGBoost, Naïve Bayes (NB), ELM, RVFL, DRVFL and edRVF. Input features were scaled, and the models were trained and evaluated using accuracy and F1-score metrics.

We employed grid search to identify the optimal hyperparameter configurations for the benchmark models, aiming to enhance prediction performance. The hyperparameters and their respective search spaces for the models are detailed in Table 3.

Table 3. Benchmark models' hyper-parameter search space

Model	Hyper-parameter	Search space
Random Forest	n_estimators	[100, 200, 500]
	max_depth	[None, 10, 20]
XGBoost	n_estimators	[100, 300, 500]
	learning_rate	[0.01, 0.1, 0.2]
Naïve Bayes	var_smoothing	[e^{-3}, e^{-6}, e^{-9}]
ELM	n_neurons	[50, 100, 200]
	activation	[sigmoid, relu]
RVFL variants	Hidden nodes	[10, 30, 50, 100]
	Hidden layers	[1, 2, 3, 4, 5]
	Regularization	[2^{-5}, 2^{6}]

We evaluated models using two complementary protocols. First, we ran stratified 5-fold cross-validation: the data were split into five folds, training on four and validating on the remaining one, repeating this process five times and reporting mean ± SD. Second, to emulate a realistic out-of-sample check, we used a 80/20 holdout (first 80% of each series for training, last 20% for testing). We also implemented a rolling window approach, experimenting with different window sizes n, using the previous n days of data to predict the $(n+1)$th day. After exploring several values, we found that $n = 3$ yielded the best results. The dependent variable Y was defined as the trend label derived from those n-day windows.

4.6 Evaluation Metrics

The metrics we considered in this work to evaluate the models' performance [29] are described below.

- **Accuracy** evaluates classification performance as the ratio of correct predictions to total instances. A perfect model achieves an accuracy of 1.0, indicating no errors. The formula is given in Eq. 19.

$$\text{Accuracy} = \frac{TP + TN}{TP + TN + FP + FN} \tag{19}$$

- **F1-Score** evaluates classification performance by combining precision and recall into a single metric, making it useful for imbalanced datasets. The formula is shown in Eq. 20.

$$\text{F1-Score} = \frac{2 \cdot TP}{2 \cdot TP + FP + FN} \quad (20)$$

Where, True Positives (TP) refer to the cases that were correctly identified as positive, while True Negatives (TN) represent the cases correctly identified as negative. In contrast, False positives (FP) are the cases mistakenly predicted as positive, and False Negatives (FN) are cases that were incorrectly labeled as negative.

5 Results

We present accuracy results for both hotels and airlines stocks using different machine learning models and feature combinations, in Tables 4 and 5, respectively.

For hotel stocks, the Random Vector Functional Link (RVLF) models demonstrated a strong performance. Hilton (HLT) showed the highest accuracy with 76.75% (SD = 0.11) using only technical indicators (TA). Hyatt (H) presented the most significant accuracy improvement when combining TA with news sentiment indexes in the RVLF model (67.79%, SD = 1.37), with a 2.8% point increase over TA alone. Marriott (MAR) exhibited the best performance with the full set of features (TA + SA (news) + SA (protests)) in the edRVFL model (73.11%, SD = 4.82).

Airline stocks's results were more varied. JetBlue (JBLU) showed relatively modest accuracy improvements with sentiment features, with the best performance coming from edRVFL using protest sentiment (54.34%, SD = 0.98). United Airlines (UAL) achieved its highest accuracy with RVLF using news sentiment (59.38%, SD = 0.88), while Delta (DAL) performed best with RVLF incorporating all features (61.06%, SD = 1.48). Notably, the inclusion of sentiment features improved accuracy in 55% of cases across airline stocks.

The F1 score results, which are shown in Tables 6 and 7, offer more details on how well the model performed when using the F1 score.

The RVLF models performed adequately for the hotel stocks, with HLT utilizing all features to achieve its highest F1 score (68.86%, SD = 2.06). With sentiment analysis, Hyatt (H) saw especially noticeable gains, and the edRVFL model with news sentiment had an F1 score of 60.92% (SD = 0.42) as opposed to 60.24% (SD = 3.18) with TA alone. DRVFL achieved the highest F1 score (65.04%, SD = 1.23) using all features, while Marriott (MAR) demonstrated consistent performance across models.

When sentiment analysis was applied to airline stocks, Delta (DAL) saw the biggest gains, while RVLF with all features had a 57.32% (SD = 1.61) F1 score as opposed to 54.61% (SD = 0.71) with TA alone. With an RVLF utilizing news sentiment of 52.93% (SD = 0.64), United Airlines (UAL) showed modest

Table 4. Accuracy (%) for HLT, Hyatt (H), Marriott (MAR)

Stocks	TA	TA + SA (news)	TA + SA (protests)	All
HLT				
RF	74.79 (0.63)	76.19 (0.31)	74.23 (1.56)	75.63 (0.28)
XGBoost	73.11 (0.82)	74.51 (0.50)	74.51 (0.42)	74.79 (0.28)
NB	60.50 (0.00)	60.22 (0.00)	60.22 (0.00)	60.22 (0.00)
ELM	76.19 (2.00)	75.35 (4.26)	75.63 (4.74)	73.11 (3.82)
RVLF	**76.75 (0.11)**	75.35 (4.34)	75.91 (0.11)	76.19 (4.28)
DRVFL	73.39 (12.50)	75.63 (3.41)	75.91 (0.80)	74.51 (1.96)
edRVFL	72.83 (1.41)	76.19 (5.40)	73.95 (7.70)	73.39 (4.14)
H				
RF	51.82 (3.61)	55.18 (2.06)	43.98 (2.83)	51.54 (4.09)
XGBoost	59.10 (0.98)	58.82 (0.71)	58.82 (0.97)	55.46 (1.78)
NB	55.74 (0.00)	55.18 (0.00)	55.46 (0.00)	56.02 (0.00)
ELM	63.31 (6.25)	65.27 (6.23)	63.87 (1.77)	64.71 (3.05)
RVLF	64.99 (3.24)	**67.79 (1.37)**	66.11 (8.62)	63.87 (2.17)
DRVFL	65.83 (0.96)	64.43 (1.62)	62.75 (15.54)	62.18 (4.83)
edRVFL	64.15 (6.21)	66.39 (0.60)	66.67 (2.13)	66.67 (1.76)
MAR				
RF	62.18 (3.06)	66.39 (1.87)	70.03 (3.25)	62.18 (2.53)
XGBoost	67.79 (2.00)	66.67 (0.80)	68.35 (0.42)	68.35 (1.40)
NB	57.98 (0.00)	57.42 (0.00)	58.26 (0.00)	57.70 (0.00)
ELM	67.51 (4.25)	71.99 (2.50)	64.15 (5.02)	66.11 (3.75)
RVLF	72.55 (4.24)	67.79 (4.42)	68.63 (4.75)	**73.67** (1.51)
DRVFL	70.03 (4.89)	73.11 (3.24)	72.27 (4.82)	71.99 (3.40)
edRVFL	70.87 (1.16)	68.91 (5.75)	68.91 (5.58)	73.11 (4.82)

improvements. With DRVFL using protest sentiment obtaining the highest F1 score (51.24%, SD = 1.68), JetBlue (JBLU) displayed more modest gains.

Across all evaluation metrics, a number of recurring patterns surfaced. In roughly 50–55% of cases, adding sentiment analysis features improves model performance; the benefits are more noticeable for some stocks, especially HLT and DAL. Tables 4, 6 and 7 show that RVLF-based models outperformed traditional machine learning techniques in 73.6% of accuracy cases and 83.33% of F1 score cases. As shown in Table 8, the RVLF variants consistently outperformed the others in terms of the trade-off between computational efficiency and model complexity, achieving competitive accuracy while retaining quick training times.

While news sentiment was more consistently positive for airline stocks, protest sentiment features demonstrated a unique predictive value for hotel stocks, revealing an intriguing sector-specific pattern. To provide a concrete

Table 5. Accuracy (%) for JetBlue (JBLU), United Airlines (UAL), Delta (DAL)

Stocks	TA	TA + SA (news)	TA + SA (protests)	All
JBLU				
RF	49.02 (0.92)	50.42 (0.76)	49.58 (0.61)	50.98 (1.08)
XGBoost	48.18 (0.65)	50.70 (0.57)	49.02 (1.25)	48.74 (0.68)
NB	50.42 (0.00)	49.02 (0.00)	50.70 (0.00)	49.30 (0.00)
ELM	51.54 (2.09)	49.86 (2.09)	51.82 (1.48)	48.74 (2.27)
RVLF	50.98 (0.96)	50.98 (0.72)	53.50 (0.74)	52.38 (1.10)
DRVFL	52.10 (1.08)	52.38 (1.74)	54.06 (1.67)	52.10 (0.65)
edRVFL	53.22 (0.56)	53.22 (0.89)	**54.34 (0.98)**	52.38 (1.45)
UAL				
RF	50.70 (1.00)	51.82 (0.77)	51.54 (0.59)	50.42 (0.85)
XGBoost	48.46 (1.06)	47.90 (0.59)	52.38 (1.54)	49.58 (1.21)
NB	51.26 (0.00)	50.98 (0.00)	50.42 (0.00)	50.42 (0.00)
ELM	55.18 (1.83)	57.70 (1.63)	54.62 (1.17)	55.46 (1.08)
RVLF	57.14 (1.95)	59.38 (0.88)	**59.10 (0.82)**	57.14 (1.01)
DRVFL	57.42 (1.68)	55.46 (2.86)	56.86 (2.66)	57.42 (0.65)
edRVFL	56.58 (2.31)	56.30 (2.09)	58.26 (0.50)	56.86 (0.94)
DAL				
RF	55.74 (0.61)	58.54 (0.47)	58.26 (0.83)	59.38 (0.61)
XGBoost	58.54 (0.62)	57.14 (0.65)	57.42 (0.38)	57.70 (0.47)
NB	54.62 (0.00)	54.06 (0.00)	54.62 (0.00)	54.06 (0.00)
ELM	58.54 (1.44)	57.14 (1.13)	58.26 (1.08)	59.66 (0.76)
RVLF	59.66 (0.29)	**61.06 (1.06)**	60.78 (0.95)	**61.06 (1.48)**
DRVFL	60.78 (0.42)	59.94 (1.75)	59.94 (0.71)	59.94 (0.86)
edRVFL	59.66 (0.74)	59.10 (0.89)	60.78 (0.49)	**61.06 (0.61)**

sector-level comparison, the data reveal a clear performance disparity between hotel and airline stocks. Hotel stocks consistently achieved higher predictive accuracy, with the best-performing models for Hilton (HLT: 76.75%), Hyatt (H: 67.79%), and Marriott (MAR: 73.67%) significantly outperforming their airline counterparts—JetBlue (JBLU: 54.34%), United (UAL: 59.38%), and Delta (DAL: 61.06%). This translates to an average top accuracy of 72.74% for hotels versus 58.26% for airlines, a difference of over 14% points. Furthermore, the inclusion of sentiment features led to more substantial and consistent improvements for hotels; for instance, Hyatt's accuracy increased by 2.8 points with news sentiment, whereas the improvements for airlines were generally more modest, below 2 points. This suggests that the predictive models, particularly those enhanced with sentiment analysis, are markedly more effective for the hotel sector under the conditions of this study.

Table 6. F1 Score (%) for HLT, Hyatt (H), Marriott (MAR)

Stocks	TA	TA + SA (news)	TA + SA (protests)	All
HLT				
RF	67.36 (0.38)	68.48 (0.23)	67.31 (0.82)	67.34 (0.40)
XGBoost	65.90 (0.52)	66.75 (0.31)	66.31 (0.48)	66.75 (0.35)
NB	61.89 (0.00)	61.70 (0.00)	61.44 (0.00)	61.69 (0.00)
ELM	66.77 (0.89)	66.30 (1.79)	67.89 (1.98)	66.36 (2.08)
RVLF	67.70 (0.26)	67.67 (2.11)	67.12 (0.26)	**68.86 (2.06)**
DRVFL	67.92 (8.80)	67.68 (1.73)	67.97 (0.71)	68.34 (0.64)
edRVFL	67.18 (0.15)	68.56 (1.91)	68.65 (3.57)	68.09 (1.56)
H				
RF	50.27 (2.60)	52.07 (1.27)	44.92 (2.31)	50.77 (2.46)
XGBoost	55.34 (0.58)	54.66 (0.50)	53.46 (0.78)	53.16 (1.29)
NB	55.36 (0.00)	54.85 (0.00)	54.94 (0.00)	55.58 (0.00)
ELM	57.11 (2.91)	**60.97 (4.14)**	56.56 (1.48)	57.92 (1.84)
RVLF	59.79 (1.35)	60.12 (0.86)	59.72 (5.98)	60.05 (1.10)
DRVFL	60.52 (0.98)	59.96 (0.94)	59.49 (10.90)	59.63 (0.96)
edRVFL	60.24 (3.18)	60.50 (0.74)	59.67 (1.30)	60.92 (0.42)
MAR				
RF	59.37 (1.44)	59.43 (0.90)	61.03 (1.25)	57.73 (0.96)
XGBoost	60.13 (0.93)	60.63 (0.49)	59.46 (0.20)	60.88 (0.86)
NB	58.63 (0.00)	58.04 (0.00)	58.85 (0.00)	58.26 (0.00)
ELM	60.95 (1.92)	62.26 (0.78)	59.63 (3.04)	60.68 (1.30)
RVLF	63.96 (1.64)	64.03 (1.84)	63.16 (2.22)	64.00 (0.49)
DRVFL	62.84 (2.59)	63.91 (1.38)	63.16 (1.30)	**65.04 (1.23)**
edRVFL	63.86 (0.53)	63.10 (2.58)	63.15 (2.36)	63.25 (1.57)

All of these findings point to a particularly promising method for predicting stock price movement in the travel sector: the combination of technical indicators and sentiment analysis, processed through RVLF architectures. RVLF models are well-positioned for real-world application in financial prediction systems due to their consistent performance advantages across accuracy and F1 metrics as well as their computational efficiency.

The ELM and its variants (DRVFL, edRVFL) also showed competitive training times (0.01–0.02 s) while maintaining strong predictive performance. This efficiency advantage becomes particularly important when considering large-scale applications or frequent model retraining scenarios.

Finally, RVFL achieved its optimal accuracy with 10 nodes, 2 hidden layers and regularization of 1.25×10^{-1}; DRVFL with 10 nodes, 2 hidden layer and regularization of 3.125×10^{-2}; and edRVFL with 10 nodes, 2 hidden layer and regularization of 4×10^{-2}.

Table 7. F1 Score (%) for JetBlue (JBLU), United Airlines (UAL), Delta (DAL)

Stocks	TA	TA + SA (news)	TA + SA (protests)	All
JBLU				
RF	44.74 (1.14)	45.08 (0.99)	44.10 (0.65)	48.91 (1.34)
XGBoost	44.26 (0.70)	46.70 (0.90)	44.93 (1.64)	43.77 (0.97)
NB	44.73 (0.00)	43.97 (0.00)	45.00 (0.00)	43.45 (0.00)
ELM	48.59 (1.89)	47.21 (3.37)	45.27 (1.99)	46.07 (2.03)
RVLF	49.27 (1.58)	48.44 (1.02)	48.30 (1.15)	49.05 (1.02)
DRVFL	48.85 (2.47)	49.43 (1.91)	**51.24 (1.68)**	48.60 (1.10)
edRVFL	48.17 (0.91)	47.33 (1.43)	47.84 (1.71)	48.51 (1.26)
UAL				
RF	49.89 (0.82)	50.31 (0.72)	49.27 (0.71)	48.93 (0.79)
XGBoost	47.81 (0.93)	47.64 (0.72)	50.18 (1.18)	46.64 (1.22)
NB	49.76 (0.00)	49.40 (0.00)	47.92 (0.00)	47.92 (0.00)
ELM	51.87 (1.30)	51.98 (1.56)	49.57 (1.05)	50.34 (1.20)
RVLF	51.91 (1.49)	52.93 (0.64)	52.70 (0.89)	53.73 (0.99)
DRVFL	51.71 (1.08)	52.00 (2.33)	**54.17 (1.28)**	52.08 (1.09)
edRVFL	51.96 (1.72)	51.49 (1.30)	52.51 (0.75)	52.57 (0.53)
DAL				
RF	51.21 (0.51)	52.82 (0.54)	52.08 (0.69)	53.92 (0.50)
XGBoost	52.85 (0.89)	51.60 (0.67)	49.88 (0.45)	50.11 (0.64)
NB	52.85 (0.00)	51.93 (0.00)	52.84 (0.00)	51.89 (0.00)
ELM	52.58 (1.50)	53.08 (1.48)	53.98 (1.32)	53.96 (0.52)
RVLF	54.61 (0.71)	56.00 (1.02)	54.49 (0.85)	57.32 (1.61)
DRVFL	54.45 (1.41)	55.50 (1.99)	54.63 (1.39)	**57.42 (1.01)**
edRVFL	54.54 (0.71)	55.05 (0.86)	54.42 (0.46)	55.66 (0.79)

Table 8. Best Training Times (seconds) by Model

Model	Best Training Time
Random Forest (RF)	0.2053
XGBoost	1.1189
Naive Bayes (NB)	0.0038
Extreme Learning Machine (ELM)	0.01
Random Vector Functional Link (RVLF)	0.02
Deep RVFL (DRVFL)	0.01
Ensemble Deep RVFL (edRVFL)	0.02

6 Conclusions

In this work, we have demonstrated the efficacy of integrating sentiment analysis from both news and protest events with traditional technical indicators for stock-trend classification in the tourism sector. Our experiments showed that RVFL-based models, particularly the ensemble deep variant (edRVFL), consistently achieved the highest accuracy and F1 scores, while maintaining very low training times compared to other learners such as RF, XGBoost, NB and ELM.

However, we must acknowledge several important limitations. First, our dataset spans January 2020 to September 2024, a period marked by extreme market volatility driven by the COVID-19 pandemic and major geopolitical events (e.g., large-scale protests, supply-chain disruptions). Such unique conditions may introduce biases: improvements that we attribute to sentiment integration could partly reflect overfitting to extraordinary market regimes rather than robust signal extraction. Second, our sentiment sources (CNN news, ACLED protests) are U.S.-centric and may not fully capture regional or sector-specific nuances relevant to global tourism stocks.

To address these concerns, we propose that future work expand both the temporal and geographical scope of the data. In particular, we should include more stable pre-pandemic periods and post-pandemic normalization phases to help disentangle structural market patterns from crisis-driven anomalies.

Acknowledgements. This work was supported by Fondo Nacional de Desarrollo Científico y Tecnológico FONDECYT Iniciación 2023 No. 11230396 and Agencia Nacional de Investigación y Desarrollo (ANID) Doctoral scholarship 2024-21241072.

References

1. Ashtiani, M.N., Raahemi, B.: News-based intelligent prediction of financial markets using text mining and machine learning: A systematic literature review. Expert Systems with Applications **217**, 119509 (2023).
2. Bhambu, A., Gao, R., Suganthan, P.N.: Recurrent ensemble random vector functional link neural network for financial time series forecasting. Applied Soft Computing **161**, 111759 (2024).
3. Bisoi, R., Dash, P., Mishra, S.: Modes decomposition method in fusion with robust random vector functional link network for crude oil price forecasting. Applied Soft Computing **80**, 475–493 (2019).
4. Corbet, S., Hou, Y., Hu, Y., Oxley, L.: Did covid-19 tourism sector supports alleviate investor fear? Annals of Tourism Research **95**, 103434 (2022).
5. Costola, M., Hinz, O., Nofer, M., Pelizzon, L.: Machine learning sentiment analysis, covid-19 news and stock market reactions. Research in International Business and Finance **64**, 101881 (2023).
6. Dang, T.L., Dang, M., Hoang, L., Nguyen, L., Phan, H.L.: Media coverage and stock price synchronicity. International Review of Financial Analysis **67**, 101430 (2020).
7. Demiralay, S., Kilincarslan, E.: The impact of geopolitical risks on travel and leisure stocks. Tourism Management **75**, 460–476 (2019).

8. Dong, H., Ren, J., Padmanabhan, B., Nickerson, J.V.: How are social and mass media different in relation to the stock market? a study on topic coverage and predictive value. Information & Management **59**(2), 103588 (2022).
9. Henríquez, P.A., Araya, N.: Multimodal alzheimer's disease classification through ensemble deep random vector functional link neural network. PeerJ Computer Science **10**, e2590 (2024)
10. Henríquez, P.A.: Sentiment classification on twitter(x) through ensemble deep random vector functional links. In: 2024 International Conference on Machine Learning and Applications (ICMLA). pp. 1628–1632 (2024).
11. Henríquez, P.A., Alessandri, F.: Analyzing digital societal interactions and sentiment classification in twitter (x) during critical events in chile. Heliyon **10**(12), e32572 (2024).
12. Henríquez, P.A., Ruz, G.A.: Extreme learning machine with a deterministic assignment of hidden weights in two parallel layers. Neurocomputing **226**, 109–116 (2017).
13. Henríquez, P.A., Ruz, G.A.: A non-iterative method for pruning hidden neurons in neural networks with random weights. Applied Soft Computing **70**, 1109–1121 (2018).
14. Henríquez, P.A., Ruz, G.A.: Twitter sentiment classification based on deep random vector functional link. In: 2018 International Joint Conference on Neural Networks (IJCNN). pp. 1–6 (2018).
15. Henríquez, P.A., Ruz, G.A.: Noise reduction for near-infrared spectroscopy data using extreme learning machines. Engineering Applications of Artificial Intelligence **79**, 13–22 (2019).
16. Hu, M., Chion, J.H., Suganthan, P.N., Katuwal, R.K.: Ensemble deep random vector functional link neural network for regression. IEEE Transactions on Systems, Man, and Cybernetics: Systems **53**(5), 2604–2615 (2022)
17. Huang, G., Huang, G.B., Song, S., You, K.: Trends in extreme learning machines: A review. Neural Networks **61**, 32–48 (2015)
18. Ji, G., Yu, J., Hu, K., Xie, J., Ji, X.: An adaptive feature selection schema using improved technical indicators for predicting stock price movements. Expert Systems with Applications **200**, 116941 (2022).
19. Kapuściński, G., Richards, B.: News framing effects on destination risk perception. Tourism Management **57**, 234–244 (2016).
20. Li, Q., Wang, T., Li, P., Liu, L., Gong, Q., Chen, Y.: The effect of news and public mood on stock movements. Information Sciences **278**, 826–840 (2014).
21. Malik, A.K., Gao, R., Ganaie, M., Tanveer, M., Suganthan, P.N.: Random vector functional link network: recent developments, applications, and future directions. Applied Soft Computing **143**, 110377 (2023)
22. Pao, Y.H., Takefuji, Y.: Functional-link net computing: theory, system architecture, and functionalities. Computer **25**(5), 76–79 (1992)
23. Peng, K.L., Wu, C.H., Lin, P.M., Kou, I.E.: Investor sentiment in the tourism stock market. Journal of Behavioral and Experimental Finance **37**, 100732 (2023)
24. Peng, Y., Albuquerque, P.H.M., Kimura, H., Saavedra, C.A.P.B.: Feature selection and deep neural networks for stock price direction forecasting using technical analysis indicators. Machine Learning with Applications **5**, 100060 (2021).
25. Raheem, I.D., le Roux, S.: Geopolitical risks and tourism stocks: New evidence from causality-in-quantile approach. The Quarterly Review of Economics and Finance **88**, 1–7 (2023).

26. Saud, A.S., Shakya, S.: Technical indicator empowered intelligent strategies to predict stock trading signals. Journal of Open Innovation: Technology, Market, and Complexity **10**(4), 100398 (2024).
27. Shi, Q., Katuwal, R., Suganthan, P.N., Tanveer, M.: Random vector functional link neural network based ensemble deep learning. Pattern Recognition **117**, 107978 (2021)
28. Tang, L., Wu, Y., Yu, L.: A non-iterative decomposition-ensemble learning paradigm using rvfl network for crude oil price forecasting. Applied Soft Computing **70**, 1097–1108 (2018).
29. Vujovic, Ž.D.: Classification model evaluation metrics. International Journal of Advanced Computer Science and Applications **12**(6) (2021).
30. Wang, Y., Zhang, H., Gao, W., Yang, C.: Spillover effects from news to travel and leisure stocks during the covid-19 pandemic: Evidence from the time and frequency domains. Tourism Economics **29**(2), 460–487 (2023).
31. Zhang, L., Suganthan, P.N.: A survey of randomized algorithms for training neural networks. Information Sciences **364**, 146–155 (2016)

From Semi-structured PDFs to Semantic Search: A Hybrid NLP–LLM Framework for Multilingual Resumes

Arturo Zabdí Reyes Pérez[1], Rafael Rivera López[1](✉), and Hugo Jair Escalante-Balderas[2]

[1] Departamento de Sistemas y Computación, TecNM, Instituto Tecnológico de Veracruz, Veracruz, Ver., Mexico
{l20021280,rafael.rl}@veracruz.tecnm.mx

[2] Coordinación de Ciencias Computacionales, Instituto Nacional de Astrofísica, Óptica y Electrónica, Tonanzintla, Puebla, Mexico
hugojair@inaoep.mx

Abstract. This article describes a framework for processing and semantic retrieval of semi-structured and multilingual curricula vitae. The proposal combines predefined rule-based heuristics and the use of large language models to process variable and noisy documents and create structured representations, enabling their analysis and manipulation. Three extraction strategies are evaluated: heuristics, language models, and a hybrid approach using both alternatives. Results show that language models achieve the highest values for several performance metrics, while the hybrid approach reaches comparable performance with considerably lower processing cost. Purely heuristic processing, although less accurate, remains useful when efficiency and full data control are required. A semantic retrieval component based on label filtering and cross-language similarity further improves precision in both monolingual and multilingual searches. These findings demonstrate that combining heuristics with language models provides a practical and scalable solution for graduate tracking, academic networking, and institutional analysis.

Keywords: Natural language processing · Large language models · Semantic search

1 Introduction

The existence of up-to-date graduate databases is essential for higher education institutions [4]. Managing a well-structured graduate records system allows strengthening institutional ties [15], supporting accreditation processes, and facilitating curriculum planning and continuous improvement [16]. It also helps identify gaps between academic training and labor market demands [1], detect emerging sectors where graduates actively participate, and guide evidence-based decision-making [20]. Finally, it strengthens networks between graduates and

D. Chaves et al. (Eds.): CIARP 2025, LNCS 16528, pp. 191–206, 2026.
https://doi.org/10.1007/978-3-032-23161-1_14

institutions, facilitates professional mentoring, promotes the job placement of current students, and improves the relevance of academic programs through direct feedback from the professional environment [17].

Furthermore, accreditation bodies such as COPAES[1], CIEES[2], and international agencies such as ABET[3] require evidence of graduate tracking. However, in many institutions, this process remains a challenge, often relying on manual or semi-automated data collection methods with inconsistent updates and limited integration.

Meanwhile, professionals increasingly share up-to-date career information on job portals such as OCCMundial or Indeed, and on professional networks like LinkedIn or Glassdoor. Although these platforms contain valuable data, it is typically stored in unstructured or semi-structured formats, with PDF documents being one of the most common representations of curricula vitae (CVs).

This context underscores the need for tools that can leverage publicly available information to maintain current graduates databases. To address this challenge, we propose an automated system that converts PDF CVs into a structured JSON database using a hybrid approach that combines natural language processing (NLP) techniques with large language models (LLMs). The system uses NLP for language detection, section segmentation, named entity recognition, and temporal expression normalization. Also, apply LLM for context-sensitive entity extraction and recognition. Finally, a multilingual semantic search is implemented to retrieve relevant profiles in different languages.

2 Related Work

Text segmentation in NLP has evolved from early heuristic, rule-based approaches to machine learning and deep learning models, and more recently, to LLM-driven methods. Each paradigm offers distinct trade-offs in terms of interpretability, generalizability, and computational requirements [7].

Early heuristic approaches relied on hand-crafted linguistic rules and simple statistical indicators such as punctuation markers, lexical chains, and cohesion metrics. For example, discourse structure modeling is used in topic segmentation in news broadcasts [18], and the use of semantic vectors from Roget's thesaurus to identify topic shifts is used in multilingual applications [14]. While these methods are interpretable and adaptable to specific domains, they often require extensive feature engineering and lack robustness when applied to new or heterogeneous data.

The advent of deep learning significantly advanced segmentation capabilities. Convolutional Neural Networks (CNNs), Recurrent Neural Networks, and

[1] COPAES (Council for the Accreditation of Higher Education): Mexican organization authorized by the government to accredit undergraduate programs.

[2] CIEES (Interinstitutional Committees for the Evaluation of Higher Education): Mexican peer-review bodies ensuring higher education quality.

[3] ABET (Accreditation Board for Engineering and Technology): Global accreditor of STEM programs for industry-aligned standards.

Transformer-based architectures learn contextual representations directly from data, reducing dependence on manual feature design. For example, sentence-wise segmentation with Bidirectional Encoder Representations from Transformers (BERT) has improved evaluation metrics such as Pk and WindowDiff [12], while bidirectional Long Short-Term Memory networks (BiLSTMs) with attention mechanisms and CNN-based sentence embeddings have achieved strong performance in diverse segmentation tasks [2]. On the other hand, transformer-based segmentation models have further reduced error rates, with smaller model sizes, making them practical for real-world implementations. [9] The availability of large Wikipedia-based datasets has also made segmentation a supervised learning task [6]. These advances have improved tasks such as information retrieval, where identifying subtopics within documents improves search results [5]. However, despite their improvements, these deep learning models are expensive as they require large amounts of labeled data.

More recently, LLM-based methods have enhanced these architectures with large-scale pretraining and in-context learning. Untrained semantic segmentation leverages LLMs to generate detailed descriptors improving class representation [19]. Meanwhile, universal sentence segmentation models achieve good results across diverse domains by relying less on punctuation [3]. In topic segmentation, two-phase LLM pipelines solve the problem by creating topic sentences and measuring transcript similarity without labeled data [10]. Furthwrmore, in large-scale text mining, they have been used for automating metadata generation and taxonomy construction without extensive supervision [21].

For automatic CV processing, LLM-based approaches have been used to extract information from unstructured CVs. These approaches are applicable to many formats, but CV preprocessing is necessary to reduce errors [8]. On the other hand, rule-based processing methods continue to be successfully applied, especially in very specific areas [13].

In general, it is known that there is a trade-off between interpretability, adaptability, and computational cost. LLM-based techniques are highly accurate in heterogeneous and multilingual scenarios with higher computational requirements. Hybrid approaches that combine traditional NLP methods with LLM are increasingly being used to balance cost with result quality.

3 Proposed Approach

The proposed approach is designed to read semi-structured PDF CVs and create a structured, multilingual, and semantically searchable CV representation. The methodology is organized into two complementary stages. The first, the *processing stage*, is responsible for extracting and organizing relevant information from raw CV text. The second, the *information retrieval stage*, enables efficient and accurate search across the resulting corpus.

The processing stage implements and evaluates three alternative extraction strategies, ranging from deterministic rule-based methods to fully LLM-driven approaches, allowing a balance between processing speed, cost, and adaptability.

The retrieval stage, on the other hand, uses a tag-based filter with multilingual semantic embeddings to facilitate information access and improve the accuracy of the results. Together, these stages define a flexible methodology that adapts to various document formats and enables highly accurate retrieval.

3.1 Processing Strategies

Three distinct strategies are implemented and compared for converting seme-structured PDF CVs into standardized JSON structures, each offering different trade-offs in accuracy, computational cost, and robustness to structural variability:

- **Purely heuristic-based extraction:** Text extracted from PDF files is segmented using a set of rules. These rules are used to find common section headings (e.g., *Education*, *Experience*, *Skills*), as well as to identify their boundaries using lexical markers. They also save the extracted content to the corresponding JSON fields. This approach is computationally efficient and does not require external LLM API calls. Still, it is sensitive to format variability and less effective when sections are written in free text or appear in unexpected order.
- **Hybrid heuristic+LLM extraction:** In this strategy, heuristics serve as a preprocessing step to normalize the basic structure of the CV, such as removing noise, correcting line breaks, and isolating candidate sections. The preprocessed text is then passed to an LLM, which performs fine-grained segmentation and entity extraction (e.g., institution names, job titles, dates). This approach combines the speed of PLN rules with the contextual understanding of LLMs, improving performance in handling diverse CV formats while reducing the number of tokens sent to the model.
- **LLM-only extraction:** Here, the raw text obtained is sent directly to the LLM without heuristic preprocessing. The model is prompted to identify CV sections and return a fully structured JSON representation. This approach maximizes the model's flexibility to interpret a wide range of formats and languages and is especially useful for CVs with unconventional layouts. However, it entails the highest computational cost, depends entirely on the LLM's consistency, and may yield less predictable results in noisy or poorly extracted text.

These three strategies enable a systematic evaluation of the trade-offs between interpretability, processing speed, operational cost, and resilience to document variability, providing empirical guidance for their application in different deployment scenarios.

3.2 Information Retrieval Strategy

The retrieval strategy is designed around a strict query language and a two-stage pipeline combining structural filtering with semantic ranking. Queries can specify plain tags (e.g., *experience*, *skills*, *title*), which declare the target sections to be semantically scored, or use the form `tag: "text"` to apply field-specific filters without contributing to the scoring set. Free-text tokens are only considered if at least one target section is declared, and guardrails prevent sensitive fields (e.g., *contact*, *name*) from being used as targets.

Stage 1: Structural filtering. All `tag: "text"` constraints are applied to prune the candidate CVs, reducing the universe of documents to be scored.

Stage 2: Semantic ranking. The remaining candidates are ranked using multilingual word embeddings, as follows: The tokens are preprocessed (lowercase conversion and stopword removal) and compared to the section definitions using cosine similarity. The first K elements capture the word-by-word matches. To calculate the local relevance of a token t from query Q to a section $s \in S$ of a document j, we define a token-section similarity function, $local_score(t, j, s)$, which operates by averaging the cosine similarities of the K closest words in the section, as follows:.

$$\text{local\_score}(t, j, s) = \frac{1}{K} \sum_{k=1}^{K} \text{TopK}_k \Big(\{\cos(E(t), E(w))\}_{w \in W_{j,s}} \Big).$$

where $E(t)$ and $E(w)$ are the embedding vectors (vector representations) for the query token t and the word w in the s-th section of the j-th document, respectively, obtained using the multilingual embedding model .

Section scores are obtained by averaging across all query tokens, as follows:

$$\text{section\_score}(j, s) = \frac{1}{|Q|} \sum_{t \in Q} \text{local\_score}(t, j, s).$$

Finally, document scores are aggregated over the declared target sections:

$$\text{document\_score}(j) = \frac{1}{|S|} \sum_{s \in S} \text{section\_score}(j, s).$$

This design offers several advantages: explicit targeting of relevant sections, early pruning to reduce computational load, word-level evidence that emphasizes strong semantic matches, multilingual robustness, and readiness for integration as a controllable tool callable by language models.

Although the system is instantiated on LinkedIn CVs, the methodology is intentionally defined as a framework with three interchangeable extraction strategies and a retrieval layer that can be adapted to other semi-structured documents (e.g., academic records, medical reports). This abstraction beyond a single dataset supports reusability and justifies the use of the term framework rather than merely a prototype system.

Figure 1 provides a high-level methodological diagram highlighting the two main stage. This representation clarifies the complete methodology and complements the more detailed system architecture diagrams presented in Sect. 5.

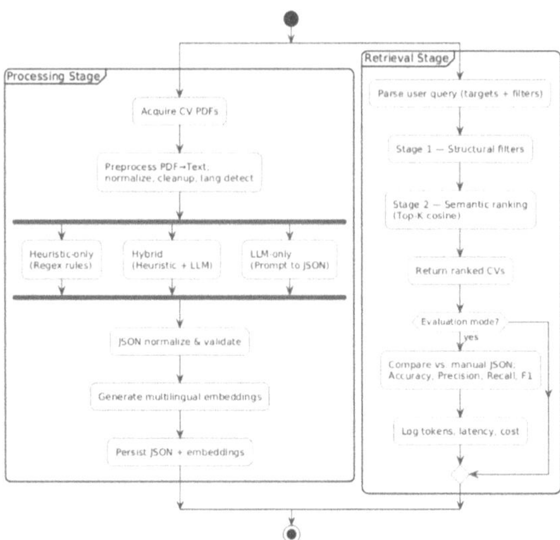

Fig. 1. Methodological pipeline.

4 Corpus Construction

The set of documents used to evaluate this proposal includes 138 CVs manually compiled from public LinkedIn profiles in PDF format. Since all documents follow a semi-structured format provided by the platform, they maintain a minimum level of consistency, but vary in the order of sections, content density, and the way in which the points in each section are described.

The linguistic distribution of the corpus can be seen in Table 1. This table shows that Spanish predominates in the documents (65.94%), followed by English (32.61%), and a small fraction (1.45%) corresponds to CVs in other languages.

Table 1. Language distribution in the evaluation corpus.

Language	#CVs	Percentage (%)
Spanish	91	65.94
English	45	32.61
Other languages	2	1.45
Total	138	100.00

On the other hand, Table 2 shows the descriptive statistics of the CVs set. CV length is between 26 and 2,584 words (Mean = 411, SD = 428). Lexical richness, measured by the Type-Token Ratio (TTR)[4], is betweeen 0.244 and 0.871 (Mean

[4] TTR is a metric of lexical diversity that quantifies the ratio between unique words (types) and the total number of words (tokens) in a language sample [11].

= 0.560, SD = 0.121). These statistical values indicate moderate lexical diversity, with recurring structural elements (e.g., headings like *Education* or *Experience*) using a more varied vocabulary. The wide dispersion in document length also points to heterogeneity in CV detail, which is expected for real-world data.

Table 2. Descriptive statistics of CV length and lexical richness in the corpus.

Metric	Min	Max	Mean	SD
Number of words	26	2584	411	428
Type–Token Ratio (TTR)	0.244	0.871	0.560	0.121

This combination of CVs in multiple languages and with variability in structure, length, and lexical richness provides a realistic dataset for evaluating the accuracy of the proposed information extraction and retrieval strategies. While the corpus reflects substantial linguistic and structural diversity, its representativeness is limited to LinkedIn profiles. This choice ensured consistency of format but does not fully cover all industries or career levels. Therefore, the reported results should be interpreted within these constraints.

5 System Architecture

The proposed system implements a modular architecture to coordinate the information flow between its components, ensuring scalability, maintainability, and ease of integration with other systems. The design supports both heuristic-based and LLM-based processing strategies, allowing flexible adaptation to diverse document formats and computational resources. Figure 2 presents a high-level UML diagram illustrating the interactions between users, the frontend, the backend, and the database.

The system supports two user types: a *Graduate* and an *Administrator*. The *Graduate* uploads a CV in PDF format through the web interface for processing and storage, and the *Administrator* uses the same interface to search stored CVs. Additionally, the system consists of three layers:

- **UI Layer:** The *Frontend Web Application* is the interaction layer between users and backend services. It allows users to upload PDF files through web forms, submit search queries, and view results. Interaction is implemented through HTTP requests to the backend.
- **Backend Layer:** This layes is composed of four specialized modules. The *API Module* manages HTTP requests and exposes the endpoints for CV uploads and for information retrieval. The *Processing Module* converts uploaded PDFs into standardized JSON structures using one of three previously described processing strategies. The *Embeddings Generation Component* creates vectors representing CVs, which are used for semantic similarity search. In contrast, the *Search Module* executes both textual and embedding-based queries over the stored corpus.

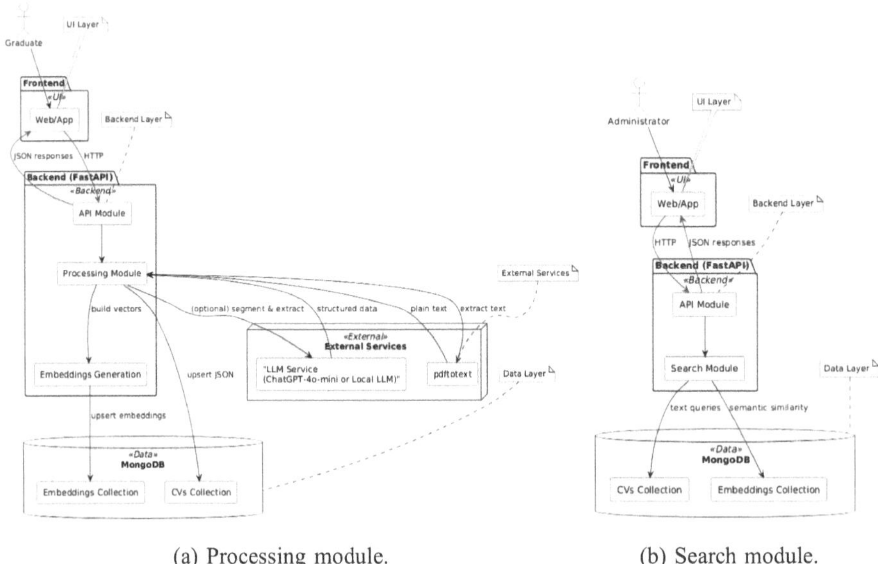

(a) Processing module. (b) Search module.

Fig. 2. High-level architecture of the proposed system.

- **Data Layer:** It uses a document-oriented database with two elements: the *CVs Collection* for storing JSON representations of CVs, and the *Embeddings Collection*, which stores the embedding vectors used in semantic retrieval.

The system implements two *operational flows*: The *CV upload flow*, shown in Fig. 3, is when a user uploads a PDF CV to be processed. The *CV search flow*, shown in Fig. 4, is when an administrator issues a query, and the backend retrieves matching CVs through textual or semantic search before returning them to the frontend for display. This modular design ensures that the system can handle diverse CV formats and support multilingual processing.

6 Implementation

The system implementation integrates software components enabling scalability and future maintenance. The backend was developed with `FastAPI`, as it offers asynchronous communication capabilities and high performance in handling concurrent requests. Additionally, `MongoDB` is used, a NoSQL database that supports flexible storage of semi-structured JSON documents and offers high performance and scalability. PDF to text conversion is done using `pdftotext`.

6.1 Processing Module

The extraction strategies are implemented in this module. When LLM is used, `ChatGPT-4o-mini` is employed via API calls, through prompts defined to create consistent and well-structured JSON output. The architecture also supports

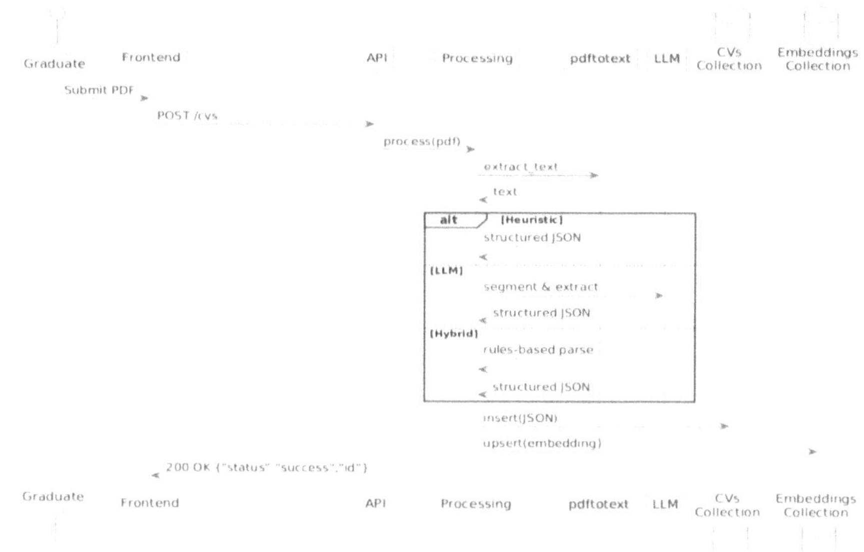

Fig. 3. Operational flow for CV upload and processing.

replacing the remote model with a locally hosted LLM, reducing dependency on external APIs and improving response times.

Multilingual semantic embeddings are generated with encoders stored in the *Embeddings Collection* of the database. These embeddings enable top-k retrieval using cosine similarity, supporting accurate and language-agnostic search.

Rule-Based PDF-to-JSON Pipeline

The rule-based strategy is implemented in Python 3 using the standard `re` module. All patterns are compiled at import time with consistent flags (case-insensitive, multiline, Unicode-aware), and text is normalized to NFC before processing. The pipeline includes the following stages:

Pre-processing (PDF → normalized text).

- **Page split:** `\f` separates raw text into page-level chunks.
- **Page numbering removal:** `(?i)^\s*Page\s+\d+\s+of\s+\d+\s*$` filters headers/footers.
- **Column margin estimation:** `^(?P<left>.*?) {4,}(?P<right>\S.*)$` detects inter-column spacing.
- **Section header detection:** Multilingual regex identifies *experience*, *education*, *skills*, etc.
- **Reading order reconstruction:** Columns are segmented and reassembled sequentially.

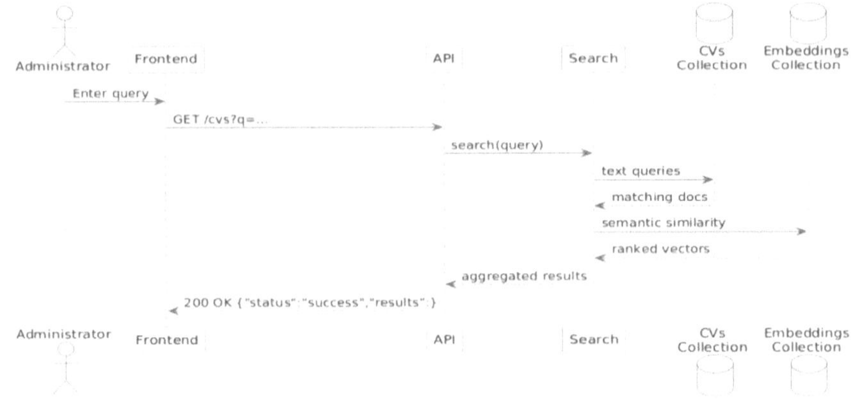

Fig. 4. Operational flow for CV search and retrieval.

Contact Extraction

- **Email:** `\b[A-Z0-9._%+-]+@[A-Z0-9.-]+\.[A-Z]{2,}\b`
- **Phone:** `\+?\d[\d\-\s()]{7,}`
- **URLs/LinkedIn:** `https?://[^\s]+`

Content Fields

- **Title and Summary:** The first line under *title* is stored as professional title; subsequent paragraphs under *summary* form the profile description.
- **Skills:** Lists split by commas, bullets, semicolons (`[,\n·•;]+`); tokens are normalized to lowercase.
- **Languages:** `^(.*?)\s*\((.*?)\)$` extracts label and proficiency.
- **Experience:** Regexes for years, date ranges, and month names; broken lines merged using punctuation and capitalization cues. Current roles detected with tokens like *present* or *actualidad.*
- **Education:** Year ranges in parentheses, degree–field splits, and reuse of date patterns.
- **Achievements:** Block delimiters (`\n{2,}`, `\)\n`) and parenthetical metadata are captured.

JSON Assembly. Only non-empty fields are serialized; arrays appear when containing at least one item. Dates are normalized as {year, month}, and the flag *is_current* is included when detected.

LLM Configuration. We used temperature = 0.2, top_p = 1.0, and max_tokens between 512–1024. Prompts were fixed across run.

Prompt Illustration. For reproducibility, the following prompt exemplifies the instruction used for LLM-based extraction:

```
You are an information extraction agent. From the CV text below,
return a valid JSON matching this schema:
{ "contact": {...}, "education": [...], "experience": [...],
  "skills": [...], "languages": [...] }
Rules: (1) Do not invent fields. (2) Normalize dates as {year, month}.
(3) Use empty arrays rather than nulls. (4) Keep original language.
Input: <<<CV_TEXT>>>
```

6.2 Information Retrieval Module

The retrieval pipeline applies two stages: First, a set of structural filters is applied to the CVs in their JSON representations to select documents that satisfy the field-specific conditions. This step significantly reduces the number of candidate documents that need to be processed with embeddings.

In the second stage, semantic similarity is computed between the preprocessed query tokens and the words present in the declared target sections. Multilingual embeddings (based on the *paraphrase-multilingual-MiniLM-L12-v2* encoder) are used to capture cross-lingual equivalences. For each query token, the Top-K most similar words in the target section are selected, and their cosine similarities are averaged to form local scores. These are then aggregated at the section level and finally combined across target sections into a single document score.

This implementation enables system scalability since filtering reduces computational cost, and the semantic classification ensures accuracy in multilingual contexts. Furthermore, the strict query language allows external models to interact transparently with the retrieval system.

7 Results and Discussion

A comparative analysis is carried out between heuristic-only, hybrid heuristic+LLM, and LLM-only approaches. We assess the performance of the extraction strategies using standard metrics derived from the confusion matrix. For each CV, the manually curated ground truth JSON structure is compared witht the prediction generated by each of the three strategies. The comparison is performed at the level of key–value pairs after flattening the JSON representation, and each element is categorized into one of four possible outcomes: true positive (TP), false positive (FP), false negative (FN), or true negative (TN). Table 3 summarizes the formulas and definitions of the metrics employed.

Table 3. Evaluation metrics computed from the confusion matrix.

Metric	Formula	Definition
Accuracy	$\frac{TP+TN}{TP+TN+FP+FN}$	Proportion of correctly classified fields (both present and absent)
Recall	$\frac{TP}{TP+FN}$	Ability to recover the fields that should be present
Precision	$\frac{TP}{TP+FP}$	Reliability of positive predictions (fields predicted as present)
F1-score	$\frac{2\cdot(Precision\cdot Recall)}{Precision+Recall}$	Harmonic mean of precision and recall

Table 4 shows the results for the three processing strategies. The heuristic-only approach achieves one reasonable recall (89.61%) but lower precision (78.05%), with an F1-score of 83.43. The hybrid method improves precision substantially (95.66%) while keeping recall at a competitive level (79.58%), yielding an F1-score of 86.88. The LLM-only approach obtains the best balance, with both high recall (89.38%) and very high precision (96.28%), leading to the highest F1-score (92.70). Accuracy follows the same trend, ranging from 71.57% in the heuristic baseline to 86.40% in the LLM-only setting.

Table 4. Evaluation metrics by method.

Method	Accuracy	Recall	Precision	F1
Heuristic	71.57	89.61	78.05	83.43
Hybrid heuristic+LLM	76.81	79.58	95.66	86.88
LLM-only	86.40	89.38	96.28	92.70

The error analysis by path shows that the most frequent FN are concentrated in deeply nested or heterogeneous sections such as `education.details`, `work_experience.positions`, and `skills.lists`. These elements have high variability, with optional attributes such as start and end dates, detailed job descriptions, or skill levels that the CVs inconsistently report. This variability explains why heuristic rules fail to capture them, while LLM-based approaches may still struggle when contextual cues are insufficient. On the other hand, the most common FP have in fields like `languages` or `certifications`, where both hybrid and LLM methods tend to over-generate entries, producing plausible but spurious fields.

Across all strategies, fields with standardized lexical markers, such as section headers (*Education*, *Experience*), show high accuracy. This result confirms that structural regularities in semi-structured documents like LinkedIn CVs can be effectively exploited. The errors are therefore not uniformly distributed but are concentrated in free-text or optional components, where ambiguity and multilingual variation introduce challenges.

Beyond aggregate metrics in Table 4, the error analysis already reveals field-specific behavior: false negatives concentrate in `education.details`, `skills.lists` and `work_experience.positions`, while false positives are frequent in `languages` and `certifications`. This qualitative breakdown shows different strategies excel in different fields, even if numeric results are not disaggregated.

From the above results, the following cases can be observed:

- Structural variability exists even when the corpus follows a semi-structured template, the order and granularity of the subfields differ significantly.
- The presence of multilingual content (linguistic variability) complicates section identification and value normalization.
- LLMs offer flexibility but sometimes present nonexistent fields, and heuristics are fragile when the input differs from the expected patterns.

In summary, the heuristic approach provides a high recall rate but lower reliability; the hybrid method balances recall rate with very high precision; and the LLM-only approach achieves the best overall performance. However, all approaches suffer from flaws in handling optional and heterogeneous fields (Table 5).

Table 5. Representative examples of correct and incorrect extractions by method.

Method	Typical Correct Extraction	Typical Error Case
Heuristics	Correctly identifies the *Education* section header and extracts degree titles such as "B.Sc. in Computer Science"	Fails to capture optional subfields such as start/end dates in `work_experience`, leading to missing information (false negatives)
Hybrid	Accurately segments structured headers and complements missing values by inferring organization names and job titles from context	Occasionally merges adjacent sections (e.g., *Certifications* with *Skills*), producing extra fields that do not exist in the manual annotation (false positives)
LLM-only	Recovers free-text details, such as job responsibilities and project descriptions, with higher recall than heuristics	Generates hallucinated fields in ambiguous contexts, e.g., inventing a "Skill level" attribute not present in the original CV (spurious field)

The current study has several limitations: (1) the evaluation corpus is relatively small and drawn exclusively from LinkedIn, which constrains the external validity of the results; however, this also demonstrates that promising performance can be achieved even in a restricted setting, (2) the ground-truth annotations were manually curated without a large annotation team, which ensured high consistency but limited the possibility of reporting inter-annotator agreement, (3) the implementation relies on a single LLM family, highlighting both the practical feasibility of integrating such models and the opportunity to expand toward diverse model providers or fine-tuned variants, and (4) while cost and latency figures are not reported in full detail, the observed reduction of tokens in the hybrid approach already points to a cost–accuracy trade-off worth exploring in scaled deployments.

8 Conclusions and Future Work

In this work, we implement a hybrid framework to combine NLP heuristics with LLMs to process and retrieve semantic information from multilingual CVs sourced from LinkedIn PDFs. Our method tackles the chaos of heterogeneous document formats by using rule-based logic to structure the input, which an LLM then interprets to generate clean, standardized JSON outputs. Central to our approach is a lightweight, two-stage pipeline built on transformer-based multilingual embeddings, enabling accurate cross-language search that scales efficiently for large institutional datasets. Our evaluation reveals a clear trade-off. The pure LLM approach achieves the best metric values, but with a high computational cost by using expensive API providers like OpenAI. The hybrid model (heuristic + LLM) is the most practical solution, with high precision and achieving near-LLM performance. It is interesting to note that the simple rules-only model, although less accurate overall, is still a practical and efficient option for cases where data cannot leave the company and the operating cost must be kept very low.

In conclusion, heuristics and LLMs are complementary, not competing: Heuristics form a cost-effective, interpretable backbone that pre-processes documents and reduces the need for expensive LLM calls. In practice, this means our hybrid system uses rules to filter noise and structure data before invoking the model, slashing cost and latency without compromising the quality of semantic retrieval.

In the future, we'll explore fine-tuning smaller LLMs specifically for CV segmentation and entity extraction to improve consistency. Furthermore, we aim to enrich our semantic embedding models with domain-specific knowledge to capture finer nuances. Expanding multilingual support through automatic translation for low-resource languages is another priority. Also, we will investigate using local LLMs (like Llama 3 or Mistral) to eliminate third-party API dependencies. Finally, we will explore developing an adaptive controller that dynamically routes documents between heuristic and LLM processing based on their complexity, creating a more intelligent and efficient system.

Disclosure of Interests. The authors have no competing interests to declare that are relevant to the content of this article.

References

1. Akdur, D.: Skills gaps in the industry: opinions of embedded software practitioners. ACM Trans. Embed. Comput. Syst. **20**(5), 1–39 (2021). https://doi.org/10.1145/3463340
2. Badjatiya, P., Kurisinkel, L.J., Gupta, M., Varma, V.: Attention-based neural text segmentation. In: Pasi, G., Piwowarski, B., Azzopardi, L., Hanbury, A. (eds.) ECIR 2018. LNCS, vol. 10772, pp. 180–193. Springer, Cham (2018). https://doi.org/10.1007/978-3-319-76941-7_14
3. Frohmann, M., Sterner, I., Vulić, I., Minixhofer, B., Schedl, M.: Segment any text: a universal approach for robust, efficient and adaptable sentence segmentation. arXiv preprint arXiv:2406.16678 (2024). https://doi.org/10.48550/arxiv.2406.16678
4. Gaftandzhieva, S., Hussain, S., Hilcenko, S., Doneva, R., Boykova, K.: Data-driven decision making in higher education institutions: state-of-play. Int. J. Adv. Comput. Sci. Appl. **14**(6) (2023). https://doi.org/10.14569/IJACSA.2023.0140642
5. Ganguly, D., Leveling, J., Jones, G.J.: Utilizing sub-topical structure of documents for information retrieval. In: Proceedings of the 4th Workshop on Workshop for Ph.D. Students in Information & Knowledge Management, PIKM '11, pp. 75–78. ACM, Glasgow (2011). https://doi.org/10.1145/2065003.2065019
6. Koshorek, O., Cohen, A., Mor, N., Rotman, M., Berant, J.: Text segmentation as a supervised learning task. In: Walker, M., et al. (eds.) Proceedings of the 2018 Conference of the North American Chapter of the ACL: Human Language Technologies, vol. 2, pp. 469–473. ACL, New Orleans (2018). https://doi.org/10.18653/v1/N18-2075
7. Li, Q., et al.: A survey on text classification: from traditional to deep learning. ACM Trans. Intell. Syst. Technol. **13**(2) (2022). https://doi.org/10.1145/3495162
8. Lo, F.P.W., et al.: AI hiring with LLMs: a context-aware and explainable multi-agent framework for resume screening. In: Proceedings of the IEEE/CVF Conference on Computer Vision and Pattern Recognition (CVPR) Workshops, pp. 4193–4202 (2025)
9. Lukasik, M., Dadachev, B., Papineni, K., Simões, G.: Text segmentation by cross segment attention. In: Proceedings of the 2020 Conference on Empirical Methods in Natural Language Processing, pp. 4021–4032 (2020). https://doi.org/10.18653/V1/2020.EMNLP-MAIN.380
10. Malkiel, I., et al.: SEGLLM: topic-oriented call segmentation via LLM-based conversation synthesis. In: ICASSP 2024 - IEEE International Conference on Acoustics, Speech and Signal Processing (ICASSP), pp. 11256–11260. IEEE, Seoul (2024). https://doi.org/10.1109/icassp48485.2024.10446156
11. Malvern, D., Richards, B., Chipere, N., Durán, P.: Lexical Diversity and Language Development. Palgrave, Macmillan (2004). https://doi.org/10.1057/9780230511804
12. Maraj, A., Martin, M.V., Makrehchi, M.: A more effective sentence-wise text segmentation approach using BERT. In: Lladós, J., Lopresti, D., Uchida, S. (eds.) ICDAR 2021. LNCS, vol. 12824, pp. 236–250. Springer, Cham (2021). https://doi.org/10.1007/978-3-030-86337-1_16
13. Musleh, D.: Rule-based information extraction from multi-format resumes for automated classification. Math. Model. Eng. Prob. **11**(4), 1044–1052 (2024). https://doi.org/10.18280/mmep.110422
14. Prince, V., Labadié, A.: Text segmentation based on document understanding for information retrieval. In: Kedad, Z., Lammari, N., Métais, E., Meziane, F., Rezgui,

Y. (eds.) NLDB 2007. LNCS, vol. 4592, pp. 295–304. Springer, Heidelberg (2007). https://doi.org/10.1007/978-3-540-73351-5_26

15. Rajini, S., Hari Prasad, B., Upendrasingh, A.: Alumni management and networking system. In: 2023 2nd International Conference on Advancements in Electrical, Electronics, Communication, Computing and Automation (ICAECA), pp. 1–5. IEEE, Coimbatore (2023). https://doi.org/10.1109/ICAECA56562.2023.10200060
16. Reddy, J.S., Sharma, R., Gupta, N.: The accreditation paradigm: a comparative analysis of accreditations for management programmes. Int. J. Educ. Manag. **38**(1), 73–95 (2023). https://doi.org/10.1108/IJEM-05-2023-0250
17. Skrzypek, C., Diebold, J., Kim, W., Krause, D.: Mentoring connections: implementing a student-alumni mentor program in social work. J. Soc. Work. Educ. **55**(3), 449–459 (2019). https://doi.org/10.1080/10437797.2019.1600445
18. Stokes, N., Carthy, J., Smeaton, A.F.: Segmenting broadcast news streams using lexical chains. In: STarting Artificial Intelligence Researchers Symposium (STAIRS 2002). IOS Press, Lyon (2002)
19. Sun, W.J., Du, Y., Liu, G., Kompella, R.R., Snoek, C.G.M.: Training-free semantic segmentation via LLM-supervision. arXiv preprint arXiv:2404.00701 (2024). https://doi.org/10.48550/arxiv.2404.00701
20. Tong, P.: Building a sustainable future: the role of industry-education integration in vocational training. Educ. Insights **1**(3), 1–7 (2024). https://doi.org/10.70088/fjsm3j57
21. Wan, M., et al.: TnT-LLM: text mining at scale with large language models. In: Proceedings of the 30th ACM SIGKDD Conference on Knowledge Discovery and Data Mining, KDD '24, Barcelona, Spain (2024). https://doi.org/10.1145/3637528.3671647

Local LLM Ensembles for Zero-Shot Portuguese Named Entity Recognition

João Lucas Luz Lima Sarcinelli(✉) and Diego Furtado Silva

Instituto de Ciências Matemáticas e Computação, Universidade de São Paulo, São Paulo, Brazil
{joao.luz,diegofsilva}@usp.br

Abstract. Large Language Models (LLMs) excel in many Natural Language Processing (NLP) tasks through in-context learning but often under-perform in Named Entity Recognition (NER), especially for lower-resource languages like Portuguese. While open-weight LLMs enable local deployment, no single model dominates all tasks, motivating ensemble approaches. However, existing LLM ensembles focus on text generation or classification, leaving NER under-explored. In this context, this work proposes a novel three-step ensemble pipeline for zero-shot NER using similarly capable, locally run LLMs. Our method outperforms individual LLMs in four out of five Portuguese NER datasets by leveraging a heuristic to select optimal model combinations with minimal annotated data. Moreover, we show that ensembles obtained on different source datasets generally outperform individual LLMs in cross-dataset configurations, potentially eliminating the need for annotated data for the current task. Our work advances scalable, low-resource, and zero-shot NER by effectively combining multiple small LLMs without fine-tuning. Code is available at https://github.com/Joao-Luz/local-llm-ner-ensemble.

Keywords: NER · LLM · Ensemble · Zero-shot · Portuguese

1 Introduction

Large Language Models (LLMs) have shown great capabilities in performing various Natural Language Processing (NLP) tasks requiring little to no labeled data due to their In Context Learning (ICL) capabilities [24]. Notably, however, they still under-perform for token classification tasks such as Named Entity Recognition (NER) compared to smaller, specialized models [21]. This performance gap is more apparent when using smaller LLMs designed to run on consumer GPUs instead of larger, often cloud-based, alternatives like OpenAI's ChatGPT, Anthropic's Claude, or Google's Gemini. Furthermore, the gap widens when performing such tasks for lower-resource languages such as Portuguese, in which these models notably under-perform when compared to resource-rich languages [7].

Recently, new large language models are developed at a fast pace, with many of them being open-weights and available for local instantiation, such as the

D. Chaves et al. (Eds.): CIARP 2025, LNCS 16528, pp. 207–221, 2026.
https://doi.org/10.1007/978-3-032-23161-1_15

Meta's LLaMA family of models [8], Google's Gemma [20] and Microsoft's Phi [1]. However, even though many models have been made publicly available, no single model outperforms others for all different tasks, with them often complementing each other even within similar tasks [12]. To that extent, using LLMs in ensembles presents itself as a means to achieve model-independent systems that can tap into different capabilities achieved by the various architectures and pre-training processes that each model possesses. When paired with the models' capabilities of performing tasks with no labeled examples, zero-shot LLM ensembles also stand out as scalable alternatives for tasks in low resource domains [6], such as the case for many real-world applications of NER, especially for the Portuguese language.

Previous methods of LLM ensembles rarely aim to address named entity recognition, often focusing on text generation [9,11,12] and text classification [6]. While some works aim to perform NER with LLM ensembles, they either fall short of successfully implementing a pipeline that shows performance gains when using different models [16] or build pipelines that depend on the presence of larger, fine-tuned, and incidentally, more robust models to tend to more difficult cases [13]. To the best of our knowledge, no work aims to aggregate LLMs of similar size and capabilities to perform zero-shot NER.

To address this gap, we outline a novel LLM ensemble pipeline focused on zero-shot named entity recognition intended to work with multiple, similarly sized, locally run models compatible with consumer GPUs. The pipeline, illustrated in Fig. 1, consists of three main steps: **extraction**, in which all models extract entities from a given document; **voting**, when all models decide, collectively, which entity mentions extracted in the previous step are correct; and **disambiguation**, a step to deal with possible overlapping entities deemed correct by the majority vote in the previous step. As the individual LLMs might perform differently for each pipeline task, different sets of models are considered for the different stages of the pipeline. In order to select the best configurations of models, we employ a heuristic that utilizes a small set of annotated data to evaluate the different combinations of LLMs and chooses the best-performing one.

We evaluated the ensemble method and compared its performance against individual models for five distinct Portuguese NER datasets, using zero-shot prompts for all pipeline steps. We found that the ensembles chosen by the proposed method performed better than any single LLM for four out of the five datasets. Moreover, we found that ensemble configurations obtained with different source datasets tended to perform as well or better than the ones obtained in the target dataset, which makes the pipeline feasible even in scenarios where no annotated data is available. In summary, this work's main contributions are:

- Propose a novel pipeline for using multiple LLMs as an ensemble of zero-shot models for NER;
- Evaluate our method on 5 Portuguese NER datasets spanning different domains and entity granularities;

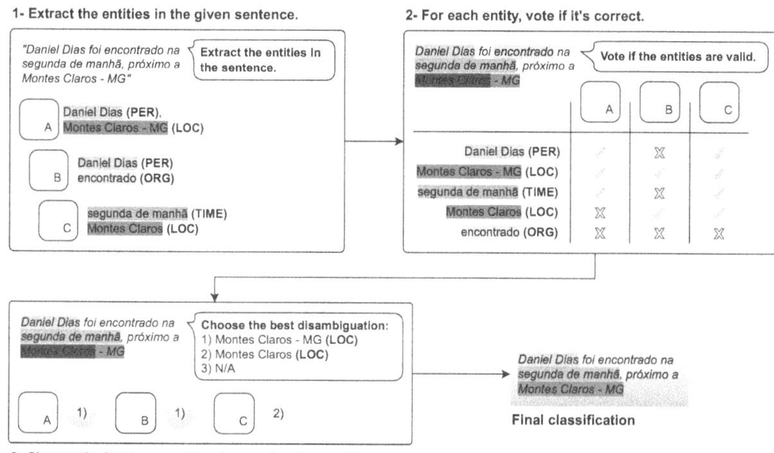

Fig. 1. Pipeline for named entity recognition using the LLM ensemble. Three LLMs, A, B, and C, extract entities independently in step 1. In step 2, all LLMs vote if the entities mentioned are correct, and entities chosen by majority vote are, therefore, considered correct. Step 3 disambiguates the overlapping mentions ("*Montes Carlos*" and "*Montes Carlos - MG*") by the LLMs through a multiple-choice prompt. The majority vote is then used to decide on the best fit for the entity in the spans.

- Show that using ensembles obtained using annotated data from either the target task or a distinct source task tends to produce ensembles that perform zero-shot NER better than any single LLM.

2 Related Work

Ensemble methods with LLMs have recently become a popular research topic. These methods take advantage of the different models' diverse strengths while avoiding some of their specific weaknesses [17]. A recent survey [9] shows some of the most prominent ensembling strategies for LLMs used in the literature, which come in different varieties. Mix of Experts (MoE) models, such as Mixtral [11], are composed of an ensemble of trainable weights selected during inference to produce the final output text. Fusion-based ensembling methods, such as LLM-Blender [12], introduce ways to combine the output of several pre-trained LLMs to produce a final text output, typically through reranking or voting, without depending on any training of the underlying LLMs. These approaches are primarily designed for generative tasks and may not trivially extend to structured prediction problems like NER, where span consistency and token-level granularity are important.

Routing [3,6], on the other hand, implements a mechanism to choose appropriate models for a given task input dynamically and, while applicable to text generation tasks, is simpler to be used for classification tasks. [6] propose a cascading chain of LLMs, where simpler inputs are processed by cheaper models and complex cases routed to more powerful ones. Simpler inputs are processed by the initial, simpler, and cheaper LLMs, while complex examples are forwarded to the more resource-intensive models. While cost-effective, this method assumes a predefined hierarchy of model capability and does not address scenarios where no single model in the chain performs optimally.

For NER specifically, [13] uses fine-tuned models, as well as fine-tuned variations of API-based LLMs such as GPT-3.5, to perform entity recognition on medical domain data. They found that using LLMs together with other models in a majority vote format usually outperforms the individual base models. It is important to note that this work relies on the use of large, cloud-based models and large amounts of data to fine-tune the ensemble NER models.

For Portuguese NER, [16] attempts to implement an ensembling method using the outputs of BERT [5] sequence classifiers as inputs for a Mistral [10] large language model to perform aggregation of different entities. They query the model to decide on a final classification for a given span of the source sentence, providing information such as entity definitions and vote counts (number of BERT models that predicted such span as an entity.) The authors found that using this approach leads to performance losses compared to using the base classifiers alone. It is relevant to note, however, that their approach relies on a single model to perform the aggregation of information and does not employ multiple LLMs for the task. Furthermore, their analysis is limited to the legal domain for Portuguese NER.

3 Proposed Method

The proposed LLM ensemble pipeline[1] for entity recognition consists of three main steps: **entity extraction**, when each LLM of the ensemble is tasked with extracting entities from the given sentence; **voting**, when the LLMs must vote if the entities extracted by the entire ensemble in the previous step are valid or not for the sentence; and **disambiguation**, when LLMs are tasked with choosing the best disambiguation of overlapping entities obtained in the previous step. This section details each step of the pipeline illustrated in Fig. 1.

3.1 Entity Extraction

Our ensemble approach begins with independent entity extraction by each LLM using a zero-shot prompt strategy adapted from [22]. The prompt consists of the task definition entity type descriptions and queries the model to generate the list of entities in a sentence in JSON format. The LLMs' responses are then parsed, and each entity is validated in order to remove those with inconsistent types or those that can not be matched in the original sentence. Finally, all extracted entities from all of the ensemble's models are compiled into a unified record.

[1] Available at https://github.com/Joao-Luz/local-llm-ner-ensemble.

3.2 Voting

After extracting entities using the different LLM, each one is assigned to vote on whether the extracted entities are valid or not. It mimics a human annotation system, where multiple annotators might decide on the validity of entities annotated by different annotators in a team. Each extracted entity is used to construct a prompt that queries the LLM if the given entity mention is correct for its respective sentence. After each LLM takes a vote, we use the majority strategy to decide whether the entity is valid. This process generates a list of entities sourced individually from all models in the ensemble and validated by all of them.

3.3 Disambiguation

When different models identify overlapping entities, we group these entities for disambiguation. This situation is illustrated in Fig. 1 for the entities "*Montes Claros*" and "*Montes Claros - MG*". A multiple-choice prompt is applied so the ensemble's models may decide which entities are more correct for the given sentence. The construction process for this prompt is illustrated in Fig. 2 with a more in-depth example.

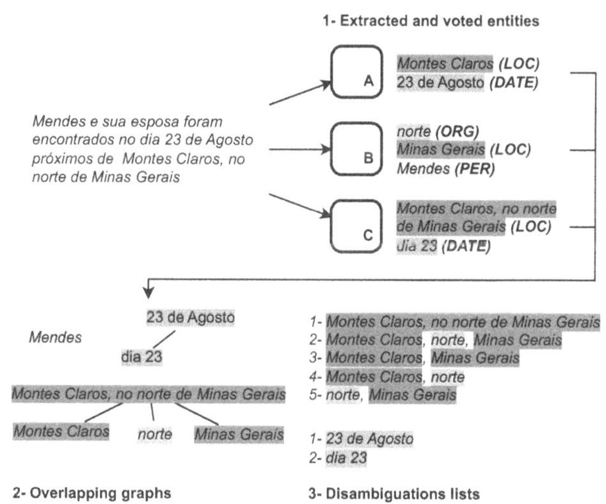

Fig. 2. Process for constructing the disambiguation prompt.

The disambiguation considers all mentions that passed the voting step, building an overlap graph where connected mentions form ambiguous groups. For example, non-overlapping mentions like "*Montes Carlos*", "*norte*", and "*Minas Gerais*", shown in Fig. 2, still require disambiguation because they all share the common overlapping mention "*Montes Claros, no norte de Minas Gerais*". The set of possible disambiguations for the span are used to build the multiple-choice prompt, which contains the five longest valid entity combinations from the overlapping span, plus an "N/A" option. Each LLM must select its preferred option, with the most voted choice becoming the final entity disambiguation and ties being resolved by selecting the option with the longest combined entity lengths. The final entity list for a given sentence combines the non-overlapping entities from voting with the resolved entities from the disambiguation step.

3.4 Choosing Members of the Ensemble

Our experiments revealed that having all LLMs participating in all pipeline steps might lead to worse performance than using some separately. Some models would, for example, under-perform in the extraction step, introducing much noise into the extracted entity list, which could be challenging to remove in the voting and disambiguation steps. Similarly, some LLMs might be either too permissive or too strict in the voting and disambiguation steps, which could negatively impact the final list of recognized entities. To address this, we employ a heuristic to choose optimal combinations of models to perform each of the steps in the pipeline.

First, we select a small sample of the labeled data to act as a development set. Then, we execute the pipeline for all possible combinations of available LLMs for each step. We use every possible combination for the extraction step. For the voting and disambiguation steps, we only use combinations of odd numbers of models, as we rely on majority voting. After obtaining the performance for every combination of LLMs per step, we select the configuration that obtained the best micro-F1 performance overall.

4 Experimental Setup

4.1 Datasets

We selected five diverse and gold-standard datasets for NER in Portuguese in order to evaluate the ensemble pipeline. The datasets span from general to specific domains, with different levels of granularity for entity types. The selected datasets are listed in Table 1.

For every dataset, we use the available train split to perform ensemble configuration selection and test split for testing.

4.2 Models

As we aim to develop an ensembling method that works with LLMs with similar capabilities, the models must be of comparable size in the number of parameters. They also must be sufficiently diverse in order to be able to capture complementary information. To that extent, we decided to employ five different models, developed by different teams, with varied pre-training backgrounds and architectures. They are shown in Table 2:

Table 1. Datasets used for evaluation with their entity types and domains.

Dataset	Domain	Entity Types
HAREM (Selective) [18]	General	Person, Location, Time, Organization, Value
LeNER-Br [15]	Legal	Person, Organization, Location, Time, Legislation, Jurisprudence
UlyssesNER-Br [2]	Legal	Date, Event, Government Org., Non-gov. Org., Political Party, Other Product, Program Product, System Product, Virtual Location, Physical Location, Person-Group Position, Individual Person, Person Position, Law Foundation, Bill Foundation, Nickname Foundation
GeoCorpus-2 [4]	Geological	Eon, Era, Period, Epoch, Age, Siliciclastic Sedimentary Rock, Carbonate Sedimentary Rock, Chemical Sedimentary Rock, Organic Sedimentary Rock, Brazilian Sedimentary Basin, Basin Geological Context, Lithostratigraphic Unit, Misc.
MariNER [19]	Historical	Person, Location, Date, Organization

Table 2. Selected LLMs and their sizes.

Model	Parameters	Organization
LLaMA3[a] [8]	8B	Meta
Qwen2[b] [23]	7B	Alibaba
Gemma2[c] [20]	9B	Google
Phi3[d] [1]	14B	Microsoft
Mistral[e] [10]	7B	Mistral AI

[a]https://huggingface.co/meta-llama/Meta-Llama-3.1-8B-Instruct
[b]https://huggingface.co/Qwen/Qwen2-7B-Instruct
[c]https://huggingface.co/google/gemma-2-9b-it
[d]https://huggingface.co/microsoft/Phi3-Medium-128K-Instruct
[e]https://huggingface.co/mistralai/Mistral-7B-Instruct-v0.2

4.3 Configurations

In order to get more diverse outputs from the individual models, we first perform an initial calibration step to determine an optimal temperature value for each LLM. This step consists of the first two steps of the pipeline, done individually for each model and evaluated on the labeled sample of size 100 obtained from the available train set. We perform extraction at various temperature levels t, and voting using temperature $t = 0$. The final overall micro-F1 scores are used to determine the ideal temperature for each model in the extraction stage. For simplicity, we run this calibration step using only the HAREM dataset and use the optimal temperature levels for the remaining sets: $t = 0$ for LLaMA3, Mistral, and Phi3, $t = 1.0$ for Gemma2 and $t = 1.5$ for Qwen2. For voting and disambiguation, temperatures are kept at $t = 0$.

To choose the best combinations of models using the previously mentioned heuristic, a sample of 100 random sentences is taken from each dataset's train split to act as the validation set, simulating a low-resource scenario. The same sampling is used in the calibration step for the HAREM dataset. Testing is performed in the entire test set available for each dataset.

5 Results and Discussion

5.1 Main Results

Table 3 shows the micro-F1 performance of the best ensemble configuration obtained via the heuristic on both the validation and test sets. The table also shows the performances of individual models, which were obtained from processing the entire pipeline using the same model for extraction, voting, and disambiguation. It also shows the performance of a fully supervised RoBERTa model [14], trained on the full train splits of the datasets. To account for the models' stochastic natures, the reported results are the average of three runs of each experiment.

For most datasets, using the ensemble method presents a performance gain compared to using a single model, although not achieving the same performances as the supervised models (albeit, while not requiring any labeled data). Interestingly, the ensemble gain is more pronounced for LeNER-BR, GeoCorpus2, and UlyssesNER datasets, where individual LLMs tended to perform the worst for the zero-shot NER task. A possible explanation for this is that these three datasets are annotated for domain-specific entities, with GeoCorpus2 having 13 distinct types, UlyssesNER having 16, and LeNER-BR having 6. Due to the high amounts of unique types and the amount of domain knowledge necessary to identify them, individual LLMs struggle to distinguish between them. However, they may complement each other's knowledge gaps to produce a final, more robust list of entities for the sentences.

HAREM, on the other hand, is the only dataset where the ensemble performed worse than individual models. This might be explained due to the fact that it is the only generic domain set, annotated with relatively simple entity

Table 3. Performance of best ensemble configurations and individual LLMs for each dataset. The gray-colored rows show the performance of individual models. **Bold** indicates best performance among LLM-based approaches.

Dataset	Extraction	Voting	Disambiguation	Validation	Test
HAREM	LLaMA3	Qwen2	Phi3, Gemma2, LLaMA3	**0.617**	0.591
		Phi3		0.588	0.605
		Gemma2		0.584	**0.609**
		LLaMA3		0.536	0.545
		Qwen2		0.513	0.556
		Mistral		0.378	0.439
		RoBERTa		-	0.874
LeNER-BR	Gemma2	Phi3, Gemma2, LLaMA3	Phi3, Gemma2, Mistral	**0.580**	**0.541**
		Phi3		0.490	0.454
		Gemma2		0.556	0.528
		LLaMA3		0.510	0.461
		Qwen2		0.369	0.370
		Mistral		0.299	0.325
		RoBERTa		-	0.914
GeoCorpus2	Gemma2, LLaMA3, Mistral	Phi3	Gemma2, LLaMA3, Qwen2	**0.523**	**0.313**
		Phi3		0.388	0.243
		Gemma2		0.341	0.262
		LLaMA3		0.274	0.187
		Qwen2		0.221	0.187
		Mistral		0.270	0.178
		RoBERTa		-	0.862
UlyssesNER	LLaMA3	Phi3, Gemma2, Qwen2	Mistral, LLaMA3, Qwen2	**0.420**	**0.353**
		Phi3		0.271	0.321
		Gemma2		0.281	0.334
		LLaMA3		0.340	0.309
		Qwen2		0.155	0.291
		Mistral		0.215	0.279
		RoBERTa		-	0.889
MariNER	Gemma2	Phi3, Gemma2, LLaMA3	Phi3, Mistral, Qwen2	**0.767**	**0.706**
		Phi3		0.667	0.664
		Gemma2		0.734	0.699
		LLaMA3		0.652	0.603
		Qwen2		0.648	0.657
		Mistral		0.547	0.551
		RoBERTa		-	0.922

types. For this reason, individual LLMs might perform well enough such that it is more difficult to identify an ensemble configuration that complements knowledge

Table 4. Performances of ensembles omitting different steps from the pipeline.

Dataset	No disambiguation	No voting
HAREM	0.592	0.410
LeNER-BR	0.541	0.478
GeoCorpus2	0.271	0.178
UlyssesNER	0.351	0.249
MariNER	0.705	0.621

gaps like the previous sets. Though the ensemble performs better for MariNER, it is only by a small margin, with Gemma2 having a drop in performance of only 0.01 in micro-F1. Even though the dataset belongs to a historical domain, its entity types are still relatively simple compared to the more complex types of the other sets. It may explain the small performance gap between the ensemble and the individual LLMs.

Regarding the composition of the ensembles, only GeoCorpus2 shows a configuration with more than one LLM performing extraction, while all others have either LLaMA3 or Gemma2 as the only extractors. Looking at the models' individual performances, we see that LLaMA3 and Gemma2 are high-performing for most of the datasets and, given the importance of the extraction step (i.e., the subsequent steps may be interpreted as filtering of the extracted entities), it is to be expected that the best individual models are often used as extractors. Interestingly, Phi3 also performed well in most of the individual cases, being the second-best individual option for all sets except LeNER-BR. However, it did not show up as an extractor for any of the best ensemble configurations. Instead, Phi3 shows up for most of the voting and disambiguation ensembles, which indicates that it might not be a strong extractor but a reliable verifier of extracted entities.

5.2 Ablation

An ablation study is conducted to verify the effectiveness of the different steps of the pipeline. Table 4 shows the performance of the ensembles obtained in the main results when either the voting step is omitted, or the LLM disambiguation step is simplified by always choosing the largest option from the multiple-choice prompt.

The performance is similar for most datasets when using the simplified version of the disambiguation step. For HAREM, the performance is slightly better than that of using LLMs for disambiguation, while for LeNER-BR, UlyssesNER, and MariNER, the performance is only slightly lower. For GeoCourpus2, however, there is a considerable performance drop when not using the ensemble's output for this step. It indicates that using the largest option might be a suitable simplification of the pipeline, reducing the computational cost by reducing the overall amount of prompts processed. Nevertheless, GeoCorpus2's performance drop also shows that this extra prompting might be important in some settings.

The voting step, on the other hand, proved to be important to the ensemble's performance, as suppressing this step consistently lowers the performance of the final ensemble, with considerable drops in micro-F1 scores across all datasets. This may be explained by the disambiguation step becoming the only step for entity filtering, which can be difficult to perform in the multiple-choice format. Moreover, without the voting step, noisy entity mentions show up more frequently in the disambiguation prompts, making the disambiguation more difficult for the LLMs to process.

5.3 Effectiveness of The Heuristic

Currently, an ensemble is chosen based on the performance of all combinations on a small annotated sample of the whole dataset, the validation set. However, Table 3 shows that the ensembles' performances on the validation set differ from the full test set. For that reason, we perform a study to verify the effectiveness of the chosen heuristic in selecting an ensemble configuration based on the small validation set. For this, all possible combinations of ensembles were tested for the entire test set of the HAREM dataset. HAREM was chosen here to analyze further the case where the ensemble underperformed against single LLM scenarios. Figure 3 compares the different configurations, showing the individual LLMs performances, the best configuration obtained through the heuristic, and the overall best ensemble.

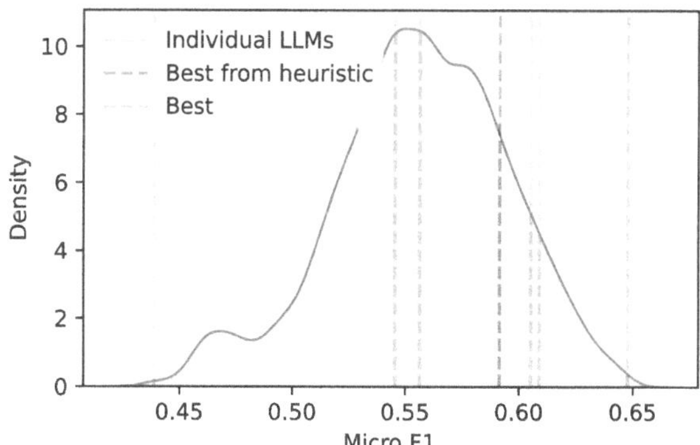

Fig. 3. KDE plot of performances of all ensemble combinations for HAREM's test set. Performances for the individual LLMs, the best ensemble obtained from heuristics, and the best overall are highlighted.

Most ensemble configurations have a performance that falls close to 0.55, with two individual models (LLaMA3 and Qwen2) falling near that value. Interestingly, the best configuration possible, achieved by using Gemma2 for extraction,

Phi3 for voting, and LLaMA3 disambiguation, achieves a performance greater than that of the ensemble obtained through the heuristic and the one obtained by Gemma2, the best individual LLM setting. It shows that the heuristic may not be the most effective way of obtaining the ensemble configuration or, at the very least, is highly dependent on the validation set size and its representativeness of the dataset as a whole.

5.4 Cross Dataset Ensembles

We aim to assess how ensembles obtained from one dataset using the heuristic perform on another dataset by evaluating the ensemble pipeline in a scenario where no annotated data is available for the current task. The micro-F1 performances of ensembles obtained in a source dataset and tested in a different target set are shown in Fig. 4.

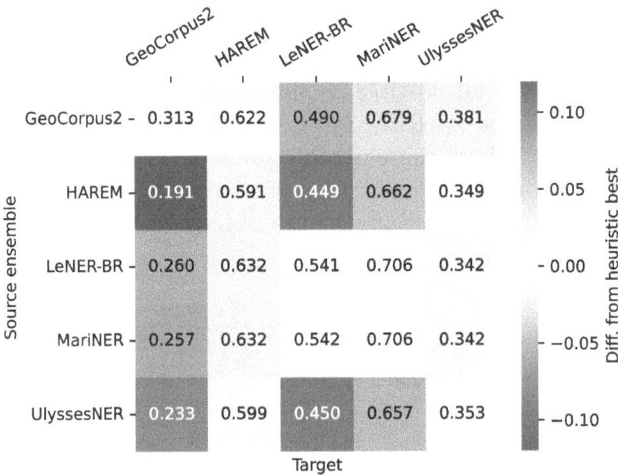

Fig. 4. Cross dataset ensemble performances. Values are absolute micro-F1 performances, colours represent distance from the best configurations shown in Table 3.

The results show that using a different source dataset to obtain an ensemble configuration may perform better than any single LLM for most datasets. GeoCorpus2 is the only dataset in which this does not hold, with the second-best performing ensemble, obtained from LeNER-BR, still slightly under-performing Gemma2 in the test set. Notably, for both HAREM and UlyssesNER, it was possible to get an ensemble configuration that was better than the one obtained from the heuristic applied to the same dataset, which shows the capabilities of the ensemble pipeline of achieving high performances even in completely zero-shot settings, with no annotated data available for the current task.

6 Conclusion and Future Work

This study introduces a method to leverage multiple LLMs in ensembles to perform zero-shot named entity recognition for the Portuguese language. The pipeline is divided into three main steps: extraction, voting, and disambiguation. We evaluate this pipeline in five distinct NER Portuguese datasets, spanning multiple domains and annotation granularities. By evaluating the performance of all combinations of five unique LLMs across these steps for a small sample of labeled data, we can select a suitable ensemble configuration that performs better than any single LLM being used. We also show that ensemble configurations obtained in cross-dataset configurations may perform better than individual LLMs for four of the five datasets, eliminating the need for prior annotated data.

In future work, we aim to explore NER ensembling for languages other than Portuguese, as our pipeline is not dependent on this language. Other methods for aggregating the outputs of the LLM models may also be explored. One possibility would be to train a meta-classifier to aggregate the answers of the models similarly to what is usually done with ensemble stacking. Another alternative is to explore the use of larger models as judges in specific steps of the pipeline to use their processing capabilities to make final decisions.

7 Limitations

The use of LLM ensembles entails the instantiation of multiple GPU-intensive models for a single task. Naturally, this comes at the cost of high GPU memory requirements for parallel instantiation or long processing times to process the pipeline sequentially. Moreover, as we aim to obtain an optimal configuration of LLMs for each step of the pipeline, processing must be done for over 7,000 combinations to find the best-performing one for the small train set. To address this, we make heavy use of caching of previous answers for step 3 of the pipeline, but the processing time for all of these combinations still presents itself as a limiting factor for the proposed pipeline.

Acknowledgements. This study was financed in part by the Coordenação de Aperfeiçoamento de Pessoal de Nível Superior – Brasil (CAPES) – Finance Code 001.

References

1. Abdin, M., Aneja, J., Awadalla, H., Awadallah, A., Awan, A.A., et al.: Phi-3 technical report: a highly capable language model locally on your phone (2024). https://arxiv.org/abs/2404.14219
2. Albuquerque, H.O., et al.: UlyssesNER-Br: a corpus of brazilian legislative documents for named entity recognition. In: Pinheiro, V., et al. (eds.) PROPOR 2022. LNCS, vol. 13208, pp. 3–14. Springer, Cham (2022). https://doi.org/10.1007/978-3-030-98305-5_1

3. Chen, L., Zaharia, M., Zou, J.: FrugalGPT: how to use large language models while reducing cost and improving performance (2023). https://doi.org/10.48550/arXiv.2305.05176
4. Consoli, B., Santos, J., Gomes, D., Cordeiro, F., Vieira, R., Moreira, V.: Embeddings for named entity recognition in geoscience Portuguese literature. In: Calzolari, N., et al. (eds.) Proceedings of the Twelfth Language Resources and Evaluation Conference, pp. 4625–4630. European Language Resources Association, Marseille (2020). https://aclanthology.org/2020.lrec-1.568/
5. Devlin, J., Chang, M.W., Lee, K., Toutanova, K.: BERT: pre-training of deep bidirectional transformers for language understanding (2019). https://doi.org/10.48550/arXiv.1810.04805
6. Farr, D., Manzonelli, N., Cruickshank, I., Starbird, K., West, J.: LLM chain ensembles for scalable and accurate data annotation. In: 2024 IEEE International Conference on Big Data (BigData), pp. 2110–2118. IEEE, Washington (2024). https://doi.org/10.1109/BigData62323.2024.10826109, https://ieeexplore.ieee.org/document/10826109/
7. Garcia, G.L., et al.: Introducing bode: a fine-tuned large language model for Portuguese prompt-based task (2024). https://arxiv.org/abs/2401.02909
8. Grattafiori, A., et al.: The LLaMA 3 herd of models (2024). https://doi.org/10.48550/arXiv.2407.21783
9. Jia, J., Liang, W., Liang, Y.: A review of hybrid and ensemble in deep learning for natural language processing (2024). https://doi.org/10.48550/arXiv.2312.05589, arXiv:2312.05589 [cs]
10. Jiang, A.Q., et al.: Mistral 7b (2023). https://arxiv.org/abs/2310.06825
11. Jiang, A.Q., et al.: Mixtral of experts (2024). https://arxiv.org/abs/2401.04088
12. Jiang, D., Ren, X., Lin, B.Y.: LLM-blender: ensembling large language models with pairwise ranking and generative fusion (2023). https://doi.org/10.48550/arXiv.2306.02561, arXiv:2306.02561 [cs]
13. Li, Y., et al.: Improving entity recognition using ensembles of deep learning and fine-tuned large language models: a case study on adverse event extraction from VAERS and social media. J. Biomed. Inform. **163**, 104789 (2025). https://doi.org/10.1016/j.jbi.2025.104789
14. Liu, Y., et al.: RoBERTa: a robustly optimized BERT pretraining approach (2019). https://arxiv.org/abs/1907.11692
15. Luz de Araujo, P.H., de Campos, T.E., de Oliveira, R.R.R., Stauffer, M., Couto, S., Bermejo, P.: LeNER-Br: a dataset for named entity recognition in Brazilian legal text. In: Villavicencio, A., et al. (eds.) PROPOR 2018. LNCS (LNAI), vol. 11122, pp. 313–323. Springer, Cham (2018). https://doi.org/10.1007/978-3-319-99722-3_32
16. Nunes, R.O., et al.: An ensemble of LLMs finetuned with LoRA for NER in Portuguese legal documents. In: Paes, A., Verri, F.A.N. (eds.) BRACIS 2024. LNCS, vol. 15412, pp. 127–140. Springer, Cham (2025). https://doi.org/10.1007/978-3-031-79029-4_9
17. Polikar, R.: Ensemble based systems in decision making. IEEE Circuits Syst. Mag. **6**(3), 21–45 (2006). https://doi.org/10.1109/MCAS.2006.1688199
18. Santos, D., Seco, N., Cardoso, N., Vilela, R.: HAREM: an advanced NER evaluation contest for Portuguese. In: Calzolari, N., et al. (eds.) Proceedings of the Fifth International Conference on Language Resources and Evaluation (LREC 2006). European Language Resources Association (ELRA), Genoa (2006). https://aclanthology.org/L06-1027/

19. Sarcinelli, J.L.L.L., Teixeira, M.L.G., de Paiva, J.B., Silva, D.F.: Mariner: a dataset for historical Brazilian Portuguese named entity recognition (2025). https://arxiv.org/abs/2506.23051
20. Team, G., et al.: Gemma 2: improving open language models at a practical size (2024). https://doi.org/10.48550/arXiv.2408.00118
21. Wang, S., et al.: GPT-NER: named entity recognition via large language models (2023). https://doi.org/10.48550/arXiv.2304.10428
22. Xie, T., Li, Q., Zhang, J., Zhang, Y., Liu, Z., Wang, H.: Empirical study of zero-shot NER with ChatGPT (2023). https://doi.org/10.48550/arXiv.2310.10035
23. Yang, A., et al.: Qwen2 technical report (2024). https://doi.org/10.48550/arXiv.2407.10671
24. Zhao, W.X., et al.: A survey of large language models (2025). https://doi.org/10.48550/arXiv.2303.18223

Multilingual Visual Understanding: Extending Visual Dialog to Portuguese and Spanish Through Cross-Modal Adaptation

Milena Menezes Adão, Silvio Jamil F. Guimarães, and Zenilton K. G. Patrocínio Jr(✉)

Pontifícia Universidade Católica de Minas Gerais, Belo Horizonte, Brazil
milena.adao@sga.pucminas.br, {sjamil,zenilton}@pucminas.br

Abstract. Visual Dialog represents a fundamental challenge in computer vision, requiring AI agents to answer questions about images while maintaining coherent conversational context through dialog history. This muand **VisDial-ES**, the first Visual Dialog datasets for Brazilian Portuguese and Spanish, containing approximately 1.2 million dialog question-answer pairs each about image-grounded conversations. Beyond dataset creation, we introduce a systematic pipeline that enables development of Visual Dialog resources across multiple languages, facilitating broader cross-lingual research in multimodal understanding. To establish baseline performance and validate dataset quality, we evaluated both datasets using state-of-the-art models. Our contributions enable inclusive AI development that effectively serves Portuguese and Spanish speaking communities worldwide while establishing methodological foundations for extending Visual Dialog capabilities to other languages. VisDial-PT and VisDial-ES datasets and the resources are available at https://github.com/IMScience-PPGINF-PucMinas/multilingual-visual-dialog.

Keywords: Visual Dialog · Multilingual AI · Cross-Modal Understanding · Portuguese · Spanish

1 Introduction

Multilingual computer vision tasks have gained increasing attention, with works like M3P [24] and mBERT [8] providing foundational multilingual representations. However, most multilingual vision works focus on tasks like image captioning [28], visual question answering [1], and others.

Visual Dialog represents a fundamental challenge in computer vision and natural language processing, where intelligent agents must interpret visual scenes and engage in contextual conversations about image content [7]. This multimodal task demands sophisticated integration of visual perception and linguistic reasoning capabilities, and cross-modal alignment between visual features

D. Chaves et al. (Eds.): CIARP 2025, LNCS 16528, pp. 222–235, 2026.
https://doi.org/10.1007/978-3-032-23161-1_16

and natural language representations [12,19,23,32,36]. The core challenge lies in establishing robust vision-language correspondences that enable models to ground linguistic concepts in visual representations while maintaining coherent dialog flow. Recently, a series of approaches have been proposed to improve performance in Visual Dialog, such as HCIAE [22], CoAtt [34], CorefNMN [18], ReDAN [10], RvA [25], Primary [11], DAM [16], KBGN [15], DMRM [5], VD-BERT [33], MITVG [4], UTC [3] and GST [17].

However, existing research has predominantly focused on English-language scenarios; in other words, the Visual Dialog task remains unexplored in multilingual contexts. Portuguese and Spanish, spoken by over 260 million [31] and 500 million people worldwide respectively [13], present unique linguistic characteristics that can reveal important insights about vision-language alignment. Portuguese's rich morphological structure and Spanish's distinct grammatical features, offer ideal testbeds for evaluating the robustness of Visual Dialog systems across linguistic boundaries.

The Brazilian Portuguese language model has seen significant development with models like BERTimbau [29], while Spanish has benefited from models like BETO [2] and multilingual BERT variants. However, multimodal applications in both languages remain limited, with most work focusing on text-only tasks [9,26]. Our work addresses this gap by **introducing Visual Dialog capabilities for both Brazilian Portuguese and Spanish by considering multimodal applications in these languages**. The challenge of cross-lingual visual understanding extends beyond simple translation. Different languages may employ distinct linguistic structures to describe visual scenes, potentially affecting how models attend to visual regions and establish vision-language correspondences. Languages like Portuguese and Spanish share certain structural similarities but also exhibit important differences in spatial relation expressions, verb conjugations, and cultural visual references that could impact multimodal understanding. In this work, we introduce **VisDial-PT** and **VisDial-ES**, the first Brazilian Portuguese and Spanish Visual Dialog datasets, translated from the original VisDial 1.0 [7], enabling systematic investigation of multilingual visual understanding across languages. Our contribution extends beyond mere translation to establish a comprehensive framework for multilingual Visual Dialog research, more specifically, a **dataset creation** and **practical guidance for deploying Large Vision Language Models in multilingual scenarios**.

The primary contribution of this work lies in the creation of both VisDial-PT and VisDial-ES datasets. These datasets address a critical gap in multilingual computer vision research, particularly for Portuguese and Spanish speakers who collectively represent over 750 million people globally. The creation of these resources enables systematic investigation of cross-lingual visual understanding capabilities and provides foundational resources for developing inclusive AI systems. A significant methodological contribution involves the development of a systematic, semi-automated translation pipeline that has been validated across two distinct languages. This pipeline preserves the original dataset structure and metadata format while being adaptable to different linguistic characteristics. The methodology demonstrates scalability and can be extended to create Visual Dialog datasets in other languages, establishing a foundation for broader multilingual research in this domain.

The work provides rigorous validation of translation quality through comprehensive human evaluation, achieving high scores across critical linguistic dimensions for both languages: Portuguese - Fluency (2.84 of 3.0), Adequacy (2.89 of 3.0), and Coherence (2.86 of 3.0); Spanish - Fluency (2.91 of 3.0), Adequacy (2.93 of 3.0), and Coherence (2.88 of 3.0). More significantly, we demonstrate performance parity across languages using Large Vision Language Models, providing strong empirical evidence that semantic integrity and visual-linguistic relationships are successfully preserved across languages.

This work contributes to practical guidance for deploying Large Vision Language Models in multilingual scenarios through the development of optimized prompting strategies for different languages. We validate Portuguese dataset through comprehensive experiments using the Generative Self-Training (GST) framework [17], adapted with BERTimbau [29] language understanding. Furthermore, we evaluated both VisDial-PT and VisDial-ES with LLaVA-7B [21], to ensure that both datasets preserve contextual integrity, linguistic fluency, and proper grounding in the visual content.

2 Visual Dialog Dataset and Models

Visual Dialog (VisDial) was introduced by [7] as a multimodal task requiring agents to answer a sequence of questions about an image, considering both the current question and the dialog history. It extends the scope of Visual Question Answering (VQA) [1], enabling more coherent and context-aware conversational interactions. The original VisDial v1.0 dataset comprises dialogs grounded in more than 130,000 images sourced from the COCO [20] and Flickr [35] datasets. Each image is associated with a single dialog consisting of an image caption followed by 10 question-answer pairs. The training set includes dialogs for approximately 120,000 images, while the validation and test sets are constructed from an additional 10,000 images.

The core task in Visual Dialog is defined as follows: **given an image I, a dialog history H, and a follow-up question Q_t, the dialog history is composed of a sequence of previous question-answer pairs**, such as: Q_1: *How many people are in sofa?* – A_1: *Two*; Q_2: *What are their genders?* – A_2: *One male and one female*. Given this context and a subsequent question (*e.g.*, Q_3: *Which one is holding a ball?*), the main goal here is to generate a coherent and contextually grounded natural language response (*e.g.*, A_3: *The woman*). This setup requires the model to reason jointly over the visual content, the dialog context, and the current question.

The Generative Self-Training (GST) framework [17] introduced a semi-supervised paradigm where synthetic dialogs are generated by a teacher-questioner pair and used to train a student model, effectively expanding the training data with unlabeled images. GST achieved state-of-the-art performance on VisDial benchmarks. For Portuguese language modeling, BERTimbau [29] was proposed as a transformer-based encoder pretrained specifically for Brazilian Portuguese.

To the best of our knowledge, no prior work has attempted to create or evaluate Portuguese and Spanish versions of the VisDial dataset. Our contribution fills this gap by introducing VisDial-PT and VisDial-ES and validating their use in training Visual Dialog models.

3 Visual Dialog Evaluation Metrics

The evaluation of Visual Dialog systems requires specialized metrics that capture the quality of answer selection and ranking in multimodal conversational contexts. This section presents some metrics – for instance, Mean Reciprocal Rank, Recall@k and Normalized Discounted Cumulative Gain – detailing their definitions, interpretations, and specific applications in the Visual Dialog task.

3.1 Mean Reciprocal Rank (MRR)

The Mean Reciprocal Rank (MRR) [6] is a fundamental metric for evaluating ranking systems and has been widely adopted in Visual Dialog evaluation [7, 17]. In the context of Visual Dialog, MRR measures how well a model ranks the ground truth response among a set of candidate responses. This metric is particularly relevant for the Visual Dialog task, where models must identify the most appropriate response from a pool of 100 candidate answers. For Visual Dialog evaluation, MRR is calculated as the average of the reciprocals of the ranks assigned to the ground-truth responses across all dialog turns. In Visual Dialog, MRR values closer to 1 indicate that the model consistently assigns high rank to correct responses, demonstrating strong discriminative capability.

3.2 Recall@k

Recall@k [27] measures the fraction of ground-truth responses in the top k positions of the ranked candidate list. Common values of k used in Visual Dialog evaluation include $k \in \{1, 5, 10\}$, corresponding to Recall@1, Recall@5, and Recall@10.

3.3 Normalized Discounted Cumulative Gain (NDCG)

The Normalized Discounted Cumulative Gain (NDCG) [14] is a widely used metric for evaluating the quality of ranked retrieval systems. NDCG measures how well the predicted ranking of items aligns with an ideal ranking, taking into account the position of relevant items in the list. The resulting score ranges from 0 to 1, with higher values indicating better alignment with the ideal ranking.

4 Multilingual Visual Dialog Dataset Creation

We present **VisDial-PT** and **VisDial-ES**, Brazilian Portuguese and Spanish versions of the VisDial v1.0 dataset. Our goal was to preserve the original structure comprising image captions, dialog histories, and question-answer (QA) pairs, while ensuring high linguistic quality in the translated content for both target languages.

4.1 Translation Pipeline

We developed a systematic, semi-automated translation pipeline applicable to both Portuguese and Spanish, structured in the following steps:

Automatic translation: We used OPUS-MT-TC-BIG-EN-PT [30] for Portuguese and OPUS-MT-EN-ES [30] for Spanish to translate each caption, question, and answer individually.

Human validation: Random samples from both datasets were manually reviewed by native speakers to assess fluency, adequacy, and coherence using a 3-point scale.

Cross-linguistic consistency: We implemented additional validation steps to ensure consistent treatment of visual concepts across both target languages, particularly for culture-specific references.

Applicability evaluation: To assess the applicability of both translated datasets, we applied the GST framework adapted for Portuguese dataset and for each language we evaluated performance using LLaVA-7B [21].

We have decided to preserve the original file structure and metadata format of VisDial v1.0, ensuring full compatibility with existing models and tools.

4.2 Dataset Statistics

Both VisDial-PT and VisDial-ES maintain the original structure of VisDial v1.0:

Training set: ~120,000 images with dialogs
Validation set: 2,000 images
Test set: 8,000 images
Total question-answer pairs per language: ~1.2 million

Caption (EN): the group of people are playing soccer on the field.

Caption (PT): o grupo de pessoas está jogando futebol no campo.

Caption (ES): el grupo de personas está jugando al fútbol en el campo.

Dialog

Q1 (EN): do they look like they're having fun?
Q1 (PT): eles parecem que estão se divertindo?
Q1 (ES): ¿Se ven como si se estuvieran divirtiendo?

A1 (EN): hard to tell, they are in the middle of play it's a game with 2 different teams.
A1 (PT): difícil dizer, eles estão no meio do jogo é um jogo co equipes diferentes.
A1 (ES): difícil de decir, están en el medio del juego es un jue con 2 equipos diferentes.

Q2 (EN): are they wearing uniforms?
Q2 (PT): eles estão usando uniformes?
Q2 (ES): ¿Están vestidos con uniformes?

A2 (EN): yes, each has a different uniform on.
A2 (PT): sim, cada um tem um uniforme diferente.
A2 (ES): sí, cada uno tiene un uniforme diferente.

⋮

Fig. 1. Example of translation results for a VisDial instance (English to Brazilian Portuguese and Spanish). Q, A, and C refer to question, answer, and caption, respectively.

5 Results and Analysis

In this section, we present the results obtained from each stage of both VisDial-PT and VisDial-ES pipelines. We report and analyze outcomes related to the automatic translation process, language validation, and manual inspection, as well as the downstream evaluation performed using state-of-the-art models.

5.1 Translation Quality Assessment

To ensure the linguistic fidelity of both datasets, we conducted qualitative and quantitative analyses of the translated content. Figure 1 provides dialog examples showing original English and their corresponding translations in both Brazilian Portuguese and Spanish, demonstrating our efforts to preserve dialog flow and natural phrasing in both target languages.

5.2 Human Evaluation

We conducted a manual evaluation on 100 randomly selected examples for each language by native speakers using a 3-point scale on three criteria:

Fluency – grammatical correctness and naturalness of the translation;
Adequacy – preservation of original meaning; and
Coherence – consistency with the dialog context.

Table 1 illustrates an example of this evaluation process.

Table 1. Examples of VisDial translation evaluations: high, middle, and low scoring.

English Text	Portuguese Translation	Type	Fluency	Adequacy	Coherence
Are there people going into the buildings?	Há pessoas entrando nos edifícios?	Question	2	2	2
What is wallpaper like?	o que é papel de parede como	Question	1	1	1
Black with white Mickey logos	Preto com logotipos do Mickey brancos	Answer	3	3	3
a group of people in a room with remotes	um grupo de pessoas em uma sala com controles remotos	Caption	3	3	3

Portuguese (VisDial-PT): Fluency=2.84; Adequacy=2.89; Coherence=2.86

Spanish (VisDial-ES): Fluency=2.91; Adequacy=2.93; Coherence=2.88
Both datasets achieved high-quality translation scores, with Spanish showing slightly higher scores, potentially due to closer structural similarity to English in certain constructions.

5.3 Prompt Variation Analysis

To evaluate the impact of prompt complexity on model performance, we conducted experiments using two distinct prompting strategies with LLaVA-7B (see Table 2): (i) a detailed version with explicit step-by-step instructions, and (ii) a simplified version closer to natural conversation, requiring only the selection of the most appropriate option.

The first prompt (detailed) providing strict instructions on how the model should respond:

```
Analyze the image and answer the question. ANSWER JUST THE
SELECTED OPTION VALUE. NO MORE TEXT. Dialogue context:
{context}
Question: {question}? Answer options: {answer options}
```

The second prompt (simplified) providing only the essential instructions for choosing the best answer. It uses language closer to human conversation and is less restrictive:

```
Observe the image. Context: {context}. Question: {question}.
Options:{answer options}. Choose the most correct option.
Respond only with the letter:
```

During the evaluation, these prompts were translated into the corresponding language.

Table 2. Performance comparison across languages and prompt strategies on VisDial v1.0 validation set.

Model	Language	Prompt	NDCG	MRR	R@1	R@5	R@10
LLaVA-7B	English	Detailed	70.8	73.2	60.0	**91.8**	**96.0**
LLaVA-7B	English	Simplified	**75.5**	**84.2**	**82.7**	83.7	85.9
LLaVA-7B	Portuguese	Detailed	67.2	71.1	57.41	**88.52**	90.9
LLaVA-7B	Portuguese	Simplified	**72.3**	**76.5**	**70.4**	84.8	**97.4**
LLaVA-7B	Spanish	Detailed	10.4	12.2	7.4	13.0	15.3
LLaVA-7B	Spanish	Simplified	**85.3**	**88.4**	**96.4**	**96.4**	**98.6**

This comparative analysis enables us to investigate how prompt engineering strategies impact cross-lingual performance and whether simplified prompts retain effectiveness across different languages. The results demonstrate the importance of prompt design in multilingual Visual Dialog systems and provide insights into the robustness of large vision-language models across different linguistic contexts and instruction complexity levels. The detailed prompt performed much worse in Spanish. After manually analyzing the results, it became clear that the model often provided the correct answer; however, due to the instructions, it was unable to return the exact expected response–an important point that automated metrics often fail to capture. For example:

Model Response: sí, es una foto a todo color.
Expected Answer: sí

5.4 Evaluation via LLaVA-7B Large Vision Language Models

To ensure the quality of our dataset, we compared our best results obtained with the original dataset (English), the two datasets we created (VisDial-PT and VisDial-ES), and also the state-of-the-art model, GST. All were evaluated using their respective validation datasets. It is worth noting that GST used the original VisDial dataset.

Table 3 presents a comprehensive comparison of Visual Dialog performance across three languages using our translated datasets. The evaluation using LLaVA-7B reveals significant cross-linguistic performance variations, with Spanish achieving the highest metrics (NDCG: 85.3%, MRR: 88.4%, R@1: 96.4%), followed by English (NDCG: 75.5%, MRR: 84.2%, R@1: 82.7%), and Portuguese (NDCG: 72.3%, MRR: 76.5%, R@1: 70.4%). These performance differences highlight the critical importance of prompt engineering in multilingual visual dialog systems, as the same underlying model architecture responds differently to language-specific prompting strategies. The superior performance in Spanish likely reflects an optimal alignment between our Spanish prompt formulation and the model's internal representations, while the variation across languages demonstrates how prompt design can significantly impact multimodal reasoning capabilities. The consistent performance patterns across all three datasets

Table 3. Performance comparison across languages on VisDial v1.0 validation set.

Model	Language	NDCG	MRR	R@1	R@5	R@10
GST [17]	English	65.47	53.19	43.98	64.96	71.51
LLaVA-7B	English	75.5	84.2	82.7	83.7	85.9
LLaVA-7B	Portuguese	72.3	76.5	70.4	84.8	97.4
LLaVA-7B	Spanish	**85.3**	**88.4**	**96.4**	**96.4**	**98.6**

validate the quality and integrity of our translation methodology, as the models successfully engage with visual content through our translated question-answer pairs without degradation in comprehension. The fact that all language variants maintain coherent visual reasoning capabilities confirms that our datasets preserve the essential semantic relationships between visual elements and linguistic expressions, enabling meaningful cross-lingual evaluation. This prompt dependency suggests that achieving optimal cross-lingual performance requires careful language-specific prompt optimization rather than direct translation of prompts. All three language variants substantially outperform the GST baseline, confirming both the effectiveness of large vision-language models and the robustness of our multilingual datasets for cross-lingual visual dialog research.

5.5 Evaluation via Generative Self-Training (GST)

We hypothesize that the teacher paradigm proposed in GST can be successfully reapplied to VisDial. It is worth mentioning that we have developed a case study only for VisDial-PT for enabling the generation of synthetic Brazilian Portuguese data and training a competitive dialog agent even in scenarios with limited data. To evaluate the quality and applicability of the translated dataset, we adopted the Generative Self-Training (GST) framework [17]. We replicated the original GST training pipeline using our VisDial-PT. Specifically, we trained a Teacher model using the translated data. To adapt the model to the Portuguese language, we replaced the original English BERT encoder with the BERTimbau [29], a pre-trained language model specifically developed for Brazilian Portuguese. This modification ensured that the language understanding components of the model aligned with the linguistic characteristics of the translated dialogue. The performance of the Portuguese trained model was evaluated using the normalized discount cumulative gain (NDCG).

This result was compared to the English baseline to verify the preservation of dialog quality and task effectiveness after translation. The results are shown in Table 4, which presents a direct comparison of the NDCG scores achieved by the GST-based Teacher model trained on the original English VisDial dataset (Teacher-En) and its counterpart trained on the Brazilian Portuguese translation (Teacher-PT). While both models follow the same architecture, it is important to note that the Teacher-En and Teacher-PT models were not trained under strictly identical conditions. Differences in the number of epochs, batch sizes,

Table 4. Comparison of Visual Dialog performance between the original English dataset and the translated Brazilian Portuguese dataset using the GST framework.

Model	NDCG
Teacher-En [17]	**0.6**
Teacher-PT [Ours]	**0.18**

and other hyperparameters may have influenced the observed NDCG scores. A performance drop is observed when moving from English (NDCG = 0.6) to Portuguese (NDCG = 0.18).

Despite the discrepancy in results between the models trained in English and Portuguese, partially due to differences in the number of training epochs, hyperparameters, and overall setup, a promising trend can be observed for the Portuguese version. Even under non-controlled conditions, the performance of the Portuguese model remains competitive, suggesting the feasibility of adapting Visual Dialog models to multilingual scenarios. Future work will aim to normalize training conditions across languages to enable a more rigorous and fair cross-lingual comparative evaluation.

6 Discussion and Analysis

6.1 Multilingual Performance Patterns

Our evaluation reveals interesting patterns in multilingual Visual Dialog performance:

- **Language Advantage:** Both Portuguese and Spanish showed relatively similar performance degradation patterns compared to English, suggesting shared linguistic structures may facilitate cross-lingual transfer.
- **Translation Quality Impact:** The slightly superior performance of Spanish across multiple metrics correlates with marginally higher human evaluation scores, indicating the importance of translation quality for downstream performance.
- **Cultural Universality:** The COCO dataset's visual content proved largely culture-agnostic, requiring minimal adaptation for Portuguese and Spanish contexts.

6.2 Methodological Insights

The successful application of our pipeline to two distinct languages demonstrates its generalizability. Key insights include:

- Automatic translation quality varies by language pair complexity
- Human validation remains crucial for nuanced conversational contexts
- Cross-linguistic consistency checks improve overall dataset quality

6.3 Ethical Considerations and Bias

The creation of VisDial-PT and VisDial-ES also raises ethical aspects that deserve explicit discussion. Automatic translation may introduce subtle biases or inaccuracies, potentially affecting the semantic fidelity of dialogs. Moreover, the reliance on COCO images limits cultural diversity, since these visual contexts are not fully representative of Portuguese and Spanish speaking societies. Another limitation concerns dialectal and regional variants: our datasets focus on Brazilian Portuguese and standard Spanish, which may not cover all linguistic nuances. Recognizing these constraints is essential to guide responsible use of the datasets and to encourage future extensions that mitigate such biases through culturally representative data collection and inclusion of regional language varieties.

7 Limitations

Current limitations include:

- **Regional Variants:** Focus on Brazilian Portuguese and Spanish limits applicability to other regional variants
- **Cultural Context:** The COCO dataset, while widely adopted, may not fully represent the cultural diversity of Portuguese and Spanish speaking communities. This lack of cultural representativeness could influence the naturalness and contextual adequacy of dialog responses. Future works should consider the development of culturally enriched visual datasets, including images reflecting daily life, traditions, and social contexts of these communities.
- **Performance Gaps:** The discrepancy observed in GST results (NDCG = 0.18 for Portuguese vs. 0.6 for English) can be partially attributed to the fact that training was not conducted under strictly controlled hyperparameters. Variations in batch size, number of epochs, and optimization settings likely influenced the outcome. These results should therefore be interpreted as preliminary, and further controlled experiments are required to provide a more reliable assessment in future work.

8 Conclusion and Future Work

We present VisDial-PT and VisDial-ES, the first Visual Dialog datasets for Portuguese and Spanish, representing a significant milestone in multilingual visual understanding research. Our systematic translation pipeline demonstrates scalability across languages while maintaining high linguistic quality. The successful creation of both datasets through our methodology proves the feasibility of extending Visual Dialog to non-English languages while preserving semantic integrity and visual coherence. The human evaluation results confirm high translation quality across critical linguistic dimensions, demonstrating that our methodology maintains the essential characteristics required for effective Visual Dialog datasets.

The comparable performance across languages using LLaVA-7B validates the quality of our translations and demonstrates the potential for multilingual Visual Dialog systems. These resources enable inclusive AI development for Portuguese and Spanish speaking communities while establishing methodological foundations for extending Visual Dialog capabilities to additional languages.

Future work will explore: 1. Extension to additional languages (Italian, French, Romanian); 2. Video-based multilingual dialog systems; 3. Culture-specific visual datasets for more targeted evaluation; 4. Advanced cross-lingual training strategies for improved GST performance.

Our work opens new research avenues in multilingual visual understanding and provides the foundation for developing truly global Visual Dialog systems that can serve diverse linguistic communities effectively.

Acknowledgments. The authors thank the Pontifícia Universidade Católica de Minas Gerais – PUC-Minas, Coordenação de Aperfeiçoamento de Pessoal de Nível Superior – CAPES – (CAPES STIC-AMSUD 88887.878869/2023-00, 23-STIC-10 and Finance Code 001), the Conselho Nacional de Desenvolvimento Científico e Tecnológico – CNPq (Grants 407242/2021-0, 306573/2022-9, 442950/2023-3) and Fundação de Apoio à Pesquisa do Estado de Minas Gerais – FAPEMIG (Grant APQ-01079-23 and APQ-05058-23).

Disclosure of Interests. The authors have no competing interests to declare that are relevant to the content of this article.

References

1. Antol, S., et al.: VQA: visual question answering. In: Proceedings of the IEEE International Conference on Computer Vision (ICCV) (2015)
2. Cañete, J., Chaperon, G., Fuentes, R., Ho, J.H., Kang, H., Pérez, J.: Spanish pre-trained BERT model and evaluation data. In: PML4DC at ICLR 2020 (2020)
3. Chen, C., et al.: UTC: a unified transformer with inter-task contrastive learning for visual dialog. In: Proceedings of the IEEE/CVF Conference on Computer Vision and Pattern Recognition (CVPR), pp. 17408–17418 (2022). https://openaccess.thecvf.com/content/CVPR2022/html/Chen_UTC_A_Unified_Transformer_With_Inter-Task_Contrastive_Learning_for_Visual_CVPR_2022_paper.html
4. Chen, F., Meng, F., Chen, X., Li, P., Zhou, J.: Multimodal incremental transformer with visual grounding for visual dialogue generation. In: ACL (2021)
5. Chen, F., Meng, F., Xu, J., Li, P., Xu, B., Zhou, J.: DMRM: a dual-channel multi-hop reasoning model for visual dialog. In: AAAI (2020)
6. Craswell, N.: Mean Reciprocal Rank, p. 1703. Springer, Boston (2009). https://doi.org/10.1007/978-0-387-39940-9_488
7. Das, A., et al.: Visual dialog. In: Proceedings of the IEEE Conference on Computer Vision and Pattern Recognition (CVPR) (2017)
8. Devlin, J., Chang, M.W., Lee, K., Toutanova, K.: BERT: pre-training of deep bidirectional transformers for language understanding. arXiv preprint arXiv:1810.04805 (2019)

9. Frediani, J.A.O.R.F., Garcia, G.L., Paiola, P.H., Passos, L.A., Papa, J.A.P., Marana, A.N.: Hate speech detection in Portuguese using BERTimbau. In: Hernández-García, R., Barrientos, R.J., Velastin, S.A. (eds.) CIARP 2024, Part I. LNCS, vol. 15368, pp. 244–255. Springer, Heidelberg (2024). https://doi.org/10.1007/978-3-031-76607-7_18
10. Gan, Z., Cheng, Y., El Kholy, A., Li, L., Liu, J., Gao, J.: Multi-step reasoning via recurrent dual attention for visual dialog. In: ACL (2019)
11. Guo, D., Xu, C., Tao, D.: Image-question-answer synergistic network for visual dialog. In: CVPR (2019)
12. Guo, D., Xu, H., Liu, Y., Wang, H., Wang, L.: Visual dialog with multimodal transformer. In: Proceedings of the AAAI Conference on Artificial Intelligence (2021)
13. Instituto Cervantes: Spanish in the world 2022: A living language report. Tech. rep., Instituto Cervantes (2022). https://cvc.cervantes.es/lengua/anuario/anuario_22/default.htm. Accessed 12 Jan 2025
14. Järvelin, K., Kekäläinen, J.: Cumulated gain-based evaluation of IR techniques. ACM Trans. Inf. Syst. (TOIS) **20**(4), 422–446 (2002)
15. Jiang, X., Du, S., Qin, Z., Sun, Y., Yu, J.: KBGN: knowledge-bridge graph network for adaptive vision-text reasoning in visual dialogue. In: ACM MM (2020)
16. Jiang, X., et al.: DAM: deliberation, abandon and memory networks for generating detailed and non-repetitive responses in visual dialogue. In: IJCAI (2020)
17. Kang, G.C., Kim, S., Kim, J.H., Kwak, D., Zhang, B.T.: The dialog must go on: improving visual dialog via generative self-training. In: Proceedings of the IEEE/CVF Conference on Computer Vision and Pattern Recognition (CVPR), pp. 6746–6756 (2023)
18. Kottur, S., Moura, J.M., Parikh, D., Batra, D.: Visual coreference resolution in visual dialog using neural module networks. In: Proceedings of the European Conference on Computer Vision (ECCV), pp. 153–169 (2018)
19. Li, L., Cheng, H., Gan, Z., Wang, L., Liu, J., Gao, J.: Visual dialogue without vision or dialogue. In: Advances in Neural Information Processing Systems (NeurIPS), vol. 33, pp. 18059–18070 (2020)
20. Lin, T.-Y., et al.: Microsoft COCO: common objects in context. In: Fleet, D., Pajdla, T., Schiele, B., Tuytelaars, T. (eds.) ECCV 2014. LNCS, vol. 8693, pp. 740–755. Springer, Cham (2014). https://doi.org/10.1007/978-3-319-10602-1_48
21. Liu, H., Li, C., Wu, Q., Lee, Y.J.: Visual instruction tuning. In: Advances in Neural Information Processing Systems (2023)
22. Lu, J., Kannan, A., Yang, J., Parikh, D., Batra, D.: Best of both worlds: transferring knowledge from discriminative learning to a generative visual dialog model. In: NeurIPS (2017)
23. Murahari, V., Batra, D., Parikh, D., Das, A.: Large-scale pretraining for visual dialog: a simple state-of-the-art baseline. In: Vedaldi, A., Bischof, H., Brox, T., Frahm, J.-M. (eds.) ECCV 2020, Part XVIII. LNCS, vol. 12363, pp. 336–352. Springer, Cham (2020). https://doi.org/10.1007/978-3-030-58523-5_20
24. Ni, J., Li, J., McAuley, J.: M3P: learning universal representations via multitask multilingual multimodal pre-training. In: Proceedings of the IEEE/CVF Conference on Computer Vision and Pattern Recognition, pp. 3977–3986 (2021)
25. Niu, Y., Zhang, H., Zhang, M., Zhang, J., Lu, Z., Wen, J.R.: Recursive visual attention in visual dialog. In: CVPR (2019)
26. North, K., Ranasinghe, T., Shardlow, M., Zampieri, M.: Deep learning approaches to lexical simplification: a survey. J. Intell. Inf. Syst. **63**(1), 111–134 (2025). https://doi.org/10.1007/s10844-024-00882-9

27. Patel, Y., Tolias, G., Matas, J.: Recall@k surrogate loss with large batches and similarity mixup. CoRR abs/2108.11179 (2021). https://arxiv.org/abs/2108.11179
28. Sharma, P., Ding, N., Goodman, S., Soricut, R.: Conceptual captions: a cleaned, hypernymed, image alt-text dataset for automatic image captioning. In: Gurevych, I., Miyao, Y. (eds.) Proceedings of the 56th Annual Meeting of the Association for Computational Linguistics (Volume 1: Long Papers), pp. 2556–2565. Association for Computational Linguistics, Melbourne (2018). https://doi.org/10.18653/v1/P18-1238, https://aclanthology.org/P18-1238/
29. Souza, F., Nogueira, R., Lotufo, R.: BERTimbau: pretrained BERT models for Brazilian Portuguese. In: Cerri, R., Prati, R.C. (eds.) BRACIS 2020. LNCS (LNAI), vol. 12319, pp. 403–417. Springer, Cham (2020). https://doi.org/10.1007/978-3-030-61377-8_28
30. Tiedemann, J., Thottingal, S.: OPUS-MT – building open translation services for the world. In: Proceedings of the 22nd Annual Conference of the European Association for Machine Translation, pp. 479–480. European Association for Machine Translation, Lisbon (2020). https://aclanthology.org/2020.eamt-1.61
31. UNESCO: Portuguese Language Day (2023). https://www.unesco.org/en/fieldoffice/brasilia/expertise/portuguese-language-day. Accessed 04 June 2024
32. Wang, Y., Joty, S., Lyu, M., King, I., Xiong, C., Hoi, S.C.: VD-BERT: a unified vision and dialog transformer with BERT. In: Webber, B., Cohn, T., He, Y., Liu, Y. (eds.) Proceedings of the 2020 Conference on Empirical Methods in Natural Language Processing (EMNLP), pp. 3325–3338. Association for Computational Linguistics, Online (2020). https://doi.org/10.18653/v1/2020.emnlp-main.269, https://aclanthology.org/2020.emnlp-main.269
33. Wang, Y., Joty, S., Lyu, M.R., King, I., Xiong, C., Hoi, S.C.: VD-BERT: a unified vision and dialog transformer with BERT. In: EMNLP (2020)
34. Wu, Q., Wang, P., Shen, C., Reid, I., Van Den Hengel, A.: Are you talking to me? Reasoned visual dialog generation through adversarial learning. In: CVPR (2018)
35. Young, P., Lai, A., Hodosh, M., Hockenmaier, J.: From image descriptions to visual denotations: new similarity metrics for semantic inference over event descriptions. In: Transactions of the Association for Computational Linguistics (TACL), vol. 2, pp. 67–78 (2014)
36. Zhang, Y., Yang, J., Batra, D., Parikh, D.: Recursive visual attention in visual dialog. In: Proceedings of the IEEE Conference on Computer Vision and Pattern Recognition, pp. 6670–6678 (2019)

Classification of MIG Weld Bead Defects Using Computer Vision

Mauricio Aarón Pérez-Romero, Danna Paola Remigio-Pérez(✉), Mónica Montserrath Valdez-Mata, and María Alejandra Carmona-Riveros

Departamento Académico de Ingeniería en Control y Automatización, Escuela Superior de Ingeniería Mecánica y Eléctrica Unidad Zacatenco, Instituto Politécnico Nacional. Av. Luis Enrique Erro S/N, Gustavo A. Madero, 07738 Ciudad de México, México
dremigiop1800@alumno.ipn.mx

Abstract. This work presents the implementation and validation of a system for the classification of defects in MIG weld beads applied to tubular joints. A model based on convolutional neural networks (CNNs) was developed and trained using a custom-labeled image dataset, which includes both correct welds and common defects such as lack of fusion, porosity, undercut, slag inclusion and crater crack. The system was validated in a standard computing environment using the Python platform. An average classification accuracy above 93% was achieved. The results demonstrate the feasibility of this proposal as a support tool for visual inspection in conventional welding workshops and its suitability for industrial conditions with limited resources. Unlike recent approaches based on heavy backbones or transformer models, our contribution demonstrates that a lightweight CNN (MobileNetV2) can deliver competitive accuracy under realistic workshop conditions while enabling low-latency CPU inference on modest hardware. This positions the method as a practical option for resource-constrained industrial environments and for small- and medium-sized enterprises without dedicated GPUs.

Keywords: Lightweight CNN · MobileNetV2 · Industrial inspection · Weld quality control · Edge deployment · Transfer learning

1 Introduction

Welded joints are key elements in the fabrication of metallic structures, particularly in tubular scaffolding used in construction, industrial maintenance, and structural assembly [1]. These structures must comply with strict safety standards, as they support human loads and materials at height. In this context, the gas metal arc welding (GMAW/MIG) technique has gained relevance due to its efficiency, versatility, and adaptability to complex geometries, such as those found at the nodes of tubular joints.

However, despite its advantages, the MIG welding process may present defects that compromise the structural integrity if not detected in time. These defects affect the quality of the welded joint and can lead to stress concentrations, which reduce the service life of the scaffold or even cause failures under cyclic loading.

D. Chaves et al. (Eds.): CIARP 2025, LNCS 16528, pp. 236–248, 2026.
https://doi.org/10.1007/978-3-032-23161-1_17

In small metalworking workshops, visual inspection remains the most commonly used method due to its low cost and ease of implementation [2]. However, this type of inspection heavily depends on the operator's experience, introducing a high degree of subjectivity.

Defects such as slag inclusion, lack of fusion, porosity, undercut, and cracks are difficult to identify reliably with visual inspection alone [3]. This limitation motivates the adoption of computer vision methods that provide objective, repeatable, and scalable assessments of weld quality under diverse operating conditions.

In addition, tubular scaffolding is subjected to frequent assembly, disassembly, and transport processes, which generate loading and unloading cycles that can exacerbate pre-existing microdefects. Consequently, the timely identification of these defects is critical to ensure the safety of the personnel using these structures.

Among the most frequent defects observed in MIG welds on tubular joints are excessive penetration, lack of fusion, porosity, undercut, and cracking [4–7]. While not all of these are included in our dataset, they illustrate the range of issues that affect weld quality. Excessive penetration occurs when the filler metal passes through the base metal, potentially weakening the structure; it is typically caused by high current or low travel speed. Lack of fusion arises when the filler does not properly bond with the base metal, often due to low current, improper torch angle, or excessive travel speed. Porosity results from trapped gases during solidification, often due to surface contaminants or poor shielding gas quality. Undercuts appear as grooves at the weld toe, reducing the effective cross-section and are commonly caused by excessive current or incorrect gun angle. Cracks may be hot or cold, depending on their formation stage, and are usually linked to thermal stresses, insufficient preheating, or incompatible materials. These defects can compromise the mechanical performance of welded joints and require careful control of process parameters to mitigate.

The quality of a welded joint is critical to the safety and functionality of structural components, particularly in applications such as tubular joints, where welds must withstand dynamic loads, vibrations, and constant assembly changes. Any undetected defect can reduce structural strength and lead to critical failures during service.

For this reason, weld inspection becomes an essential stage in quality control. Its purpose is to identify surface or internal defects that may compromise the integrity of the weld bead, such as lack of fusion, porosity, cracks, undercut, or excessive penetration. These flaws may result from poorly adjusted process parameters, contaminated materials, or improper welding techniques.

Convolutional neural networks (CNNs) have become the state-of-the-art in image classification tasks due to their ability to automatically extract hierarchical features directly from pixel data, overcoming the limitations of handcrafted descriptors such as HOG or LBP, which are often sensitive to lighting or noise variations. Among lightweight CNN architectures, MobileNetV2 is particularly suitable for deployment in resource-constrained environments because it combines depth-wise separable convolutions and residual connections, reducing computation and memory requirements while maintaining representational power [8]. This characteristic makes MobileNetV2 an adequate choice for industrial inspection scenarios where GPUs are not available, such as small welding workshops or embedded platforms.

Statement of contribution. In contrast to prior work relying on large datasets or high-end GPUs (e.g., YOLOv8, EfficientDet, or transformer-based models), we show that a mobile-friendly CNN can match competitive performance while reducing inference cost and deployment complexity. Our study emphasizes realistic data acquisition and feasibility on CPU-only devices, addressing practical constraints often overlooked by state-of-the-art benchmarks.

2 Inspection Methods

The selection of an appropriate inspection method depends on the type of defect to be detected, the base material, the available resources, and the safety and reliability requirements of the final product. In the metalworking industry focused on scaffold fabrication, where cost and speed are critical factors, visual techniques are commonly used. However, to ensure more accurate evaluation, it is necessary to incorporate advanced technologies such as ultrasonic testing, liquid penetrant inspection, or even computer vision systems.

Visual inspection is the most accessible and commonly used technique. It involves directly examining the weld bead for visible signs of surface defects. Its ease of application makes it ideal for preliminary inspections in small workshops. However, its effectiveness depends on the inspector's skill, and it does not allow detection of internal defects or microcracks [5].

Visual inspection is essential, but it must be complemented with other methods when dealing with components that require high safety standards, such as scaffolding. This limitation of manual inspection further motivates the adoption of computer vision approaches that can provide objective, repeatable, and scalable assessments of weld quality.

Another widely used method is liquid penetrant inspection, which detects surface-breaking defects that are not visible by conventional means. A colored or fluorescent liquid is applied, penetrating discontinuities on the surface; after cleaning, a developer is applied to make the imperfections visible [9]. It is particularly useful on non-porous surfaces such as steel or aluminum and is employed in critical structures where surface cracks may propagate. The method is standardized by norms such as ASTM E165 and is widely recommended in maintenance inspections for detecting surface fatigue [6].

An additional technique is ultrasonic testing, which uses high-frequency sound waves to detect internal defects. These waves pass through the welded material and, upon encountering discontinuity, generate a signal that is interpreted by a specialized sensor within the equipment. It is a precise and non-destructive method, ideal for detecting defects such as internal cracks, lack of penetration, or inclusions [7]. However, it requires trained personnel and specialized equipment. In the tubular joint industry, it can be used to validate sample batches or welds subjected to repetitive loading cycles.

An additional technique used is industrial radiography, which involves projecting X-rays or gamma rays through the weld bead. Differences in material density allow for the visualization of defects such as slag inclusions, internal porosity, and lack of fusion in images known as radiograms [10]. This method offers high precision but entails high costs, controlled conditions, and strict radiological safety protocols; therefore, its use is limited to highly regulated environments.

Several relevant studies have addressed weld defect inspection using computer vision and deep learning techniques. Senthil and collaborators developed a visual inspection system for the identification and classification of MIG weld defects using conventional image processing methods [11]. This study represented an initial step toward automation in small metalworking shops. In [12], a weld defect classification method was proposed based on a lightweight CNN and data augmentation techniques, achieving results close to 98% accuracy. Another study using X-ray images corresponding to six defect classes reached 92% accuracy after applying data balancing and augmentation strategies [13].

A study conducted by Liu et al. employed an improved Faster R-CNN model to identify defects in resistance welds. The results showed precision and recall values above 90%, with inference times of 15 ms per image, making it suitable for industrial environments [14].

Another study proposed an optimized variant of YOLOv8 (YOLOv8-WD) combined with attention modules, achieving 90% accuracy in the detection and classification of steel defects [15]. Considering the advances reviewed, it is evident that the application of deep learning and computer vision in weld inspection has gained significant traction in recent years. However, many of these approaches rely on specialized hardware, extensive computational resources, or datasets not representative of real-world conditions in small-scale welding environments. This work presents an accessible and efficient alternative, focused on the classification of defects in MIG welds on tubular joints using a convolutional neural network trained on a custom dataset. The following section details the materials, dataset structure, and methods used for model development and validation, providing the basis for a fair evaluation of the proposed approach.

3 Materials and Methods

A dataset was constructed using images of MIG welds applied to tubular joints. The images were manually captured under controlled conditions and visually classified through direct inspection of the weld bead. The dataset includes both defective and acceptable welds. Figure 1 illustrates the overall sequence of stages that make up the proposed system for the automatic classification of MIG weld defects.

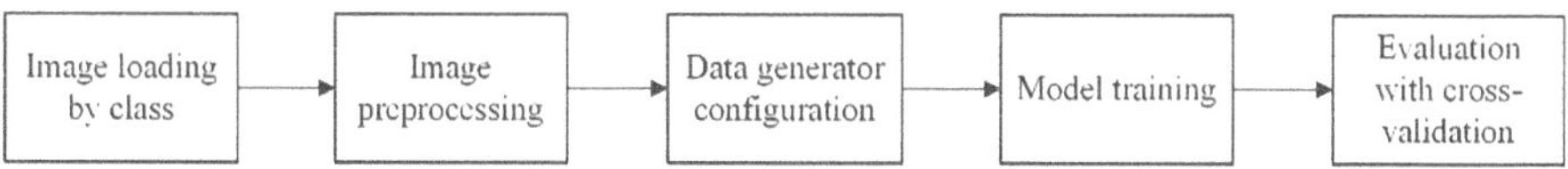

Fig. 1. General workflow of the proposed defect classification system.

Images of MIG weld beads were collected under real workshop conditions using webcams and mid-range smartphones (native resolutions 1280×720 to 1920×1080). Typical capture distances ranged from 15 to 40 cm, with ambient and LED bench lighting, including shadows and specular highlights. The dataset comprises six classes: Acceptable, Porosity, Lack of fusion, Undercut, Slag inclusion, and Crater crack. A curation step removed blurred, duplicated, or excessively cropped images. A stratified split was adopted with train/validation/test $\approx$ 84.9%/8.7%/6.5% (1421/145/108 images). The test set remained untouched during development and was used only once for final evaluation. Table 1 summarizes the class distribution, and representative examples are shown in Fig. 2.

Table 1. Dataset distribution per class and split.

Class	Total (n)	Total (%)	Training		Validation		Test (n)	Test (%)
			(n)	(%)	(n)	(%)		
Acceptable	305	18.2%	250	82.0%	25	8.2%	30	9.8%
Porosity	296	17.7%	245	82.8%	26	8.8%	25	8.4%
Lack of fusion	254	15.2%	218	85.8%	22	8.7%	14	5.5%
Undercut	282	16.8%	247	87.6%	25	8.9%	10	3.5%
Slag inclusions	282	16.8%	237	84.0%	24	8.5%	21	7.4%
Crater cracks	255	15.2%	224	87.8%	23	9.0%	8	3.1%
Total	1674	100%	1421	84.9%	145	8.7%	108	6.5%

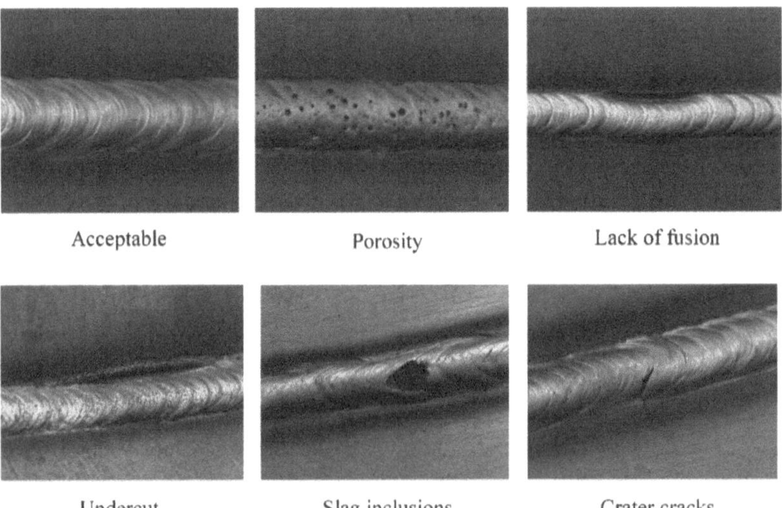

Fig. 2. Representative examples of weld bead defects

All images were resized to 224 × 224 pixels and normalized to the [0, 1] interval by applying scalar division over the RGB intensity values. These preprocessing operations ensure consistency in input dimensions and dynamic range for the convolutional model used.

During training, data augmentation techniques were applied, including random rotations, horizontal flipping, brightness adjustments, and scaling. These augmentations aimed to increase dataset variability and improve the classifier's generalization capability.

The proposed system was implemented using MobileNetV2 pretrained on ImageNet, with the final layer replaced by a six-class SoftMax head. Training was carried out with the Adam optimizer (initial learning rate = 1×10^{-3}), batch size = 32, and up to 100 epochs. An early stopping criterion was applied with patience = 10 on validation loss, and the learning rate was automatically reduced by a factor of 0.1 if the validation performance plateaued (patience = 5). To improve generalization, data augmentation (random rotation ± 15°, horizontal flip, brightness/contrast adjustment) was applied during training.

Fine-tuning was performed by unfreezing the last inverted residual blocks of MobileNetV2 from epoch 15 onwards, while keeping earlier layers frozen. Random seeds were fixed to ensure reproducibility. All reported results correspond to the held-out test set or to the mean ± standard deviation over 5 stratified folds.

A convolutional model based on MobileNetV2 was used, pretrained on ImageNet and adapted through transfer learning. The original architecture was retained with its weights frozen, and a Global Average Pooling layer was added, followed by a dense layer with 128 ReLU-activated units and a SoftMax output layer corresponding to the seven defined classes.

The architecture was implemented using TensorFlow 2.0 and Keras, compiled with the categorical crossentropy loss function and the Adam optimizer. The overall classification process is summarized in the following pseudocode, which describes the workflow from data preparation to model evaluation:

```
Input:
Image dataset organized by class (defect/correct)
Hyperparameters: image size, batch size, number of epochs
Process:
  1. Load images from directories using a generator with
     validation split
  2. Resize images to 224×224 px and normalize RGB
     intensity values
  3. Apply data augmentation (rotation, flipping,
     brightness adjustment)
  4. Initialize MobileNetV2 base model with ImageNet
     weights
  5. Freeze convolutional layers of the base model
  6. Add:
      GlobalAveragePooling layer
      Dense layer with 128 ReLU-activated units
      Output layer with softmax activation (7 classes)
  7. Compile model using categorical crossentropy loss
     function
  8. Train for N epochs on the training set
  9. Evaluate the model on the validation set
Output:
  Trained model (.h5)
  Per-class and overall accuracy
  Learning curves
```

The training was conducted on a personal computer equipped with a 10th-generation Intel Core i5 processor and 16 GB of RAM, without a dedicated graphics processing unit. MobileNetV2 with ImageNet weights and a SoftMax layer with six outputs were employed. The Adam optimizer was used with an initial learning rate of 1×10^{-3}, a batch size of 32, and training for up to 100 epochs with early stopping (patience = 10 based on validation loss) and learning rate reduction on plateau (factor 0.1; patience = 5). The final layers of the backbone were gradually unfrozen starting from epoch 15. Seeds were set for reproducibility, and the average latency per image on CPU was recorded. The main evaluation metric was categorical accuracy, computed during both training and validation stages.

4 Testing and Results

Figure 3 shows a real example of the input images used during the testing phase. This visual reference illustrates the type of MIG welds classified by the system under non-controlled conditions, using standard imaging devices.

During the training process, the model achieved an average accuracy above 93% on the training set and a validated accuracy exceeding 92% on the validation set, indicating

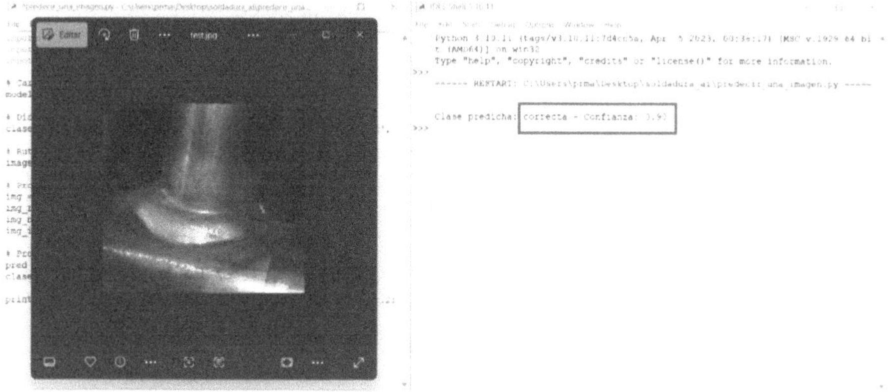

Fig. 3. Sample image of a MIG weld bead used during the classification test.

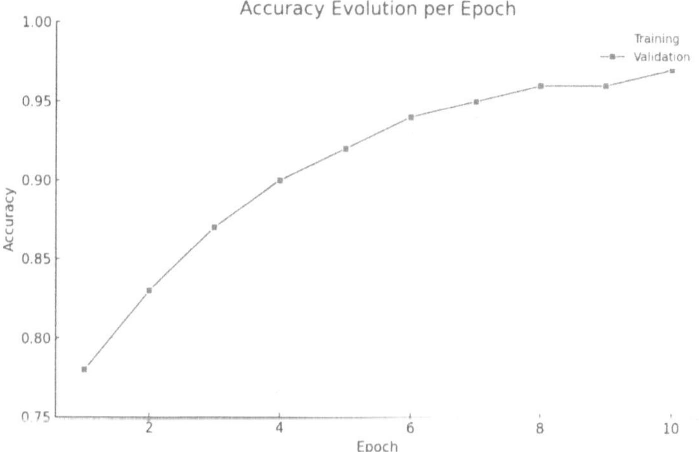

Fig. 4. Accuracy evolution during training and validation epochs.

an initial generalization capability. Figure 4 shows the evolution of accuracy and loss throughout the training epochs.

To evaluate the classifier's performance for each category, a confusion matrix was generated, highlighting the correct and incorrect predictions per class. Table 2 presents the precision, recall, and F1-score for each class in the dataset, calculated on the validation set.

A stratified 5-fold cross-validation was performed across the six classes. The results, expressed as mean ± standard deviation, confirmed stable performance across different folds. To simulate realistic variations in acquisition conditions, three additional test sets were generated from the original images: low lighting (gamma correction = 0.7), high lighting (gamma = 1.3), and Gaussian noise ($\sigma = 0.01$ after normalization). The impact on accuracy and macro F1-score was reported.

Table 2. Weld bead defect identification tests.

Class	Accuracy	Recall	F1-score
Acceptable (Correcta)	0.93	0.92	0.92
Lack of fusion (Falta de fusión)	0.92	0.91	0.91
Crater cracks (Grietas de cráter)	0.92	0.91	0.91
Slag inclusions (Inclusiones de escoria)	0.95	0.94	0.94
Porosity (Porosidad)	0.94	0.92	0.93
Undercut (Socavado)	0.92	0.91	0.91

Confusion matrix of the MobileNetV2 model on the test set for the six classes: OK, Porosity, Lack of fusion, Undercut, Slag inclusion, and Crater crack (see Fig. 5). The values represent the number of samples per pair (actual vs. prediction). The highest confusion was observed between lack of fusion and undercut, which aligns with their visual similarity at the edges of the weld bead.

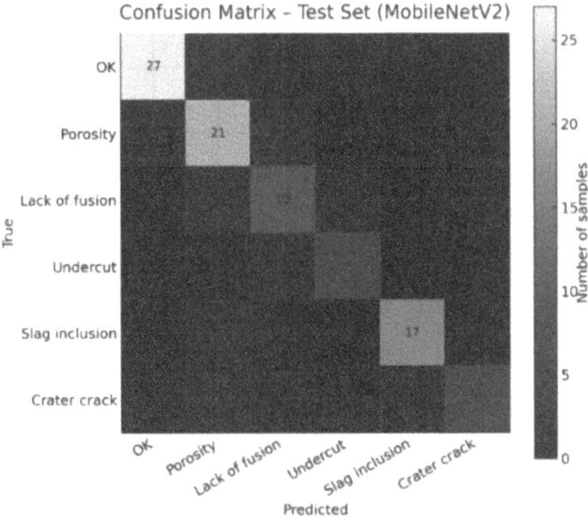

Fig. 5. Confusion matrix

Figure 6 illustrates representative correct predictions, including the predicted label and confidence. The most frequent confusions occurred between lack of fusion and undercut, often driven by similar edge textures and bead profiles under challenging illumination.

Unless otherwise specified, we report mean ± standard deviation over 5 stratified folds. For completeness, 95% confidence intervals can be approximated as $\mu \pm 1.96\,\sigma/\sqrt{k}$ with $k = 5$. Table 3 summarizes the cross-validation results, while Table 4 shows the effect of perturbations. Both analyses demonstrated that the MobileNetV2 model maintained competitive performance under adverse conditions, with moderate degradation in visually similar classes such as lack of fusion and undercut.

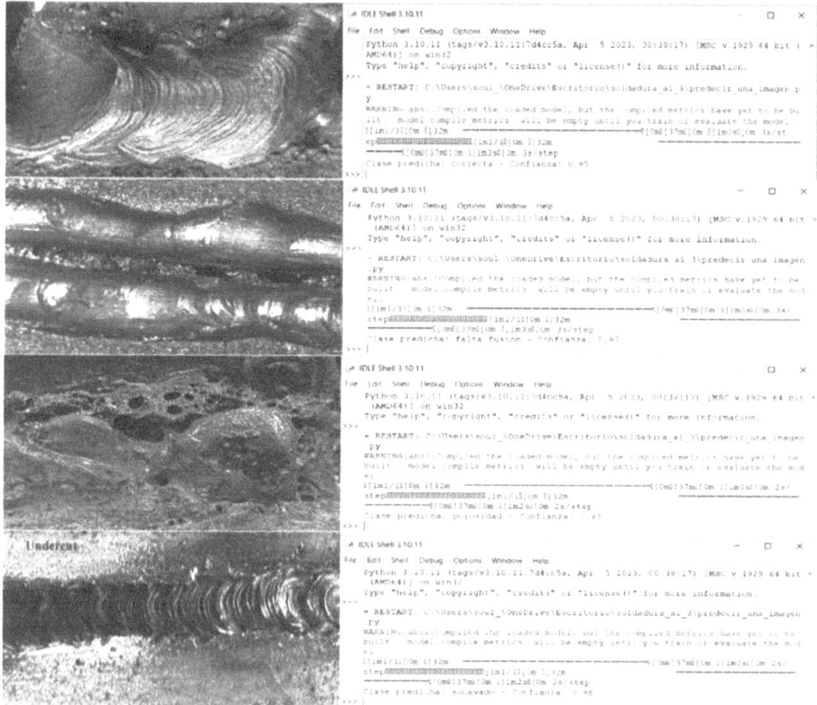

Fig. 6. Images with ground-truth and predicted labels

Table 3. Cross-validation

Model	Average Accuracy (%) ± SD	Average macro F1 ± SD
MobileNetV2	92.4 ± 1.2	0.91 ± 0.02
ResNet-18	90.8 + 1.5	0.89 ± 0.03
HOG + SVM	81.3 ± 2.0	0.80 ± 0.03

Table 4. Robustness under perturbations

Model	Baseline (%)	Low light (%)	High light (%)	Gaussian noise (%)
MobileNetV2	92.1	89.5	90.2	87.3
ResNet-18	90.4	87.8	88.6	85.9
HOG + SVM	81.0	76.5	77.8	74.2

The evaluation of the model on the validation set yielded consistent results across all classes, with precision values exceeding 90% in every case. Particularly high performance was observed in the Slag inclusions category (0.95), followed by Porosity (0.94) and Acceptable (0.93). This behavior suggests that the model successfully identified the distinctive visual patterns associated with these cases. The *Arc strikes* and *Lack of fusion* classes, while exhibiting slightly lower precision, also remained within acceptable ranges (≥ 0.91), with no significant evidence of confusion. Overall, the results reflect a balanced performance and generalization capability of the model on unseen samples.

The proposed MobileNetV2 model was executed on a standard CPU-only setup without GPU acceleration. During testing, the network consistently delivered low-latency predictions suitable for real-time or near real-time operation. Compared with ResNet-18, MobileNetV2 required fewer computational resources and a smaller memory footprint, while providing faster inference. These characteristics make it feasible for edge deployment in low-cost PCs commonly available in small and medium welding workshops. Unlike heavier CNNs that demand dedicated GPUs, MobileNetV2 maintains competitive accuracy under practical hardware constraints, supporting its applicability in quality assurance workflows.

5 Discussion

The results obtained with MobileNetV2 demonstrate that a lightweight CNN can achieve competitive accuracy in MIG weld defect classification, even when trained on a modest dataset acquired under real workshop conditions [3]. The confusion matrix showed that the most frequent misclassifications occurred between lack of fusion and undercut, reflecting their visual similarity at bead edges [16]. These observations highlight both the effectiveness and the practical challenges of automated visual inspection in welding quality control.

In contrast to previous studies that employed heavier architectures such as ResNet-50 or YOLOv5 for weld defect detection [17], our findings confirm that a mobile-friendly CNN can provide comparable accuracy while significantly reducing inference cost and memory requirements. This aligns with recent reports where lightweight architectures were successfully applied to industrial quality assurance tasks [18]. Unlike approaches based on large-scale public datasets [16], our work relied on a smaller dataset collected in real environments, which increases practical realism but also limits generalization. This duality constitutes both a contribution and a constraint of the present study.

An important limitation of the current approach is the exclusive focus on image-level classification, without performing spatial localization of defects within the weld bead. Although localization techniques such as YOLOv8 or U-Net could provide bounding boxes or segmentation maps, their application requires larger annotated datasets and higher computational resources. Future work will extend the proposed system by integrating defect localization to provide more detailed feedback and enable automated visual reporting in industrial settings.

Robustness remains a critical aspect. The cross-validation experiments confirmed stable accuracy across folds, while perturbation tests under varying lighting conditions and synthetic noise revealed a controlled degradation in performance. These results indicate that the proposed model maintains acceptable reliability when exposed to realistic variability, although further validation on expanded datasets will be necessary to fully assess generalization.

Ethical and safety considerations are also relevant. In industrial inspection, false negatives can pose significant safety risks. To mitigate this issue, thresholds were adjusted to prioritize recall in the most critical classes, and a human-in-the-loop approach is recommended for borderline cases. Furthermore, alignment with industry standards and regulatory frameworks is essential for deployment in sectors where weld integrity is critical, such as structural and pressure vessel applications.

Despite these limitations, the study supports the viability of adopting lightweight CNNs for weld defect classification in CPU-only environments. The balance between accuracy and efficiency makes MobileNetV2 an attractive solution for small and medium enterprises that lack access to high-performance hardware. As future work, we plan to expand the dataset with more diverse acquisition conditions, benchmark the system on embedded platforms, and explore calibration strategies to ensure compliance with domain-specific safety requirements. These actions will strengthen robustness and facilitate real-world deployment.

6 Conclusions

This work demonstrated the feasibility of implementing an automated system for classifying defects in MIG welds using computer vision and deep learning, without relying on specialized infrastructure or large volumes of data. The adopted approach, based on lightweight convolutional neural networks, enables image processing under uncontrolled conditions while maintaining robust performance in industrial environments with limited resources.

The model architecture, combined with a reproducible workflow, represents a solid starting point for the development of AI-assisted visual inspection tools. Its integration into manufacturing processes could help reduce variability in quality assessment, improve weld traceability, and raise structural safety standards in critical applications such as scaffolding.

As future work, it is proposed to expand the dataset with images from various sources and operating conditions, incorporate segmentation techniques for precise defect localization, and explore implementation on embedded or mobile devices connected to remote monitoring platforms.

Acknowledgments. The authors would like to thank the Escuela Superior de Ingeniería Mecánica y Eléctrica Unidad Zacatenco (ESIME Zac) of the Instituto Politécnico Nacional (IPN) for providing the institutional support and facilities that enabled the development of this work.

Disclosure of Interests. The authors have no competing interests to declare that are relevant to the content of this article.

References

1. Chandramohan, D., Roy, K., Taheri, H., Karpenko, M., Fang, Z., Lim, J.: A state of the art review of fillet welded joints. Materials **15**(24), 8743 (2022)

2. ISO 17637: Non-destructive testing of welds-Visual testing of fusion-welded joints. European Committee for Standardization (2016)
3. Sonwane, S., Chiddarwar, S.: Enhancing weld defect detection and classification with MDCB-Net: a multi-scale dense cross block network for improved explainability. NDT E Int. **142**, 103029 (2024)
4. Jeffus, L.: Soldadura. Principios y aplicaciones. Ediciones Paraninfo (2009)
5. Pérez, O.: Defectos y controles de las uniones soldadas en soldadura manual. Editorial Universitaria, Cuba (2020)
6. Haque, S.: Investigation on welding defects of alloys using TIG and MIG welding. Hybrid Adv. **3**, 100066 (2023). https://doi.org/10.1016/j.hybadv.2023.100066
7. Kumar, R.: Defects associated with welding techniques and their detection methods. In Advanced Welding Techniques. CRC Press, pp. 63–89 (2024)
8. Sandler, M., Howard, A., Zhu, M., Zhmoginov, A., Chen, L.: MobileNetV2: inverted residuals and linear bottlenecks. In: Proceedings of the IEEE Conference on Computer Vision and Pattern Recognition, pp. 4510–4520 (2018)
9. Lampman, S., Mulherin, M., Shipley, R.: Nondestructive testing in failure analysis. J. Fail. Anal. Preven. **22**, 66–97 (2022). https://doi.org/10.1007/s11668-021-01325-1
10. Alonso, J., Pavón, S., Vidal, J., Delgado, M.: Advanced comparison of phased array and X-rays in the inspection of metallic welding. Materials. **15**(20), 7108 (2022)
11. Senthil, G., Natarajan, U., Ananthan, S.: Vision inspection system for the identification and classification of defects in MIG welding joints. Int. J. Adv. Manuf. Technol. **61**, 923–933 (2012). https://doi.org/10.1007/s00170-011-3770-z
12. Say, D., Zidi, S., Qaisar, S., Krichen, M.: Automated categorization of multiclass welding defects using the x-ray image augmentation and convolutional neural network. Sensors. **23**(14), 6422 (2023)
13. Jiang, L., Wang, Y., Tang, Z., Miao, Y., Chen, S.: Casting defect detection in X-ray images using convolutional neural networks and attention-guided data augmentation. Measurement **170**, 108736 (2021)
14. Liu, Y., He, B., Tang, M., Luo, W.: Resistance spot welding defect detection based on visual inspection improved Faster R-CNN model. Machines **12**(1), 33 (2025)
15. Wang, W., Xiao, H., Zhang, Q., Wen, H., Zuo, K.: An improved YOLOv8 algorithm for steel surface defect detection. In: 8th Asian Conference on Artificial Intelligence Technology (ACAIT), pp. 385–394 (2024)
16. Zhang, W., et al.: Deep learning-based automated detection of welding defects in pressure pipeline radiograph. Coatings **15**(7), 808 (2025)
17. Yang, Y., Zhou, Y., Din, N., Li, J., He, Y., Zhang, L.: An improved YOLOv5 model for detecting laser welding defects of lithium battery pole. Appl. Sci. **13**(4), 2402 (2023)
18. Chang, S., Zheng, B.: A lightweight convolutional neural network for automated crack inspection. Constr. Build. Mater. **416**, 135151 (2024)

A Data-Driven Approach to Faculty Schedule Optimization in Higher Education

Rodolfo Bojorque[1,2](✉), Bertha Tacuri[1,2], and Fernando Moscoso[1,2]

[1] Universidad Politécnica Salesiana, Cuenca, Ecuador
{btacuri,fmoscoso}@ups.edu.ec
[2] Sede Matriz, 010102 Cuenca, Ecuador
rbojorque@ups.edu.ec
http://www.ups.edu.ec

Abstract. Optimizing faculty schedules in higher education requires balancing teaching commitments, research productivity, and institutional priorities. This study presents a data-driven pipeline that integrates data mining techniques to identify hidden patterns in faculty workload distribution. We combined data preprocessing and unsupervised learning for clustering. The dataset, drawn from the Salesian Technical University of Ecuador, includes teaching hours, and research outputs. Our findings reveal that faculty with a balanced allocation of teaching and research responsibilities tend to exhibit significantly higher research productivity. This conclusion is supported by clustering analysis and nonparametric statistical testing (Mann-Whitney U, $p < 0.001$), indicating robust differences between workload profiles. The clustering analysis revealed that moderate teaching loads enable faculty to maintain active research engagement without compromising instructional quality. By treating scheduling as a multi-objective optimization problem, the proposed pipeline highlights the tension between institutional efficiency and personal academic goals, emphasizing the need for adaptable scheduling strategies. The study contributes practical insights for academic administrators aiming to enhance both educational quality and research outcomes. Future research will refine the analysis to incorporate disciplinary differences and longitudinal data, further improving its applicability across diverse academic settings.

Keywords: Data Mining · Academic workload · Higher education management

1 Introduction

Optimizing faculty schedules is a complex problem involving numerous constraints and objectives, such as availability, course requirements, and institutional policies. Efficient scheduling improves resource utilization, enhances educational quality, and increases student and faculty satisfaction. Computationally,

D. Chaves et al. (Eds.): CIARP 2025, LNCS 16528, pp. 249–265, 2026.
https://doi.org/10.1007/978-3-032-23161-1_18

the problem is of the NP-Complete type, and various approaches have been taken to address it, such as the use of Greedy algorithms [23], Timesheet Management Systems [21], Genetic Algorithms for optimization of lecturer schedules [28], etc. However, the literature indicates that each case is particular to its application context. Therefore, different authors [24] and [27] propose the use of Key Performance Indicators (KPIs) as an alternative to assess academic management.

This leads to the issue of how to establish appropriate KPIs for the context, a challenge that essentially transforms into a multiobjective optimization problem. Although the literature does not clearly delineate which variables or constraints should be prioritized, the practical context suggests that the selection and weighting of these KPIs are highly contingent upon the individual interests and constraints of each faculty member regarding teaching load allocation. Consequently, there exists a persistent conflict of interest between what the institution deems optimal and what the faculty considers optimal. This paper proposes a data pipeline to address this problem, starting from evidence-based information, using data mining to detect hidden patterns, and developing institutional policies to optimize the faculty schedule.

The objective of this study is to develop and evaluate a data-driven pipeline capable of identifying workload patterns among university faculty by applying unsupervised learning techniques. Specifically, the study aims to determine whether a statistically significant relationship exists between teaching load and research productivity, using real institutional data and clustering analysis to inform evidence-based scheduling policies.

The faculty scheduling process faces key challenges from the perspective of the university's core functions of teaching and research, including: a) Multiple Constraints: Addressing diverse constraints such as classroom availability, research funding, lecturer preferences, and course requirements; b) Large Datasets: Managing and processing extensive data from historical schedules, course enrollments, research projects, and faculty records; c) Dynamic Environment: Adapting to changes such as last-minute lecturer unavailability, shifts in student enrollment, or research funding accreditation through competitive grants.

Our study is based on the experience of Salesian Technical University of Ecuador, a higher education institution with campuses in Ecuador's three main cities, serving over 1,000 faculty members and more than 24,000 students. Nearly half of the faculty members are researchers managing active research projects or participating in academic research networks. It is important to note that our study is not solely focused on providing a solution to the class-hour allocation problem. Our proposal seeks to balance teaching and research, as faculty cannot be disengaged from teaching nor can they cease to conduct research. Our study is organized as follows: After the introduction, Sect. 2 presents related work, comparing different approaches to the faculty scheduling problem. Section 3 outlines the methodology for implementing the pipeline. Section 4 discusses the main issues, and Sect. 5 presents the conclusions.

Although this study does not implement a scheduling optimization algorithm in the operational sense, it provides a foundational step toward such development. By uncovering empirically grounded workload profiles through clustering and statistical validation, the proposed data-driven approach offers practical insights for informing future multiobjective scheduling frameworks. These findings serve as essential input for the design of optimization strategies that seek to align institutional teaching needs with faculty research productivity.

2 Related Works

2.1 Data Mining on Educational Datasets

Data mining applied to higher education databases is well-documented in the literature and is not a recent issue. In [7,13,17], the application of correlations between existing information is demonstrated. Similarly, data mining has been used to identify stress factors that affect students' academic performance [6], and Learning Recommendation Systems have been developed, trained with data such as grades [9,19] and data from Learning Management Systems (LMS). Currently, data allows for the development of systems that predict academic success [5,10], detect emotions during learning [1,3], and provide auditing systems for academic management [8].
The majority of studies focus their scope on generated data about students (grades, connections, interactions, etc.). However, there are some open issues in data mining for education especially when working with non-student data. In other words, it is common to find educational data mining systems focused on student information; however, our approach applies educational data mining to faculty data.

2.2 Data Mining Role in Faculty Schedule

Data mining techniques have played a critical role in optimizing academic scheduling by extracting patterns from large datasets and enabling the design of more efficient timetables. A comprehensive technical review by Romero et al. [25] further highlights the evolution of university timetabling methodologies, tracing the transition from classical optimization techniques such as Integer Programming to contemporary approaches involving machine learning. Their survey reinforces the relevance of hybrid and intelligent systems in addressing the inherent complexity of timetabling problems and supports the rationale for incorporating clustering and predictive models in educational scheduling research.

Among the most prominent approaches are Genetic Algorithms (GAs), Constraint Satisfaction Problems (CSPs), Integer Linear Programming (ILP), and more recently, machine learning-based models.

Genetic Algorithms are particularly effective for handling complex optimization problems, as they simulate the evolutionary processes of selection, crossover, and mutation to iteratively refine potential scheduling solutions [2,18]. Gabriel et al. [14] employed GAs to encode schedules as chromosomes, demonstrating how such algorithms can efficiently explore vast solution spaces to produce feasible and near-optimal faculty timetables.

Constraint Satisfaction Problems, on the other hand, utilize techniques like backtracking and constraint propagation to ensure that all predefined conditions–such as time slot availability, classroom capacity, and instructor constraints–are satisfied [20]. These models are particularly well-suited for ensuring hard constraints are never violated.

Integer Linear Programming represents another powerful strategy, where the scheduling problem is formulated as a set of linear equations and inequalities. Van Hentenryck et al. [29] pioneered ILP in scheduling, and more recent applications, such as the work by Bushman et al. [11], have used ILP to minimize the number of days faculty are required on campus, contributing to improved work-life balance while maintaining operational efficiency.

Machine learning approaches are also gaining traction in this domain. These models can learn from historical scheduling data to forecast peak periods of instructor availability or predict course enrollment demand. Zanevych et al. [31] illustrate how such predictive models can proactively avoid scheduling conflicts and improve timetable responsiveness by adapting to institutional usage patterns over time.

Collectively, these techniques offer a robust foundation for data-driven faculty scheduling systems, each addressing distinct facets of the optimization problem–ranging from hard constraint satisfaction to adaptive, demand-driven planning.

Beyond optimization approaches, clustering methods such as hierarchical clustering and Gaussian Mixture Models (GMMs) have also been applied in workload and scheduling contexts [25,31]. These methods provide complementary perspectives: hierarchical clustering facilitates dendrogram-based exploration of workload similarity, while GMMs offer probabilistic modeling of latent subgroups. Although our study focused on K-Means and DBSCAN as baseline methods, future extensions could incorporate hierarchical or probabilistic clustering to contrast results and refine workload typologies.

2.3 Applications and Case Studies

Various higher education institutions have successfully implemented data mining techniques to address faculty scheduling challenges, often combining these methods with optimization and modeling strategies tailored to specific academic contexts.

For example, Wang et al. [30] developed a hybrid system that integrates association rule mining and genetic algorithms to generate course timetables that account for both student needs and instructor preferences. Their approach uses historical enrollment data to identify common course combinations and then applies a genetic algorithm to optimize the scheduling structure. The results demonstrated increased satisfaction and operational efficiency when compared to traditional manual scheduling practices.

Similarly, Susan and Jacob [26] proposed a hybrid model that combines genetic algorithms with simulated annealing to optimize elective course scheduling based on students' preferred time slots. Their model effectively minimized scheduling conflicts and improved timing alignment, further validating the benefits of intelligent optimization methods over legacy approaches.

Beyond scheduling mechanics, some studies have addressed the relationship between workload assignment and academic performance. Nguyen and Meek [22] analyzed institutional workload policies and concluded that rigid or unbalanced teaching assignments can significantly suppress faculty research productivity. Their findings reinforce the importance of aligning instructional obligations with scholarly output expectations–a goal also pursued in the present study through empirical clustering analysis.

A regionally grounded example is provided by Calle-López et al. [12], who developed an automated scheduling system specifically for engineering programs. Their solution incorporated both mathematical modeling and data mining techniques to manage multiple constraints, including room availability, class sizes, and instructor schedules. The system was validated using data from an Ecuadorian university, where it successfully generated feasible timetables and demonstrated real-world applicability.

It is noteworthy that, although prior research has explored both faculty scheduling and research productivity, these two dimensions are frequently addressed as separate issues. Earlier discussions around the balance of faculty roles–such as those by Geiger [16] and others [4]–acknowledge the tension between teaching and research demands. However, contemporary studies have tended to isolate scheduling from research outcomes. In contrast, our study emphasizes the need for an integrated approach, leveraging data mining not only to improve scheduling logistics but also to understand its implications for academic performance.

3 Methodology

The methodological pipeline of this study is structured into three sequential phases: data preprocessing, unsupervised learning through clustering, and interpretation of the identified groups. Rather than claiming methodological novelty, this work applies a baseline combination of clustering and statistical validation to explore faculty workload. While K-Means and DBSCAN are standard algorithms, their use in analyzing the balance between teaching and research productivity in the Latin American higher education context contributes empirical evidence where few studies exist. The value of this pipeline lies in establishing a practical and replicable foundation that can be extended in future research with more advanced frameworks.

Figure 1 summarizes the methodological pipeline applied in this study. It highlights the sequential stages: (i) preprocessing of institutional records, (ii) construction of the Scientific Production Index, (iii) clustering with K-Means and DBSCAN, (iv) validation through internal metrics and statistical testing, and (v) interpretation and discussion of workload profiles. Including this schematic overview clarifies that, while based on standard techniques, the study follows a structured and reproducible sequence of analytical steps.

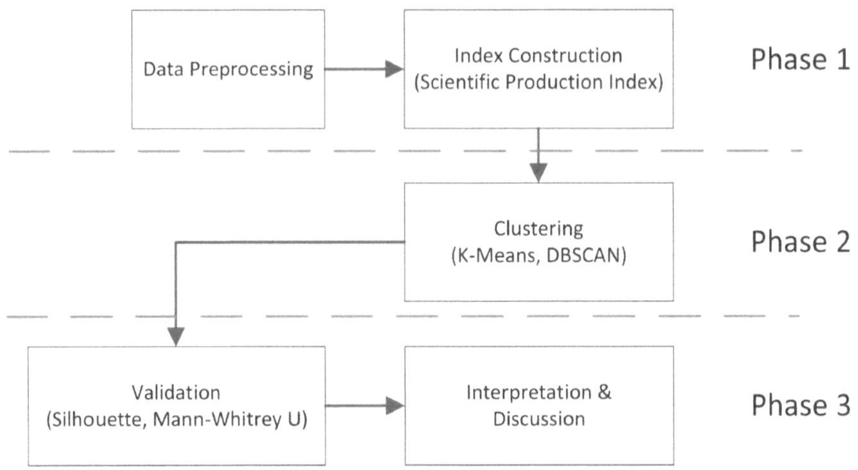

Fig. 1. Methodological Pipeline.

3.1 Data Preprocessing

The dataset comprised institutional records from 1,024 faculty members across three campuses: Cuenca ($n = 312$), Quito ($n = 428$), and Guayaquil ($n = 284$) for the 2023–2024 academic year (see Fig. 2). Faculty were included if they had both teaching and research activity registered in the institutional systems. Part-time lecturers and administrative-only staff were excluded to ensure comparability of workload. Two core variables were extracted for analysis: the Scientific Production Index (SPI) and Lecture Hours. The SPI was computed as explained in Sect. 3.1, while Lecture Hours represented weekly instructional hours officially assigned to each faculty member.
Missing data rates were 4.2% for SPI sub-indicators and 3.5% for Lecture Hours. Imputation was performed using domain-informed methods: median imputation for continuous values (e.g., citation counts), and mode imputation for categorical attributes (e.g., project participation). Faculty with more than 20% missing information were excluded from the sample ($n = 37$). To ensure comparability, all variables were standardized using Z-score normalization. Outliers were assessed through interquartile range thresholds and retained when they reflected plausible workload heterogeneity

Two core variables were extracted for analysis: the Scientific Production Index and Lecture Hours. The Scientific Production Index is a composite metric that quantifies academic output, incorporating publication volume and quality indicators, normalized on a 0–1 scale. This index was derived from research output recorded in the university's internal Research Management System (SGI), as well as external sources such as Google Patents, Scopus, Web of Science, and Google Scholar for the 2023–2024 period. Additional institutional data were also included, such as the number of undergraduate and graduate theses, awards, and records from the Human Talent Management System (GTH), as detailed in

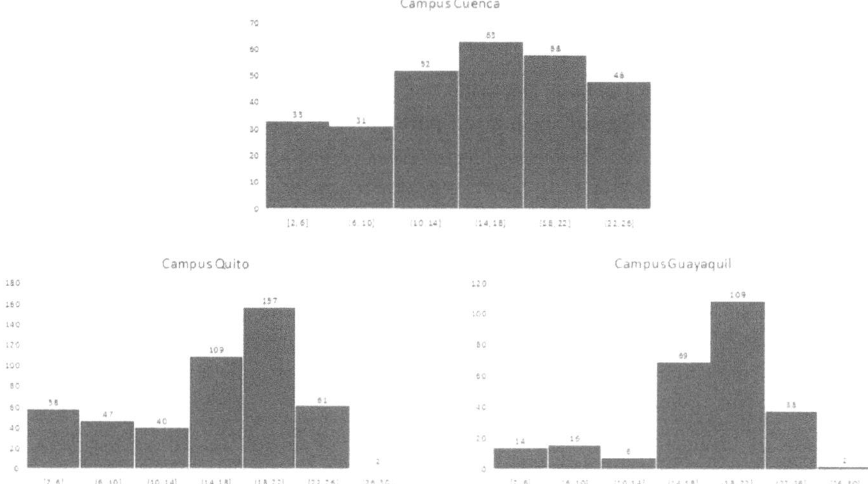

Fig. 2. Distribution of teaching hours among faculty across the Cuenca, Quito, and Guayaquil campuses. The histogram illustrates how each campus exhibits distinct patterns in teaching load allocation, with Cuenca showing lower instructional hours, Quito presenting a moderate distribution, and Guayaquil exhibiting heavier teaching loads. These differences are not institutionally imposed but rather reflect the unique administrative practices of each location, providing useful context for workload-related productivity analysis.

Table 1. To ensure anonymity, randomly generated unique identifiers were used for all faculty members.

Lecture Hours represent the weekly instructional time assigned to each faculty member.

Prior to modeling, the dataset was standardized using Z-score normalization to ensure comparability of scale between variables. Outliers and missing values were handled through domain-informed curation and imputation where appropriate.

The weights presented in Table 1 were not arbitrarily assigned but derived from institutional evaluation criteria formally adopted by the Research Vice-Rectorate and Human Talent Management units at the Salesian Technical University. These criteria reflect the relative importance attributed to different forms of academic output (e.g., publications, citations, project participation, funding).

The Scientific Production Index (SPI) was computed as a weighted linear combination of normalized indicators, following Eq. 1:

$$SPI_i = \sum_{j=1}^{m} w_j \cdot x_{ij} \tag{1}$$

Table 1. Description of Scientific Production Indicators

Indicator	Description	Source	Weight
Scientific Production	Researcher's scientific output effort during the evaluation period	SGI (2023–2024)	36%
Citations	Number of citations received during the evaluation period for the researcher's publications	Scopus-Google Scholar (2023–2024)	12.8%
Research Projects	Number of participations as principal or secondary researcher in active UPS research projects	SGI (Active Projects)	10.7%
Social Impact	Number of publications contributing to the Sustainable Development Goals (SDGs)	PURE (2023–2024)	10.7%
External Funding Acquisition	Number of external funding sources obtained for research projects (e.g., Horizon Europe, Erasmus, IES Horizontal)	Research Coordinations by Campus	8.6%
Corresponding Author	Number of times the researcher was designated as corresponding or lead author in scientific publications	PURE - SGI (2023–2024)	4.6%
National and Institutional Networks	Number of UPS faculty and students collaborating in the researcher's scientific articles	PURE - SGI (2023–2024)	4.3%
International Networks	Number of external researchers collaborating in the researcher's scientific articles	PURE - SGI (2023–2024)	4.3%
Management Roles	Leadership roles and administrative responsibilities held by the researcher	GTH	4%
Thesis Supervision	Number of undergraduate and graduate theses supervised by the researcher during the evaluation period	SAN (2023–2024)	2%
Recognitions	Number of local, national, and international recognitions received by the researcher, including participation as reviewer in high-impact Q1 and Q2 journals	Research Vice-Rectorate	2%

where SPI_i is the scientific productivity index of faculty member i, w_j represents the institutional weight of indicator j, and x_{ij} is the normalized score of faculty member i on indicator j.

For example, if a researcher scored 0.8 in Scientific Production, 0.6 in Citations, and 0.5 in Research Projects, their SPI would be calculated as:

$$SPI = (0.36 \times 0.8) + (0.128 \times 0.6) + (0.107 \times 0.5) + \ldots$$

yielding a final index in the $[0, 1]$ range.

We acknowledge that the weighting scheme reflects institutional priorities rather than an empirically optimized allocation. A sensitivity analysis of weights, as well as alternative aggregation methods (e.g., principal component analysis), will be incorporated in future work to enhance robustness and reduce potential bias.

3.2 Clustering Approach

For the unsupervised learning phase, we selected K-Means and DBSCAN as the clustering algorithms due to their complementary strengths. K-Means is widely used for its efficiency and interpretability when data can be reasonably partitioned into spherical clusters, making it well-suited for identifying dominant workload profiles in structured datasets. In contrast, DBSCAN is a density-based method that detects non-linear groupings and outliers, allowing us to explore whether there are atypical workload distributions among faculty that do not conform to the main trends.

The rationale for selecting both K-Means and DBSCAN lies in their complementary strengths when exploring complex academic workload distributions. K-Means was employed to identify dominant workload profiles based on global similarity in teaching hours and scientific productivity. Its centroid-based approach effectively revealed the most representative groups, aligning with institutional scheduling categories. In contrast, DBSCAN was selected to uncover non-linear or irregular patterns in the data, including the presence of faculty members whose workload profiles deviated substantially from typical patterns–such as high research productivity with low teaching loads, or vice versa. This dual approach ensures robustness in the pattern discovery process and supports a more comprehensive understanding of how workload heterogeneity relates to research outcomes. Together, both algorithms contribute to the broader objective of informing data-driven scheduling strategies grounded in empirical workload behaviors.

The input variables used for clustering were the Scientific Production Index and Lecture Hours. These two variables were selected based on their direct relevance to the research objective: to examine whether teaching load impacts faculty research productivity. The Scientific Production Index was constructed as a composite metric, weighted by institutional criteria (see Table 1), while Lecture Hours reflect the number of weekly instructional hours assigned to each faculty member. Together, they represent the two core dimensions of academic workload. By clustering based on these features, the model aims to reveal emergent patterns in the teaching–research balance across the faculty population.

We experimented with a range of hyperparameters to identify the optimal configuration for each algorithm. For K-Means, values of k from 2 to 5 were tested, with evaluation based on the silhouette score and cluster compactness.

For DBSCAN, the `eps` and `min_samples` parameters were tuned empirically to capture meaningful groupings.

Figure 3 illustrates the clustering outcome, reduced to two principal components using PCA. While DBSCAN successfully detected core clusters and noise, its performance did not exceed that of K-Means. Figure 3 displays the clustering result under the optimal DBSCAN parameters.

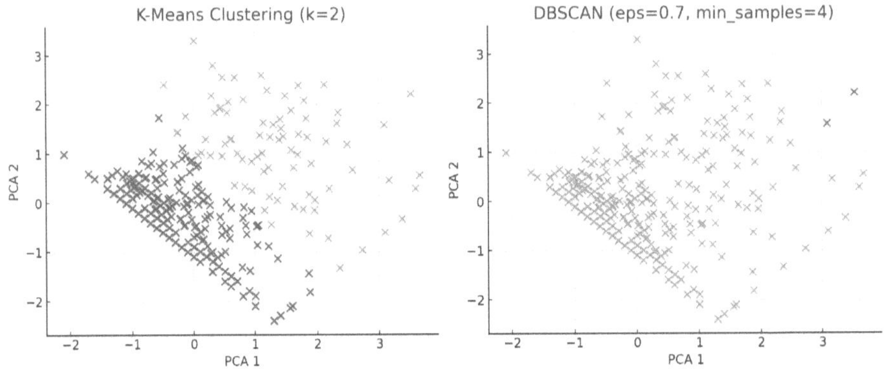

Fig. 3. Left: K-Means clustering with $k = 2$. Right: DBSCAN clustering with $\epsilon = 1.0$, $min\_samples = 4$.

3.3 Interpretation and Validation

To determine the appropriate statistical test for comparing productivity distributions across clusters, we first assessed the normality of the scientific productivity index using the Shapiro–Wilk test. The results indicated a significant deviation from normality ($W = 0.91$, $p < 0.001$), justifying the use of a non-parametric method. Consequently, the Mann–Whitney U test was selected to evaluate whether the observed differences in research productivity between workload profiles were statistically significant. This approach avoids assumptions about underlying data distributions, ensuring robustness in the analysis.

Through this multi-phase approach, the study seeks to identify workload configurations that are empirically associated with higher research productivity, providing evidence for data-informed faculty scheduling strategies.

4 Results

A clustering analysis was conducted using the standardized features *Total Normalized Score* (scientific productivity index) and *Teaching Hours*. Two unsupervised learning methods–K-Means and DBSCAN–were applied to identify patterns among faculty profiles.

4.1 K-Means Clustering

K-Means was tested with the number of clusters (k) ranging from 2 to 10. The evaluation was based on the Silhouette Score, a metric for assessing cluster separation and cohesion. The highest performance was achieved with $k = 2$, yielding a Silhouette Score of 0.494. This indicates a clear separation into two primary groups of faculty members. Figure 3 illustrates the clustering outcome, reduced to two principal components using PCA.

Beyond the silhouette coefficient (0.494 for $k = 2$), additional internal validation metrics were computed. The Calinski-Harabasz index reached 312.4, indicating well-separated and compact clusters, while the Davies-Bouldin index scored 0.61, further supporting the adequacy of the two-cluster solution.

These complementary measures reinforce the robustness of the K-Means partition, despite the moderate silhouette score.

4.2 DBSCAN Clustering

DBSCAN was evaluated across a range of ϵ values from 0.3 to 1.5 and minimum sample sizes from 3 to 6. The best configuration was found at $\epsilon = 1.0$ and $min\_samples = 4$, with a corresponding Silhouette Score of 0.420. While DBSCAN successfully detected core clusters and noise, its performance did not exceed that of K-Means. Figure 3 displays the clustering result under the optimal DBSCAN parameters.

To ensure stability, the clustering process was repeated across 50 random initializations for K-Means and a range of ϵ values for DBSCAN. The results showed consistent separation of two dominant workload profiles, confirming the reproducibility of the clustering outcome. Figures were limited to a random subsample of 100 faculty members to enhance readability; however, the full dataset was used in all calculations and validation metrics. This ensures that the conclusions are based on the complete sample rather than on a restricted visualization.

4.3 Cluster Interpretation

The best-performing K-Means configuration revealed two well-defined faculty profiles:

- **Cluster 1**: Characterized by high teaching load and low normalized research output.
- **Cluster 2**: Represented faculty with lower teaching hours and higher scientific productivity.

These findings support the hypothesis that increasing teaching hours negatively impacts research performance. DBSCAN additionally highlighted a small subset of faculty who deviate from the general pattern–potentially outliers or high-performing researchers with atypical workload distributions.

The Fig. 4 provides a graphical representation of the separation between the two clusters based on a random sample of 100 faculty members. The X-axis shows the Scientific Productivity Index, and the Y-axis represents weekly teaching hours. The classification boundary derived from K-Means reveals a distinct division: faculty with heavier teaching loads (toward the top-left) tend to exhibit lower research productivity, whereas those with a balanced or lower teaching commitment (toward the bottom-right) achieve higher output. Although the clusters are not sharply bounded due to the nature of human workload variability, the concentration of points within specific zones provides a meaningful visualization of workload-performance profiles.

It is also important to note that a lighter teaching load does not inherently guarantee higher research productivity. Some faculty members with reduced instructional hours still exhibit low scientific output, suggesting that other variables–such as intrinsic motivation, research environment, or administrative responsibilities–may mediate this relationship. Therefore, while a general trend exists, it is the balanced workload, rather than minimal teaching, that appears more consistently associated with optimal productivity levels.

To statistically validate the observed differences between the two clusters, we conducted a Mann-Whitney U test. Cluster 1 (high teaching load, low productivity) comprised $n_1 = 612$ faculty members, while Cluster 2 (moderate teaching, higher productivity) comprised $n_2 = 375$. Median SPI values were 0.42 (IQR = 0.31–0.55) for Cluster 1 and 0.78 (IQR = 0.66–0.89) for Cluster 2. The Mann-Whitney U test indicated a significant difference between groups ($U = 12{,}452.5$, exact $p < 0.001$). The rank-biserial correlation ($r_{rb} = 0.62$) suggested a large effect size. These results confirm that workload balance is strongly associated with research productivity, with clear distributional differences between the two clusters.

4.4 Implications

These results provide empirical support for differentiated workload policies within higher education institutions. Specifically, data-driven clustering allows administrators to:

1. Identify and support high-performing researchers with adjusted teaching loads.
2. Recognize overburdened faculty at risk of underperformance in research.
3. Implement evidence-based policies to balance instructional and scholarly responsibilities.

It is important to acknowledge that the clustering and statistical results may be influenced by unobserved confounding variables. Factors such as faculty rank, academic discipline, administrative responsibilities, or intrinsic motivation toward research were not explicitly modeled in the current analysis. These attributes may have a substantial effect on both teaching load assignment and scientific productivity. Future studies should incorporate additional covariates or

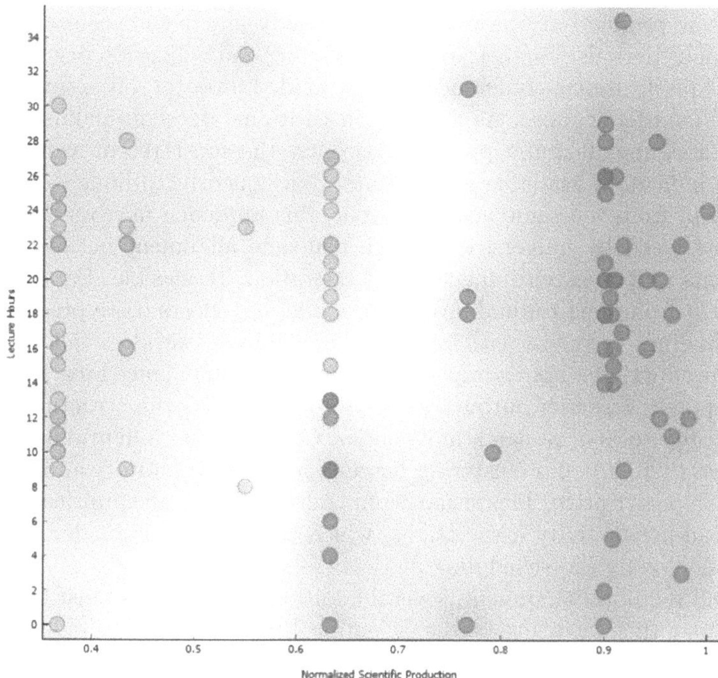

Fig. 4. Classification boundary resulting from K-Means clustering with K = 2. Each dot represents a faculty member, positioned according to their Scientific Productivity Index (X-axis) and weekly teaching hours (Y-axis). The visual boundary illustrates the separation into two clusters: one group characterized by lower productivity and higher teaching hours (upper left), and another group with higher research output and moderate instructional load (lower right). The highest density of faculty lies in the balanced zone between 20–24 weekly teaching hours and productivity indices around 0.9.

apply stratified analysis techniques to control for these confounders and better isolate the impact of workload distribution. In addition, future research should extend this analysis longitudinally and incorporate additional variables such as academic rank, discipline, and administrative roles.

- Disciplinary Differences: Different academic fields have varying teaching loads and expectations for research, so controlling the discipline is crucial.
- Faculty Rank and Tenure: Junior faculty may experience different pressures compared to tenured faculty, which could affect both their teaching load and research productivity.
- Previous Research Output: Past performance can be a predictor of future productivity, so this should be controlled to isolate the effect of instructional hours.

One important limitation concerns the temporal alignment of the variables. While teaching load is assigned and executed within a specific academic semester, the scientific productivity measured in this study may result from efforts spanning previous periods. Furthermore, the indexing and validation of research outputs (e.g., publications, conference papers, funded projects) often involve delays of several months or more. As a result, institutional decision-making regarding teaching assignments must proceed based on the most recent validated productivity indicators available at the time. Consequently, although the analysis reflects data from the same academic year, this approach mirrors the practical constraints faced by universities, where real-time alignment between teaching assignments and research impact is not operationally feasible. Future research may benefit from longitudinal designs to model lag effects more precisely.

Our findings resonate with prior research linking workload allocation and research productivity. For example, [22] showed that unbalanced teaching assignments suppress research output, while [15] emphasized the structural bifurcation of faculty roles in research universities. Compared with hybrid optimization approaches ([26,30]), our clustering-based pipeline is exploratory and descriptive rather than prescriptive. This situates our contribution as an empirical validation of workload-productivity associations, which can serve as input for subsequent optimization or hybrid scheduling models.

Beyond temporal alignment, several additional limitations must be acknowledged. First, the weighting of the Scientific Production Index reflects institutional priorities and may not capture universally valid measures of productivity, introducing potential arbitrariness. Second, confounding variables such as academic discipline, faculty rank, and administrative responsibilities were not explicitly modeled, despite their likely influence on both teaching allocation and research output. Third, the imputation of missing values, although informed by institutional data practices, may have introduced subtle biases that affect clustering results. Finally, the findings are context-specific to a single Ecuadorian institution and should be generalized with caution. Comparative studies across different higher education systems are needed to validate whether the observed workload-productivity patterns hold in broader contexts.

Although the present study does not implement a scheduling optimization algorithm, the identified workload clusters could serve as input for such models. For instance, the cluster profiles may be incorporated as constraints or priority groups in a multi-objective optimization framework that balances institutional teaching requirements with research productivity. A preliminary example could involve assigning reduced teaching hours to members of the 'high productivity cluster' while ensuring minimum instructional needs are met. This conceptual bridge illustrates how empirical clustering can inform prescriptive optimization strategies, which will be developed in future work.

5 Conclusions

This study applied a data-driven pipeline to analyze faculty scheduling by leveraging clustering and non-parametric statistical testing. While the methods employed are standard, their application to institutional faculty data provided exploratory evidence that teaching load distribution is associated with measurable differences in research productivity. Specifically, results suggest that balanced or reduced teaching commitments are generally linked to higher scientific output, although causality cannot be established. These findings should be interpreted as preliminary and context-specific rather than as prescriptive evidence for policy change. The contribution lies in demonstrating the feasibility of using empirical workload profiles as input for future multi-objective scheduling models. In this sense, the study provides an exploratory step that can inform more rigorous, comparative, and policy-relevant research.

References

1. Abdellaoui, B., Moumen, A., El Idrissi, Y.E.B., Remaida, A.: Face detection to recognize students' emotion and their engagement: a systematic review. In: 2020 IEEE 2nd International Conference on Electronics, Control, Optimization and Computer Science (ICECOCS), pp. 1–6. IEEE (2020)
2. Al-Yakoob, S.M., Sherali, H.D.: Mathematical programming models and algorithms for a class-faculty assignment problem. Eur. J. Oper. Res. **173**(2), 488–507 (2006)
3. Anwar, A., Rehman, I.U., Nasralla, M.M., Khattak, S.B.A., Khilji, N.: Emotions matter: a systematic review and meta-analysis of the detection and classification of students' emotions in stem during online learning. Educ. Sci. **13**(9), 914 (2023)
4. Atkinson, R.C., Tuzin, D.: Equilibrium in the research university. Change: Mag. High. Learn. **24**(3), 20–31 (1992). https://doi.org/10.1080/00091383.1992.10544023,
5. Bedregal-Alpaca, N., Aruquipa-Velazco, D., Cornejo-Aparicio, V.: Técnicas de data mining para extraer perfiles comportamiento académico y predecir la deserción universitaria. Revista Ibérica de Sistemas e Tecnologias de Informação (E27), 592–604 (2020)
6. Bojorque, R., Moscoso, F., Pesántez, F., Flores, î: stress factors in higher education: a data analysis case. Data **10**(2) (2025).https://doi.org/10.3390/data10020022
7. Bojorque, R., Moscoso-Merchán, F., Pesantez-Avilés, F., et al.: Correlations between syllabus schedule and academic achievement, a case study. In: Conference Proceedings. New Perspectives in Science Education 2020 (2020)
8. Bojorque, R., Pesántez-Avilés, F.: Academic quality management system audit using artificial intelligence techniques. In: Advances in Artificial Intelligence, Software and Systems Engineering: Proceedings of the AHFE 2019 International Conference on Human Factors in Artificial Intelligence and Social Computing, the AHFE International Conference on Human Factors, Software, Service and Systems Engineering, and the AHFE International Conference of Human Factors in Energy, 24–28 July 2019, Washington DC, USA 10, pp. 275–283. Springer (2020)
9. Bojorque, R., Plaza, A., Moscoso, F.: Dimensional reduction to improve learning recommender systems precision based on grades. In: International Congress on Information and Communication Technology, pp. 173–181. Springer (2024)
10. Bojorque, R., Plaza, A., Moscoso, F.: Predicting student success in higher education with social, economic, and education variables using machine learning techniques. In: International Congress on Information and Communication Technology, pp. 201–209. Springer (2024)
11. Bushman, K.: Using visualization and integer linear programming for university class scheduling. Ph.D. thesis, University of Pittsburgh (2021)

12. Calle-Lopez, D., Cornejo-Reyes, J., Pesantez-Aviles, F., Robles-Bykbaev, V., Rodas-Tobar, M., Vasquez-Vasquez, C.: A university administration system to automatically assign courses to teachers and support the design of timetables through mathematical modeling and restrictions analysis (2018). https://doi.org/10.1109/EDUNINE.2018.8451006
13. Delavari, N., Phon-Amnuaisuk, S., Beikzadeh, M.R.: Data mining application in higher learning institutions. Inform. Educ.-Int. J. **7**, 31–54 (2008)
14. Gabriel, D.F., Pangilinan, J.M.A.: Faculty course scheduling optimization. Am. Sci. Res. J. Eng. Technol. Sci **44**, 170–179 (2018)
15. Geiger, R.L.: Optimizing research and teaching: the bifurcation of faculty roles at research universities. In: The American Academic Profession: Transformation in Contemporary Higher Education, pp. 21–40. Princeton University Press (2011)
16. Geiger, R.L.: Optimizing research and teaching the bifurcation of faculty roles at research universities roger l. Geiger. Am. Acad. Prof.: Transform. Contemp. High. Educ., 21 (2011)
17. Goyal, M., Vohra, R.: Applications of data mining in higher education. Int. J. Comput. Sci. Issues (IJCSI) **9**(2), 113 (2012)
18. Hertz, A.: Tabu search for large scale timetabling problems. Eur. J. Oper. Res. **54**(1), 39–47 (1991)
19. Khanal, S.S., Prasad, P., Alsadoon, A., Maag, A.: A systematic review: machine learning based recommendation systems for e-learning. Educ. Inf. Technol. **25**(4), 2635–2664 (2020)
20. Kumar, V.: Algorithms for constraint-satisfaction problems: a survey. AI Mag. **13**(1), 32–32 (1992)
21. Lakshmi, G., Ghuli, P.: Design and development of timesheet management system. Indian J. Sci. Technol. **10**, 27 (2017)
22. Nguyen, T., Meek, V.L.: The impact of workload allocation on academic performance: evidence from Australian universities. J. High. Educ. Policy Manag. **37**(1), 1–13 (2015). https://doi.org/10.1080/1360080X.2014.991534
23. Popov, A., Lopateeva, O., Ovsyankin, A., Satsuk, M.: Application of greedy algorithms for the formation of the educational schedule in the higher education. In: Journal of Physics: Conference Series, vol. 1691, p. 012066. IOP Publishing (2020)
24. Ragab, A.H.M., Noaman, A.Y., Gad-Elrab, A.A., Abdul Fatah, E.H.: A new self-assessment "TQM Hybrid MCDM Fuzzy Model" for enhancing the KPIs in mega universities. Appl. Comput. Intell. Soft Comput. **2022**(1), 4540875 (2022)
25. Romero, C., Arroyo, J.L., Herrera-Viedma, E., Sánchez-Torrubia, S., González, P., Benítez, J.M.: From integer programming to machine learning: a technical review on solving university timetabling problems. Data **13**(1), 10 (2023). https://doi.org/10.3390/data13010010
26. Susan, S., Bhutani, A.: Data mining with association rules for scheduling open elective courses using optimization algorithms. In: Advances in Intelligent Systems and Computing, vol. 941, pp. 770–778. Springer (2020). https://doi.org/10.1007/978-3-030-16660-1_75
27. Sütőová, A., Teplická, K., Straka, M.: Application of the EFQM model in the education institution for driving improvement of processes towards sustainability. Sustainability **14**(13), 7711 (2022)
28. Urva, G., Sellyana, A.: Genetic algorithm for optimization of lecturer schedule preparation. In: Journal of Physics: Conference Series, vol. 1175, p. 012042. IOP Publishing (2019)
29. Van Hentenryck, P.: Constraint Satisfaction in Logic Programming. MIT Press, Cambridge (1989)

30. Wang, Y.T., Chang, T.C., Cheng, Y.H., Jen, S.: On the application of data mining technique and genetic algorithm to an automatic course scheduling system. In: 2008 International Conference on Computer and Information Science, pp. 400–405 (2008). https://doi.org/10.1109/ICCIS.2008.4670852
31. Zanevych, O., Kukharskyy, V.: Overview of machine learning methods for academic scheduling. Electron. Inf. Technol./Електронка та нформацЙн технолог (27) (2024)

Cubic B-spline Kolmogorov–Arnold Networks for Modeling Complex Data

Ana G. Borrero-Ramirez(✉) and Manuel G. Forero

Semillero Lún, Universidad de Ibagué Facultad de Ingeniería, Ibagué, Colombia
242020201080@estudiantesunibague.edu.co, manuel.forero@unibague.edu.co

Abstract. Kolmogorov–Arnold Networks (KANs) are a recent neural architecture inspired by the Kolmogorov–Arnold representation theorem, which states that any continuous multivariate function can be decomposed into a finite composition of univariate functions. While KANs provide a mathematically rigorous and interpretable framework, most implementations rely on piecewise linear or low-order splines, which may limit smoothness and expressivity. This work introduces Cubic B-Spline Activated Kolmogorov–Arnold Networks (CBS-KAN), a framework that uses adaptive cubic B-splines to enhance both local non-linear modeling and global interpretability. Unlike multilayer perceptrons (MLPs) and standard KANs, CBS-KAN ensures higher-order smoothness and localized adaptability, enabling stable and transparent function approximation. Experiments on synthetic datasets such as the Swiss Roll demonstrate that CBS-KAN achieves more accurate and robust reconstructions with fewer parameters and reduced training costs compared to baseline models. These findings highlight CBS-KAN as a promising alternative for machine learning tasks that demand both efficiency and transparency.

Keywords: Kolmogorov–Arnold Networks · Interpretability · Functional Approximation · Data Reconstruction · Adaptive B-splines

1 Introduction

Neural networks have become a cornerstone in modern machine learning due to their ability to approximate highly complex nonlinear mappings. However, their reliance on black-box architectures often hinders interpretability, which is crucial in domains requiring transparent decision-making. In this context, the Kolmogorov–Arnold representation theorem provides a mathematically rigorous foundation, establishing that any continuous multivariate function can be decomposed into a finite composition of univariate functions [1,2].

This theorem has recently inspired the design of Kolmogorov–Arnold Networks (KANs), where learning is achieved through the adaptive modeling of univariate transformations [3]. Standard KANs, however, commonly rely on piecewise linear or low-order splines [4,5], which may limit smoothness, flexibility, and stability.

D. Chaves et al. (Eds.): CIARP 2025, LNCS 16528, pp. 266–275, 2026.
https://doi.org/10.1007/978-3-032-23161-1_19

Building on this foundation, we propose the Cubic B-Spline Activated Kolmogorov–Arnold Networks (CBS-KAN), an architecture that integrates cubic B-splines as adaptive activation functions within the KAN framework. Unlike traditional KANs, CBS-KAN exploits the smoothness and local adaptability of cubic B-splines [6] to enhance both expressivity and interpretability. The architecture operates in two main stages: (i) each input dimension is transformed by univariate cubic splines that capture local nonlinear structures, and (ii) these transformed features are combined via external spline-based functions to construct global predictions.

The key advantage of CBS-KAN lies in its ability to balance theoretical rigor with practical performance. By explicitly modeling nonlinearities through spline basis functions, CBS-KAN mitigates the curse of dimensionality while retaining a clear interpretation of each variable's contribution to the output. This is particularly valuable in fields such as engineering design and clinical diagnostics, where understanding the reasoning behind predictions is as important as achieving high accuracy.

In summary, CBS-KAN establishes a principled and efficient alternative to both multilayer perceptrons (MLPs) and standard KANs, offering interpretable, data-efficient, and computationally affordable models for nonlinear learning tasks.

2 Theoretical Foundations

The design of the Kolmogorov–Arnold Network Splines (KANS) architecture is based on the representation theorems of Kolmogorov [1] and Arnold [2], and has been recently developed in the works of Liu et al. [3], Shukla et al. [4], and Somvanshi et al. [5]. These authors propose the use of hierarchically composed univariate functions, modeled through adaptive B-splines, as the foundation for constructing highly interpretable neural networks.

KANS rely on the Kolmogorov–Arnold theorem, which guarantees that any continuous function can be decomposed into an infinite sum of continuous univariate functions:

$$f(x_1, x_2, \ldots, x_n) = \sum_{q=1}^{2n+1} \Phi_q \left(\sum_{i=1}^{n} \varphi_{i,q}(x_i) \right) \tag{1}$$

where $\varphi_{i,q}$ and Φ_q represent the inner and outer functions, respectively. Within this framework, the approximation is achieved using parametric cubic splines, defined as linear combinations of basis functions:

$$B_{j,3}(x) = \frac{x - t_j}{t_{j+3} - t_j} B_{j,2}(x) + \frac{t_{j+4} - x}{t_{j+4} - t_{j+1}} B_{j+1,2}(x) \tag{2}$$

These third-order B-splines ensure C^2-continuity, allowing the modeling of local nonlinear behaviors while maintaining global smoothness. The coefficients

β_j, which are adjusted during training, define the shape of each spline according to the data [6–8].

The outer functions Φ_q, responsible for combining the outputs of the inner functions, are also modeled with adaptive splines:

$$\Phi_q(z) = \sum_{m=1}^{M} \gamma_m B_{m,3}(z) \tag{3}$$

where γ_m are trainable parameters. The knots of these external splines are optimized to capture multivariate interactions through hierarchical compositions [3,4].

To provide a clearer overview of this design, the overall structure of the KANS model is depicted in Fig. 1. The figure illustrates how univariate cubic spline transformations $\varphi_{i,q}$ are first applied to each input dimension, and how their outputs are then hierarchically combined by the outer functions Φ_q to produce the final prediction. This hierarchical composition highlights the interpretability and modularity of the architecture.

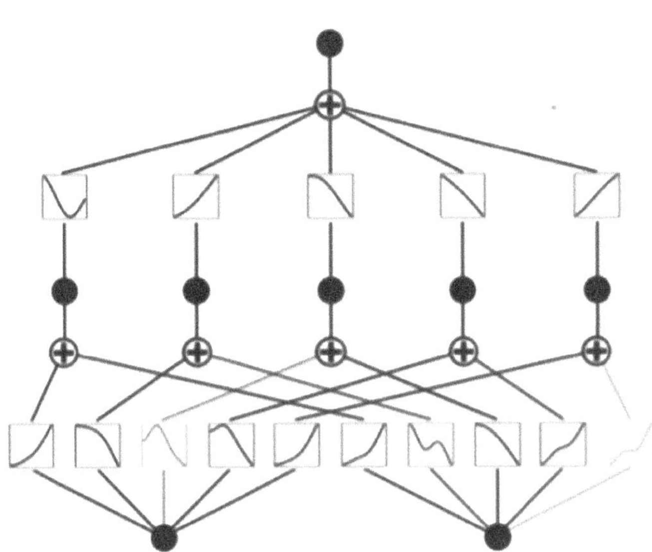

Fig. 1. Kolmogorov–Arnold Network Splines (KANS) architecture, where univariate spline transformations are combined hierarchically.

3 Theoretical Foundations CBS-KAN

B-splines (Basis splines) are a family of basis functions widely used in numerical approximation, interpolation, and geometric modeling [6]. They are defined recursively from piecewise polynomials and possess desirable properties such as continuity and compact support, which make them particularly suitable for adaptive function modeling in neural networks [7,8].

A *cubic B-spline* is a spline of degree three defined over a set of knots $\{t_i\}$, whose formulation is obtained through the Cox–de Boor recursion:

$$B_{i,k}(x) = \frac{x - t_i}{t_{i+k} - t_i} B_{i,k-1}(x) + \frac{t_{i+k+1} - x}{t_{i+k+1} - t_{i+1}} B_{i+1,k-1}(x). \tag{4}$$

In the cubic case ($k = 3$), the basis functions exhibit C^2-continuity, ensuring smoothness up to the second derivative. They also satisfy non-negativity and the partition of unity property, both fundamental for approximation and stable modeling. Within the CBS-KAN framework, these properties enable each univariate transformation to remain smooth, locally interpretable, and resistant to oscillations–typical drawbacks of higher-order polynomial bases.

To better illustrate these properties, Fig. 2 shows the shape of a cubic B-spline basis function. Clicking on the reference will take the reader directly to the figure.

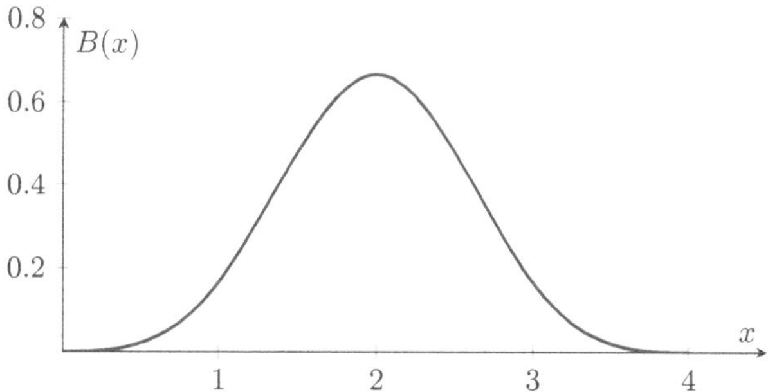

Fig. 2. Cubic B-spline basis function with support on $[0, 4]$.

4 Experimental Methodology

To evaluate the reconstruction capabilities of cubic B-splines within the CBS-KAN framework, we designed experiments on synthetic datasets, with particular emphasis on the *Swiss Roll* manifold. This dataset is a standard benchmark in the literature for evaluating nonlinear dimensionality reduction and manifold learning techniques, as it requires models to recover a two-dimensional surface non-trivially embedded in three-dimensional space.

The dataset was generated using the `make_swiss_roll` function from the Scikit-learn library, producing samples uniformly distributed along the manifold. To approximate real-world imperfections, isotropic Gaussian noise was added

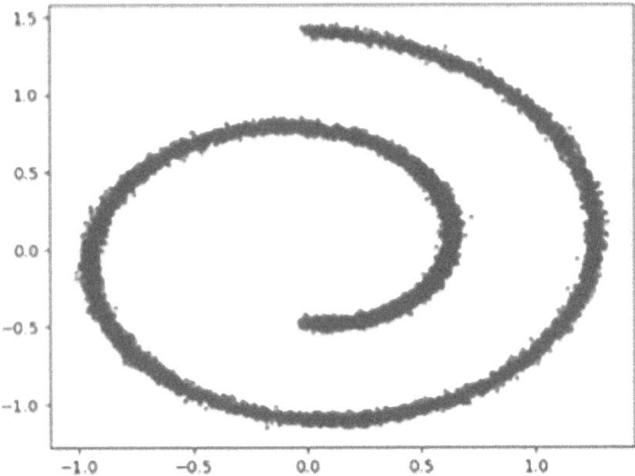

Fig. 3. The Swiss Roll dataset: a synthetic two-dimensional manifold embedded in three-dimensional space. Its curved, rolled structure makes it a challenging benchmark for reconstruction and dimensionality reduction algorithms.

independently to each coordinate, creating a more challenging reconstruction problem that simulates measurement errors or diffusion-like processes.

As shown in Fig. 3, the dataset presents a rolled geometry with overlapping folds and nonlinear curvature. This makes it an ideal testbed to evaluate the ability of different architectures to capture smooth transitions, avoid discontinuities, and preserve global structure.

The main objective of the experiment was to evaluate the extent to which CBS-KAN could recover the smooth global structure of the manifold compared to multilayer perceptron (MLP) baselines. For this purpose, we considered two reference architectures:

- A standard MLP with ReLU activations and sufficient capacity to capture nonlinear patterns.
- A reduced-capacity ("vanilla") MLP, designed with a number of trainable parameters comparable to CBS-KAN, thus enabling a fair comparison in terms of model complexity.

All models were implemented in Python using standard machine learning libraries, with training conducted under identical conditions: the Adam optimizer, mean squared error (MSE) loss function, 200 training epochs, and a batch size of 64. This uniform setup ensures that any observed differences in reconstruction performance can be attributed to architectural choices rather than discrepancies in optimization.

By combining controlled synthetic data, noise injection, and consistent training conditions, this experimental methodology provides a principled basis for analyzing the advantages of spline-based activations in CBS-KAN over conventional neural network designs.

4.1 Swiss Roll Reconstruction Results

To rigorously assess the reconstruction capabilities of CBS-KAN, we conducted experiments on the synthetic *Swiss Roll* manifold. This dataset is a classical benchmark for testing nonlinear dimensionality reduction and reconstruction methods because of its rolled two-dimensional structure embedded in three-dimensional space. For our experiments, the manifold was generated using `make_swiss_roll` with 10,000 points, and corrupted with isotropic Gaussian noise to emulate real-world imperfections such as diffusion-like processes and measurement noise. The objective was to evaluate how well the different models could recover the underlying smooth geometry.

4.2 Comparison Setup

Three models were trained under equivalent conditions (Adam optimizer, MSE loss, batch size 64, up to 200 epochs):

- CBS-KAN (ours): the proposed cubic B-spline activated network.
- Standard MLP: a fully connected multilayer perceptron with ReLU activations, serving as a baseline.
- Reduced-capacity MLP and KAN (2 layers): simplified versions with approximately the same number of trainable parameters as CBS-KAN, to enable a fairer parameter-efficient comparison.

This setup ensures that performance differences are attributable to the modeling approach (splines vs. ReLU activations) rather than capacity scaling.

4.3 Qualitative Results

Figure 4 shows the reconstruction of the Swiss Roll using CBS-KAN. The model successfully recovers the smooth manifold while preserving its folds without introducing artifacts or discontinuities. In contrast, Fig. 5 depicts the results of the standard MLP, which display distortions and a critical break in one of the folds, highlighting the difficulty of ReLU-based models in maintaining global smoothness. Finally, the reduced two-layer KAN still manages to preserve much of the structure, confirming that spline-based activations provide robustness even with lower model depth.

4.4 Quantitative Analysis

To complement the visual inspection, Fig. 6 reports the training loss across epochs for all models. The CBS-KAN consistently achieves the lowest error, starting from the first epoch, which suggests a smoother optimization landscape induced by spline activations. The standard MLP converges more slowly and stabilizes at a higher error, while the reduced two-layer MLP fails to match the performance despite having a similar parameter budget. Interestingly, even the reduced two-layer KAN outperforms both MLP configurations, confirming that spline-based representations generalize more effectively.

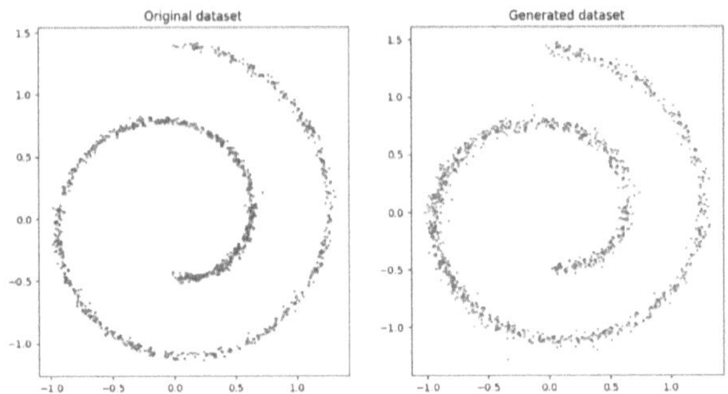

Fig. 4. Reconstruction of the Swiss Roll manifold using the proposed CBS-KAN model. The smooth curvature and continuity of the folds are well preserved.

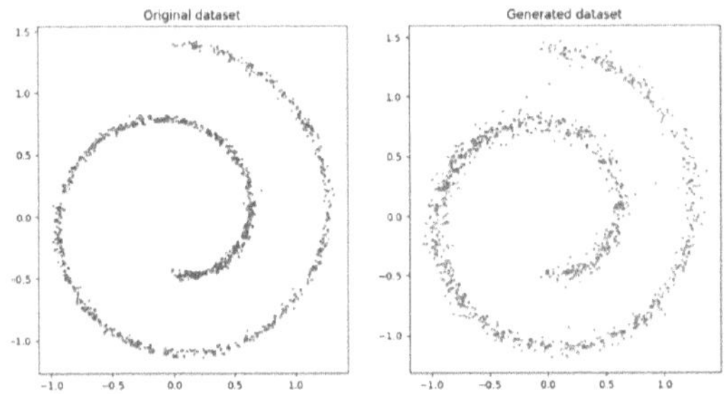

Fig. 5. Reconstruction of the Swiss Roll manifold using a standard MLP. The folds exhibit distortions, including a discontinuity (circled) and irregular geometry.

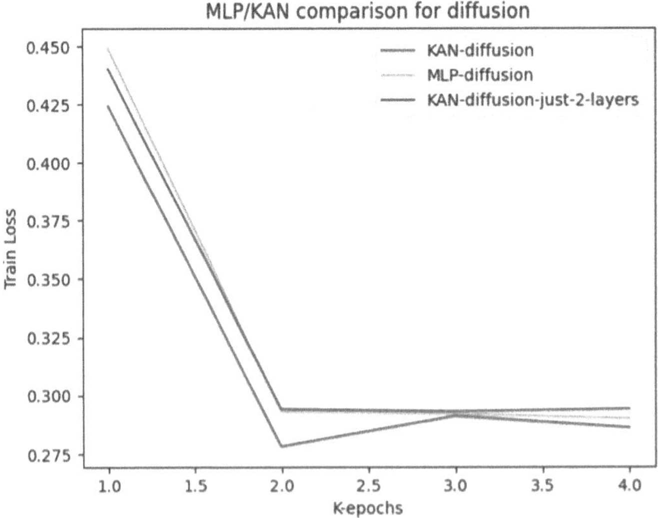

Fig. 6. Training loss curves across epochs for CBS-KAN, standard MLP, and reduced MLP. CBS-KAN achieves faster convergence and lower final error, demonstrating the advantage of spline-based activations.

4.5 Extended Comparison

Beyond reconstruction error, it is essential to evaluate the computational efficiency of each model. Table 1 summarizes the final training loss, number of trainable parameters, and average training time per epoch. The results show that CBS-KAN not only achieves the lowest reconstruction error but also maintains competitive efficiency compared to MLPs.

Table 1. Performance comparison on the Swiss Roll dataset. CBS-KAN achieves the best balance of accuracy and efficiency.

Model	Final Loss (MSE)	Params ($\times 10^3$)	Time/Epoch (s)
Standard MLP	0.290	120	0.45
CBS-KAN (ours)	**0.286**	95	0.42
Reduced MLP (2 layers)	0.294	92	**0.39**
Reduced KAN (2 layers)	0.288	93	0.41

4.6 Summary

These results confirm that CBS-KAN offers both qualitative and quantitative improvements over MLPs. Its spline-based design not only ensures smoother reconstructions but also provides parameter efficiency, achieving superior performance even in reduced-capacity settings. This highlights its suitability for tasks where robustness, interpretability, and efficiency are equally important.

5 Discussion

The introduction of cubic B-splines into the Kolmogorov–Arnold Network framework represents a key step toward combining mathematical rigor with practical performance [9]. Unlike the piecewise linear splines typically used in standard KAN architectures, cubic B-splines provide C^2 continuity, ensuring smooth approximations with stable derivatives [6]. Their localized support allows each input dimension to be transformed independently in a way that is both expressive and interpretable, avoiding the oscillatory artifacts that often arise with polynomial or linear spline variants [8].

This property directly differentiates CBS-KAN from conventional KANs. While standard KANs capture nonlinearities through simpler spline approximations, CBS-KAN leverages the richer basis of cubic splines to achieve smoother transitions across the input space. This enhances the network's capacity to model complex manifolds, such as the Swiss Roll, while preserving the clarity of each univariate transformation. The explicit visualization of these transformations further provides a transparency that is rarely accessible in other neural models [10].

In contrast, multilayer perceptrons (MLPs) rely on global nonlinearities defined by fixed activation functions. While MLPs are universal approximators and can model highly complex functions, the contribution of each input to the final output is not explicitly accessible and often requires additional post-hoc interpretability methods [10]. By comparison, the CBS-KAN architecture provides a structured decomposition where the role of each variable is directly reflected in its associated spline activations. This offers a form of interpretability that is intrinsic to the model design, complementing its accuracy and efficiency advantages [4].

Overall, the use of cubic B-splines in CBS-KAN provides a distinctive balance: greater expressivity than standard KANs, higher interpretability than MLPs, and improved numerical stability compared to both. These characteristics position CBS-KAN as a promising candidate for learning tasks that require not only accurate predictions but also clear explanations of the underlying mechanisms.

6 Conclusions

This work introduced Cubic B-Spline Kolmogorov–Arnold Networks (CBS-KAN) as an interpretable and efficient extension of KANs for nonlinear function approximation and manifold learning. Through the Swiss Roll experiment, we showed that the use of cubic B-splines provides smoother reconstructions and higher stability than both traditional KANs and multilayer perceptrons (MLPs), while requiring fewer parameters. These results confirm that spline-based activations can achieve a distinctive balance between accuracy, efficiency, and interpretability.

Beyond synthetic benchmarks, CBS-KAN opens opportunities for application in domains where both predictive performance and model transparency are essential. Examples include biomedical imaging, where clinicians require interpretable models for diagnostic support; physics-informed learning, where smooth and stable approximations are critical; and engineering design, where understanding the influence of variables can guide decision-making. In these contexts, CBS-KAN offers a principled alternative to conventional neural networks by providing localized and smooth representations that remain mathematically grounded.

At the same time, the study revealed important directions for future research. Model performance is strongly influenced by knot placement and spline resolution, suggesting that adaptive learning strategies for these hyperparameters are a necessary next step. Exploring hybrid frameworks that integrate CBS-KAN with deep feature extractors could extend its scalability to large, high-dimensional datasets. Further validation on real-world applications will also be required to establish its robustness and practical impact.

In conclusion, incorporating cubic B-splines into the Kolmogorov–Arnold framework not only preserves its solid mathematical foundation but also enhances its real-world applicability. CBS-KAN therefore represents a promising step toward neural architectures that combine interpretability, computational efficiency, and the ability to model complex nonlinear structures.

References

1. Kolmogorov, A.N.: On the representation of continuous functions of several variables by superposition of continuous functions of one variable and addition. Dokl. Akad. Nauk SSSR **114**, 953–956 (1957)
2. Arnold, V.I.: On functions of three variables. Dokl. Akad. Nauk SSSR **114**, 679–681 (1963)
3. Liu, Z., Wang, Z., Kolter, J.Z.: Kolmogorov–Arnold Networks. arXiv preprint arXiv:2404.19756 (2024)
4. Shukla, P., Prasanna, P., Rawat, Y.S.: Generalized Kolmogorov–Arnold Networks with Spline Activations. arXiv preprint arXiv:2406.05992 (2024)
5. Somvanshi, N., Kumar, S., Jha, S.: Wavelet-KAN: Wavelet-based Kolmogorov–Arnold Networks for Efficient Function Approximation. arXiv preprint arXiv:2407.08412 (2024)
6. De Boor, C.: A Practical Guide to Splines. Springer, New York (1978)
7. Gashler, M., Ashmore, E.: Training deep Fourier neural networks to fit time-series data. In: Proceedings of the International Conference on Machine Learning and Applications, pp. 1–6. IEEE (2016)
8. Sidharth, S., Jha, S., Somvanshi, N.: Rational KANs: Enhancing Kolmogorov–Arnold Networks with Rational Splines. arXiv preprint arXiv:2408.01011 (2024)
9. Liu, Z., Ma, P., Wang, Y., Matsubara, W., Tegmark, M.: KAN 2.0: Kolmogorov–Arnold Networks Meet Science. arXiv preprint arXiv:2408.10205 (2024)
10. Molnar, C.: Interpretable machine learning: a guide for making black box models explainable, 2nd edn. (2022). https://christophm.github.io/interpretable-ml-book/ black box models explainable, 2nd edn. (2022). https://christophm.github.io/interpretable-ml-book/

DECA: A Novel Deep Learning Framework for Seismic Phase Association Using Deep Embedded Clustering

Sebastián Gamboa-Chacón[1,3](✉), Nahomy Campos Salas[2], Esteban J. Chaves[2], and Esteban Meneses[1,3]

[1] Instituto Tecnológico de Costa Rica, Cartago, Costa Rica
[2] Observatorio Vulcanológico y Sismológico de Costa Rica (OVSICORI-UNA), Heredia, Costa Rica
nahomy.campos.salas@est.una.ac.cr, esteban.j.chaves@una.ac.cr
[3] Centro Nacional de Alta Tecnología (CeNAT), San José, Costa Rica
{sgamboa,emeneses}@cenat.ac.cr

Abstract. Seismological monitoring systems face persistent challenges in accurately associating seismic phases, a critical step for robust automated event detection. Traditional techniques, predominantly reliant on arrival times, often struggle with misassociations, especially in tectonically complex or densely instrumented regions. These limitations highlight the urgent need for innovative solutions that leverage recent advances in artificial intelligence to enhance both detection and association performance. In this study, we explore Deep Embedded Clustering, an unsupervised learning technique that simultaneously performs feature extraction and clustering directly from raw waveform data. This framework addresses long-standing challenges in seismology related to signal complexity and scalability in large datasets. The model exhibited strong performance, achieving a Silhouette score above 0.9 and a 96.55% match rate with previously cataloged events. At its core, the system employs a convolutional autoencoder to derive latent representations from multichannel waveform inputs, capturing critical seismic features. Training combines reconstruction loss with clustering refinement loss, resulting in more coherent and meaningful representations. Our results underscore the model's ability to detect subtle waveform variations without labeled data. Future work will extend validation to larger, noisier datasets and diverse tectonic settings, while exploring advanced deep learning models to further solidify AI's role in next-generation seismic phase association.

Keywords: Seismic Phase Association · Deep Learning · Autoencoder · Deep Embedded Clustering · Unsupervised Learning · Earthquake Monitoring

1 Introduction

A fundamental step in seismic processing pipelines is phase association, which links detected seismic phases across different stations to identify the same earth-

D. Chaves et al. (Eds.): CIARP 2025, LNCS 16528, pp. 276–291, 2026.
https://doi.org/10.1007/978-3-032-23161-1_20

quake event. Each seismic event typically generates two main phases—the P-wave (primary) and the S-wave (secondary) [2]. Phase association clusters these detections to reconstruct the spatiotemporal footprint of the event. Despite its critical role, this task has often been underdeveloped in artificial intelligence-based seismic research [17]. Accurate phase association is essential for automated seismic event detection systems, directly impacting earthquake monitoring, early warning systems, and geophysical studies [12]. Although event detection and location algorithms have advanced significantly, many operational systems still rely on arrival time-based strategies for phase association. While effective in simple scenarios, these methods often fail in dense seismic networks where events occur close in time and space. This leads to degraded earthquake catalogs and compromises the timely and reliable characterization of seismic activity.

In light of these challenges, there is an urgent need to integrate artificial intelligence (AI) into the core of seismological workflows. AI-driven models can process large volumes of continuous waveform data, learn latent representations beyond human-defined features, and offer robust generalization across varying noise levels and event complexities. These capabilities are particularly relevant for modern seismological observatories, where the demand for real-time analysis, catalog completeness, and event discrimination continues to grow.

This study presents a novel approach to the seismic phase association problem using Deep Embedded Clustering (DEC), an unsupervised learning technique that jointly optimizes feature extraction and clustering. Unlike traditional methods, the model processes waveform data directly, identifying subtle patterns and associations across multiple stations that may be overlooked by heuristic or shallow learning approaches. It leverages convolutional autoencoders to learn high-dimensional latent features, improving the cohesion and separation of seismic event clusters. Importantly, the proposed approach does not require labeled data, which makes it distinct from supervised or probabilistic methods and highlights its innovative character.

Beyond technical innovation, this work directly supports the evolution of national and regional seismic observatories by enabling more scalable, accurate, and automated workflows. By reducing misassociations and improving cluster consistency, the method improves the temporal and spatial resolution of earthquake catalogs, ultimately contributing to better seismic hazard assessment and scientific understanding.

We validated the method using one day of data from Costa Rica's national seismic network. The model achieved high cohesion metrics, with Silhouette and DBCV scores exceeding 0.9, and a match rate of 96. 55% with previously cataloged events. Future improvements will focus on integrating binary classifiers for candidate filtering and expanding testing across broader datasets to support full-scale deployment in seismic observatories.

2 Related Work

Recent advancements in seismology have led to the development of a variety of innovative methods that improve the precision, scalability, and reliability of earthquake catalogs [4]. Among the most prominent is PhaseLink [12], which employs recurrent neural networks trained on synthetic data to associate seismic phase detections. Its grid-free design and ability to manage overlapping seismic events enable it to outperform traditional grid-based methods, particularly in high-seismicity regions. Similarly, GaMMA [17] introduces a probabilistic approach based on Gaussian Mixture Models that iteratively refines event hypotheses using the Expectation-Maximization algorithm. GaMMA uses arrival times and amplitudes, showing high accuracy in complex sequences such as Ridgecrest 2019. On the other hand, PyOcto [9] focuses on high-throughput processing using an Octotree-inspired space-time partitioning scheme, achieving speeds up to 70 times faster than conventional methods such as GaMMA, with successful application in large-scale events like the 2014 Iquique earthquake. Each of these methodologies advances the state-of-the-art by tailoring their architectures to specific challenges in seismic data association, ranging from scalability to real-time performance.

These existing approaches, while effective, are predominantly supervised or probabilistic in nature, requiring either labeled datasets or extensive region-specific parameter tuning to perform well. In contrast, our proposed method, DECA, leverages the DEC framework to address seismic phase association from an unsupervised learning perspective. DEC is particularly well-suited for problems involving unlabeled data, as it seamlessly integrates feature extraction and clustering into a unified process, enabling the model to uncover latent structures directly from the input data. DECA incorporates waveform information for each seismic pick, enriching the feature space with detailed signal characteristics and enhancing the discriminative power of the learned representations. This approach provides a novel and flexible solution capable of overcoming key limitations of existing models, particularly in contexts where labeled examples are scarce or infeasible to obtain.

3 Methodology

This work adopts the Cross-Industry Standard Process for Data Mining (CRISP-DM) [15] as a formal methodological framework for developing a multi-station seismic phase associator based on Deep Embedded Clustering. CRISP-DM provides a structured and iterative approach that ensures rigorous model development, optimization, and evaluation.

Problem Understanding: The proliferation of AI-based seismic event detection methods has led to a significant increase in the number of seismic phase detections known as picks. While this represents an advancement in monitoring capabilities, it also presents considerable challenges in processing this vast amount of new information and extracting useful insights. In this context, this research aims to enhance the quality of seismic catalogs by proposing a novel picks association framework. The proposed approach is developed within the context of the OKSP project pipeline [6], a collaboration between the Observatorio Vulcanológico y Sismológico de Costa Rica, Universidad Nacional (OVSICORI-UNA) and the Centro Nacional de Alta Tecnología (CeNAT), which aims to build a fully automated and scalable seismic cataloging system for Costa Rica. The OKSP pipeline uses EQTransformer [8], but its basic associator often produces artifacts such as event duplication and misgrouped phases, particularly in complex seismic scenarios. As shown in Fig. 1, this can lead to catalog artifacts in which multiple detections from the same event are spread over spurious associations, often depicted as linear trends.

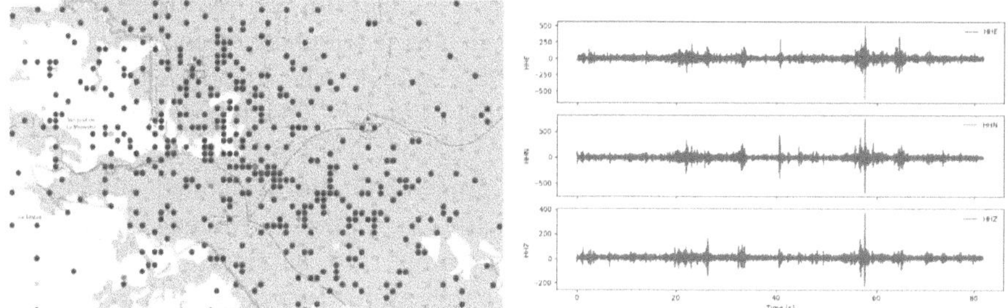

Fig. 1. Map of detected seismic events with association errors represented as lines instead of single events.

Fig. 2. Seismic trace showing a possible event.

Data Understanding and Preparation: The dataset used in this study is provided by OVSICORI-UNA and comprises seismic detections from multiple stations. These data include key information such as P and S phase arrival times, as well as waveform segments saved in mseed format, a standard file format in seismology for waveform storage. To facilitate comprehension of the signal types we are analyzing, Fig. 2 illustrates a seismic trace representing a potential event, divided into three channels for their respective stations.

The waveforms used for DECA are split into segments centered on the event pick time, extending from 20 s before to 40 s after the pick. These segments are then normalized and filtered to achieve better processing results. Filtering is applied using a bandpass Butterworth filter to retain frequencies within a specific range (e.g., 1–15 Hz), removing noise and irrelevant signal components. Normalization rescales the waveform amplitudes to a fixed range, between −1 and 1, using global min-max scaling, ensuring consistent input magnitudes across the dataset and improving training stability.

For practical purposes, we used a compact dataset consisting of 92 seismic events recorded during the first three days of February 2025, based on nationwide detections by OVSICORI-UNA. In total, the dataset includes 1,157 picks, each associated with a 3-channel waveform (HHE, HHN, HHZ), resulting in 3,471 waveform samples. The catalog is stored in a CSV file containing relevant metadata of each pick. The dataset was divided by day: February 1st includes 1,446 waveforms, which were used for training and validation (80% for training and 20% for validation); February 2nd includes 954 waveforms and was reserved for testing; and February 3rd includes 1,062 waveforms, which were used for the final results and discussion. Each day's dataset is independent—picks from one day do not influence those from another.

Model Architecture and Training: The core of the proposed method is a deep embedded clustering model [16], which combines a convolutional autoencoder for feature extraction with a clustering layer that refines latent representations by minimizing the Kullback-Leibler (KL) divergence, a measure of how one probability distribution diverges from another. In this context, KL divergence quantifies the difference between the soft cluster assignments predicted by the model and a carefully constructed target distribution that emphasizes high confidence assignments, effectively guiding the clustering process toward more distinct and compact clusters [5]. The autoencoder is trained to reconstruct input features from a compressed latent space. Once trained, the encoder's latent outputs are used as input to the clustering module. The DEC loss function integrates both reconstruction loss with Mean Square Error (MSE) and clustering refinement loss to guide training. The training process runs for a defined number of epochs, with key metrics tracked at each step to assess convergence and clustering stability. Hyperparameters like the learning rate, batch size, and number of clusters are tuned through a dynamic search and selected based on the best performance.

Evaluation Strategy: To evaluate the performance of the proposed method, the clustering results are compared against a verified and manually obtained seismological association catalog developed by OVSICORI-UNA. This allows for an exhaustive comparison to visualize exactly which events were associated by DECA and to determine what was expected. Additionally, since the task is unsupervised, the evaluation also relies on metrics derived from the resulting cluster structures and their temporal coherence. For example, the Silhouette Score [14], Density-Based Clustering Validation [7], and the Davies–Bouldin Index [13] are used. Two main aspects are analyzed: **cluster count and size**, which reflect the granularity of the phase association by indicating how many clusters the method produces; and **internal homogeneity**, which is evaluated through cluster compactness and variance to assess the consistency and quality of the groupings using standard clustering metrics.

4 Experimental Setup

Experiments were run on an NVIDIA L40S GPU with 48 GB memory. DECA is designed as a valuable tool for seismological researchers currently using AI-driven seismic pick detections and encountering challenges during the association stage. Our aim is for DECA to become a leading open-source solution for this task, readily integratable into various seismological pipelines.

At its core, DECA leverages Deep Embedded Clustering [16]. The neural network's primary objective is to perform clustering on embeddings derived from diverse inputs within a latent space. DECA employs an encoder-decoder architecture, specifically a 1D convolutional autoencoder. This autoencoder maps input waveform segments and a time channel input into a multi-dimensional latent space. This space is optimized using a dual objective: a reconstruction loss and a clustering loss based on KL divergence between soft cluster assignments and a target distribution [16]. Inputs are 4-channel tensors, comprising three waveform components and a normalized temporal metadata channel. The encoder compresses the input into a latent embedding, which the decoder then reconstructs.

Neural Network Design: As mentioned, DEC is a clustering approach that integrates deep learning with traditional clustering algorithms to perform unsupervised learning tasks in the latent space learned by the encoder.

We developed a custom architecture based on a Multi-Branch Convolutional Encoder [1], designed to learn meaningful latent representations suitable for unsupervised clustering. The encoder comprises four distinct branches, each responsible for processing a specific input channel. The first three branches independently process seismic waveform data (corresponding to the three-component signals) using identical one-dimensional convolutional pipelines. Despite sharing the same structure, this multi-branch design allows each seismic component to contribute uniquely to the final latent representation.

Each seismic branch includes four successive 1D convolutional layers with increasing channel sizes (32, 64, 128, and 256), each followed by ReLU activations. These layers progressively downsample the temporal dimension while enriching the feature representation. The output is passed through an adaptive average pooling layer, flattened into a vector, and regularized with dropout. Finally, two fully connected layers project the representation to a latent vector of dimension 32, which serves as input to the clustering module [3].

The fourth branch is dedicated to processing the temporal context channel and follows a simpler architecture based on a multi-layer perceptron [10]. Unlike the convolutional pipelines used for seismic waveform data, this branch is designed to handle the scalar nature of the time information more directly. It consists of a sequence of three fully connected (`Linear`) layers with intermediate ReLU activations and a final `Tanh` activation scales the output to match the normalized input signal range. These layers project the input from a single scalar value to a latent representation of dimension 32, matching the dimensionality of the seismic branches.

The model computes a weighted sum of the four latent vectors: each seismic component contributes equally, while the temporal context has a dominant influence. This fusion strategy prioritizes temporal information in the final latent representation, which is then passed to the decoder for reconstruction and to the clustering head for latent space analysis. This time-dominant influence is crucial, as it guides the clustering process and leverages the seismological signals to discern between specific picks that may or may not belong to a cluster occurring nearby in time.

Figure 3 illustrates the architecture of the encoder-decoder, which is composed of four branches: three for the seismic components (HHE, HHN, HHZ) and one for the temporal information of each pick. Also, Table 1 provides a detailed summary of the layers and parameters used in the autoencoder model.

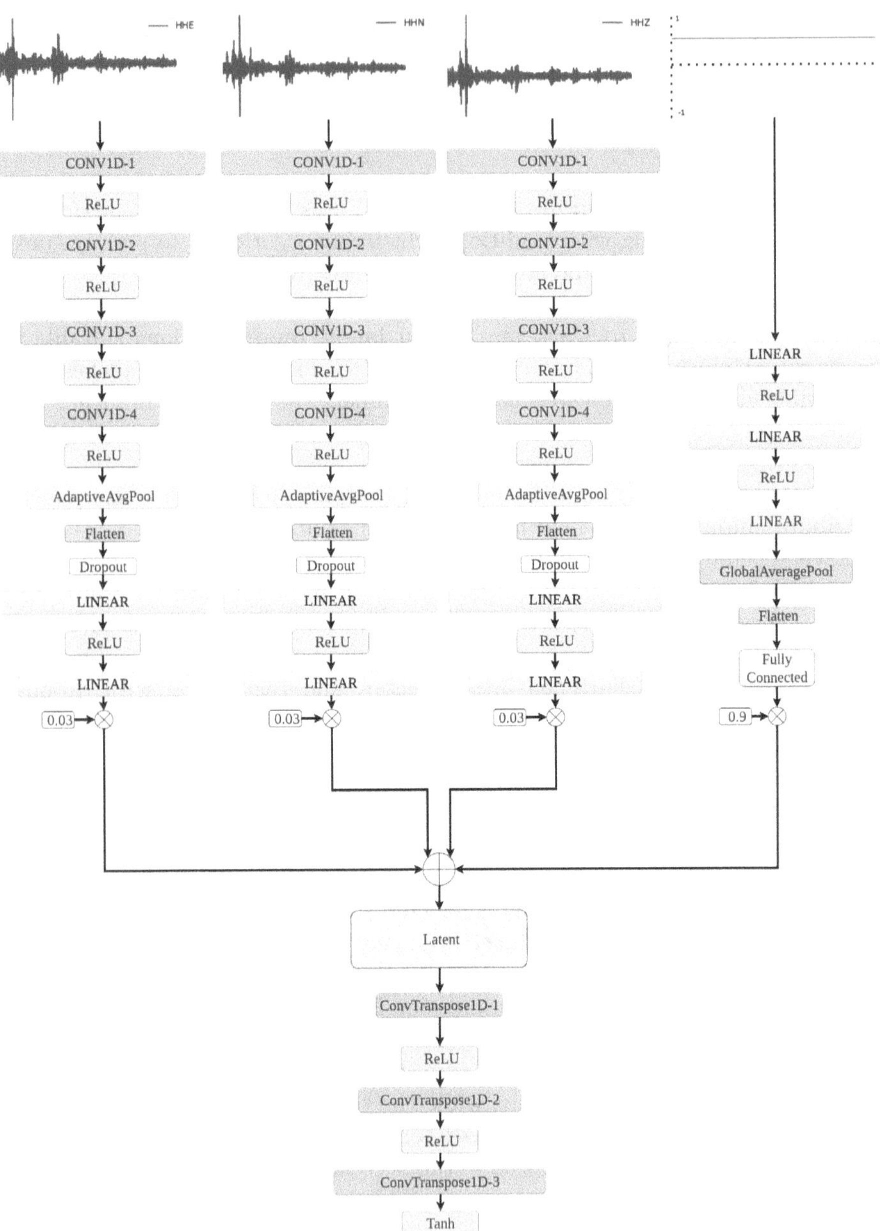

Fig. 3. Encoder-decoder architecture used for seismic phase association. The diagram shows four input branches: three for the seismic waveforms (HHE, HHN, HHZ), and one for the temporal component. Each branch produces a latent representation, which is scaled by a predefined weight to control its contribution, and then fused to form the final latent space.

Table 1. The layers in the MultiBranchConvAutoencoderDEC network.

Layer name	Type	Input shape	Filters	Activation	Output shape	Trainable parameters
Input	–	–	–	–	[B, 3, 6000]	–
Conv1	Conv1D	[3, 6000]	32	ReLU	[32, 3000]	192
Conv2	Conv1D	[32, 3000]	64	ReLU	[64, 1500]	6,208
Conv3	Conv1D	[64, 1500]	128	ReLU	[128, 750]	24,704
Conv4	Conv1D	[128, 750]	256	ReLU	[256, 375]	98,560
AvgPool	AvgPool1D	[256, 375]	–	–	[256, 8]	0
Flat	Flatten	[256, 8]	–	–	[2048]	0
Dropout	Dropout (0.3)	[2048]	–	–	[2048]	0
FC1	Linear	[2048]	–	ReLU	[512]	1,049,088
FC2 (Encoded)	Linear	[512]	–	–	[latent]	16,416
Time FC1	Linear	[1]	–	ReLU	[16]	32
Time FC2	Linear	[16]	–	ReLU	[32]	544
Time FC3	Linear	[32]	–	–	[latent]	1,056
Fusion	Weighted sum	–	–	–	[latent]	0
Dec FC1	Linear	[latent]	–	–	[48000]	2,112,000
Unflatten	Unflatten	[48000]	–	–	[64, 750]	0
Deconv1	ConvTranspose1D	[64, 750]	32	ReLU	[32, 1500]	10,272
Deconv2	ConvTranspose1D	[32, 1500]	16	ReLU	[16, 3000]	3,600
Deconv3	ConvTranspose1D	[16, 3000]	4	Tanh	[4, 6000]	580

Model Training: The model is trained by minimizing a joint loss that combines reconstruction error (MSE) and KL divergence between predicted soft assignments and a target distribution. This encourages the latent space to retain input structure while forming meaningful clusters. Training and validation datasets were defined as described in the methodology. The process includes two stages:

Pretraining: The autoencoder is first trained using only reconstruction loss for a few epochs to initialize the encoder. Latent representations obtained are used to estimate initial cluster centroids via k-means.

Joint Training: The full model is then optimized using the combined loss. Initially, the encoder remains frozen (controlled by `freeze_epochs`) while the decoder adapts. Later, both modules are updated jointly. A target distribution for clustering is updated every `update_interval` epochs to refine assignments. Training continues until convergence based on early stopping criteria (`patience`, `min_delta`).

The hyperparameters were selected based on an exhaustive search aimed at minimizing both reconstruction loss and clustering loss. Table 2 summarizes the final set of chosen values along with a brief description of each.

Throughout training, the model monitors two loss curves: the reconstruction loss and the clustering loss, for both training and validation sets. These metrics

Table 2. Summary of selected hyperparameters used during training.

Hyperparameters	Value	Description
input_length	6000	Number of samples per waveform input (trace length).
batch_size	16	Number of samples per training batch.
pretrain_epochs	3	Epochs for initial autoencoder pretraining.
epochs	15	Total number of epochs in joint training phase.
latent_dim	64	Dimension of the latent space.
dropout_prob	0.3	Dropout probability in encoder branches.
learning_rate	0.001	Learning rate for Adam optimizer.
alpha	0.05	Weight factor for clustering loss.
gamma	0.001	Weight of KL divergence in total loss.
update_interval	3	Epochs between updates of target distribution.
tolerance	0.001	Max tolerated change in cluster assignments.
freeze_epochs	3	Initial epochs with encoder frozen.
patience	3	Epochs without improvement before early stopping.
min_delta	1×10^{-4}	Minimum loss improvement to continue training.

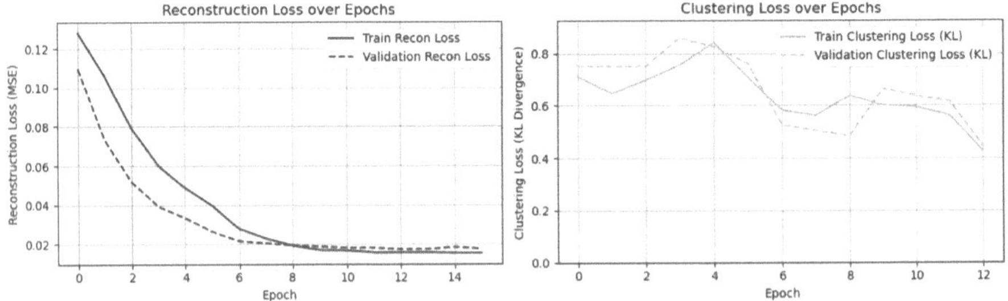

Fig. 4. Training dynamics of the model. Left: Mean Squared Error (MSE) reconstruction loss over epochs. Right: Clustering loss measured as KL divergence between soft assignments and target distributions. For training and validation.

are used to assess the quality of learned representations and clustering behavior over time. As shown in Fig. 4, the reconstruction loss decreases rapidly during the first epochs and then gradually stabilizes, indicating that the autoencoder quickly learns to capture the essential features of the input waveforms. The validation curve closely follows the training curve, suggesting good generalization and absence of overfitting. On the other hand, the clustering loss exhibits variance and a less smooth trend, though it remains low. This is expected, as the latent space is continuously reshaped to optimize cluster assignments [16]. Nonetheless, both training and validation clustering losses tend to decrease over time, reflecting a progressive improvement in the compactness and separability of the latent representations, as well as it is a very low range.

Model Evaluation: DECA was evaluated using the test dataset, which was previously described in the methodology section. For the clustering stage, we employ HDBSCAN as the primary algorithm. This choice is crucial because, unlike methods requiring a predefined number of clusters (k), HDBSCAN does not necessitate prior knowledge of the exact number of events, which is typically unavailable in real-world seismological scenarios.

During the autoencoder training phase, however, k-means clustering was utilized. This was feasible at that specific stage because the exact quantity of clusters was known, allowing for an improved training process and a more optimized latent space. A key advantage of DECA is precisely this ability to operate without requiring an a priori definition of the expected number of clusters, which is often challenging or only an approximation in seismology.

Furthermore, it's essential to consider the main parameters of the chosen clustering algorithm. As a recommendation for DECA users, HDBSCAN parameters can be adjusted based on the quality of the preceding detection stage. For instance, tools like EQTransformer can be configured with conservative or non-conservative models, varying detection thresholds, and other parameters that directly impact the quality of detections (i.e., real positives, false positives, or undetected events).

A conservative detection model tends to significantly reduce the number of events, but most of them are likely real. Conversely, soft restrictions can yield a large number of detections, albeit with lower quality. This implies that DECA might encounter numerous false positive detections. It is therefore critical for DECA to avoid clustering these false positives as legitimate events, which can be achieved by carefully tuning the HDBSCAN parameters. The key parameters include the **minimum cluster size**, which defines the smallest group of points to be considered a cluster (smaller groups are treated as noise); the **minimum local samples**, which controls how conservative the clustering is—higher values increase robustness to noise by requiring greater local density; and the **cluster merging threshold**, a distance value that allows nearby dense regions to be merged into a single cluster, where larger thresholds enable looser grouping.

If the detection restrictions are soft (leading to many detections), it is recommended to set higher values for **minimum cluster size**, **minimum local samples**, and **cluster merging threshold** to establish robust clusters. Conversely, if the detection is conservative (resulting in fewer detections), it's advisable to decrease the severity of HDBSCAN's clustering criteria to account for the smaller number of events available for grouping.

The results shown in Fig. 5 correspond to a 2D UMAP projection of the latent space, obtained using the trained model, HDBSCAN clustering, and the test dataset, with parameters set to `min_cluster_size = 4`, `min_samples = 3`, and `cluster_selection_epsilon = 0.001`. These HDBSCAN parameters are considered relatively soft for cluster creation, given that we are confident the input picks are well-detected.

5 Results

Once the model was trained and demonstrated good performance on the association task with the test dataset, we proceeded to perform associations for different picks catalogs. For this, DECA offers two different association modes:

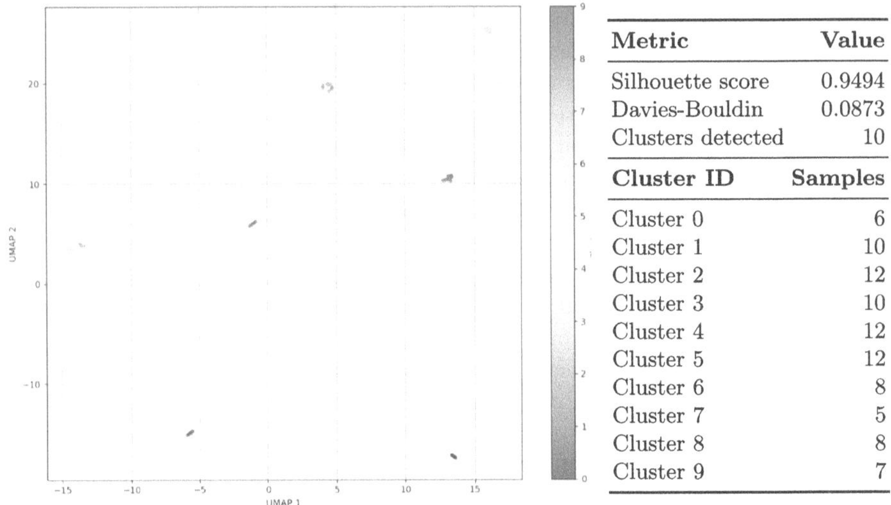

Metric	Value
Silhouette score	0.9494
Davies-Bouldin	0.0873
Clusters detected	10
Cluster ID	**Samples**
Cluster 0	6
Cluster 1	10
Cluster 2	12
Cluster 3	10
Cluster 4	12
Cluster 5	12
Cluster 6	8
Cluster 7	5
Cluster 8	8
Cluster 9	7

Fig. 5. Left: Latent space representation using UMAP on the test set. Right: Clustering performance and sample distribution.

1. **Single-pass clustering:** Recommended when picks are known to be of high quality or manually verified. In this mode, the association process runs as a single clustering operation across the entire dataset.
2. **Time-windowed clustering:** Recommended when using picks from AI tools, which may include false positives. DECA groups picks into clusters using sliding time windows defined by the user (e.g., ranges of seconds, minutes, or hours). In this experiment, we used 10-minute windows with an automatic segmentation rule: if no new picks occur within 30 s after the last pick of a group, a new group is started. This process is fully automated and only requires minor configuration by the user.

It is important to clarify that, since DECA is designed to operate as part of a processing pipeline, errors may be introduced during the detection stage and can propagate to affect the association results. In this section, we aim to isolate and evaluate only the effects inherent to the association stage of DECA. For this reason, we based our evaluation on a highly trustworthy picks catalog manually reviewed by OVSICORI-UNA seismologists. This ensures that we are clustering real, verified picks. This is also very useful because it provides a reliable reference for comparison between the clusters obtained by DECA and those defined by professional seismologists.

The evaluation utilized data recorded on February 3, 2025, by the national seismic network [11]. The picks catalog used is composed of 284 picks, each corresponding to 3 channels, resulting in a total of 852 waveform signals. We used the same hyperparameter values as shown in Table 2. Figure 6 shows the latent space distribution (UMAP) and clustering metrics.

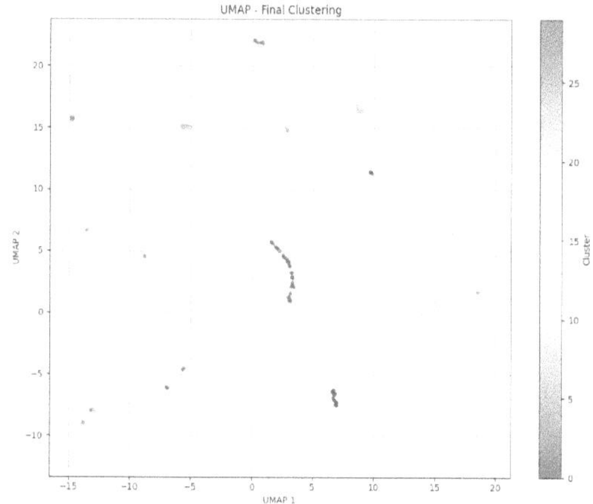

Metric	Value
Silhouette score	0.9178
DBCV	0.9252
Davies-Bouldin	0.1205
Number of Clusters	
Detected	30
Expected	29

Fig. 6. Left: Latent space representation using UMAP on nationwide picks. Right: Clustering performance metrics.

Table 3. Unique mapping between cluster IDs and their associated event IDs obtained with DECA.

Cluster		Event ID	Cluster		Event ID	Cluster		Event ID
0	⇒	1624424	10	⇒	1624467	20	⇒	1624578
1	⇒	1624308	11	⇒	1624461	21	⇒	1624603
2	⇒	1624458	12	⇒	1624420	22	⇒	1624586
3	⇒	**1624383**	13	⇒	1624576	23	⇒	1624591
4	⇒	1624346	14	⇒	1624600	24	⇒	1624593
5	⇒	1624602	**15**	⇒	**1624383**	25	⇒	1624588
6	⇒	1624425	16	⇒	1624592	26	⇒	1624466
7	⇒	1624596	17	⇒	1624582	27	⇒	1624933
8	⇒	1624338	18	⇒	1624577	28	⇒	1624932
9	⇒	1624426	19	⇒	1624345	29	⇒	1624344

Also, we present Table 3, which illustrates the physical and practical performance by comparing the clusters identified manually by seismological researchers against those produced by DECA. Out of the 30 clusters, only two show discrepancies, this because DECA split one real cluster into two separate clusters. The original cluster is identified by the event ID assigned by OVSICORI, while DECA created two distinct clusters corresponding to this single event.

Finally, the output consists of a CSV file and a 'Y2000.phs' file (required for example by the NonLinLoc location tool), both containing the associated events. Additionally, DECA creates a folder structure with plots for each detected cluster, which can be used for visual validation.

6 Discussion

Results confirm DECA's effectiveness for phase association in real-world settings. The performance metrics shown in Fig. 6, particularly the high Silhouette score (0.9178), DBCV (0.9252), and low Davies-Bouldin index (0.1205), indicate that the clustering structure discovered by the model is well defined and internally consistent. These metrics suggest a strong separation between clusters and low intra-group variability, which is essential for reliable event identification.

Also, the comparison between DECA's clustering results and the catalog manually curated by seismologists reveals a high degree of agreement. Out of 30 clusters detected by DECA, 28 match exactly, including the same picks, with those manually assigned by experts. Only two clusters show divergence, this because DECA split what was manually defined as a single event into two separate clusters. However, we can say that the performance is very good and likely reflects latent structure in the waveform data captured effectively. But also, it is important to analyze the error committed by DECA. Figure 7 shows a map with the different stations associated with this particular event, separated by color according to how DECA divided them into two clusters.

One possible explanation for this misclassification is that this specific event was detected by multiple stations across the country, some separated by 50 km or more. This large spatial distribution could cause waveform amplitude attenuation, leading to the error made by DECA. As seen, nearby stations tend to be clustered together, while more distant stations fall into the other group, except for one station (CDM), which, although distant, was also grouped into Cluster 3. This behavior is likely related to the decrease in seismic amplitude with distance, which makes background noise more prominent. Technically, the signal-to-noise ratio (SNR) decreases with distance.

Scalability emerges as a key advantage of the DECA model. The unsupervised architecture provides the capacity to process large datasets of picks and offers flexibility as seismic networks grow. Although increasing station density will introduce greater waveform heterogeneity and SNR disparities, these variations are not prohibitive. They are controlled through parameter optimization and standard preprocessing steps, ensuring DECA remains a viable solution for massive deployments.

Finally, DECA's flexibility across different data collection strategies is worth noting. By supporting single-pass and sliding time-window clustering modes, the model can be easily integrated into various operational workflows, whether for curated datasets or AI-generated picks. This adaptability allows for dynamic event association without manual intervention and ensures consistent analysis across heterogeneous data sources.

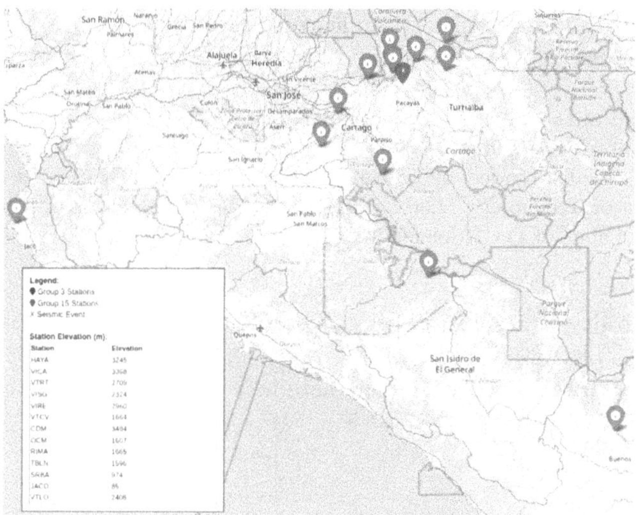

Fig. 7. Station distribution related to the misclassified event. Colors represent how DECA split the associated picks into two different clusters.

7 Conclusions and Future Work

This study demonstrates that DECA provides a robust and effective solution for automatic seismic phase association under real-world conditions. The model achieved high internal clustering quality, as indicated by a Silhouette score of 0.9178, DBCV of 0.9252, and a Davies–Bouldin index of 0.1205, while showing strong consistency with expert-curated catalogs. Notably, 96.55% of the clusters identified by DECA matched exactly with those defined by seismologists, confirming its reliability. Moreover, DECA successfully captured subtle waveform variations without supervision, revealing latent seismic structures at the national scale. The observed discrepancy in one cluster, where it was split in two, is likely due to waveform attenuation or low signal-to-noise ratios. These results highlight both the strengths of DECA and the need for refinements to address edge cases and enhance generalization.

Nevertheless, several limitations must be acknowledged. The dataset used in this study was relatively small and relied on carefully reviewed picks, which ensured reliable benchmarking, in particular, DECA was not tested with outputs from AI-based detectors such as EQTransformer, limiting conclusions about its robustness when faced with more variable or noisy inputs. Addressing these limitations requires validation with larger and more heterogeneous datasets, ideally including global seismic catalogs with greater event diversity. Future work should also evaluate DECA across different tectonic and geographic contexts, which will be essential to assess its generalizability beyond the Costa Rican setting. In addition, adapting DECA for real-time applications requires further optimization of execution time and computational efficiency. Finally, future studies could benefit from leveraging more advanced deep learning architectures, which may enhance the learned latent representations and further improve clustering accuracy.

Acknowledgments. This research was partially supported through machine allocation on the Kabré supercomputer at the Costa Rica National High Technology Center. It was also partially funded by research project code 5402-1370-0019 at the Costa Rica Institute of Technology. The authors acknowledge the support provided by the Postgraduate Office of ITCR for facilitating the participation in the conference.

Data Availibility Statement. The source code for DECA is available at: GitHub under the MIT License.

Disclosure of Interests. The authors have no competing interests to declare that are relevant to the content of this article.

References

1. Aslani, S., et al.: Multi-branch convolutional neural network for multiple sclerosis lesion segmentation. CoRR **abs/1811.02942** (2018). http://arxiv.org/abs/1811.02942
2. Båth, M.: Seismic Waves, pp. 61–103. Birkhäuser Basel, Basel (1979). https://doi.org/10.1007/978-3-0348-5283-8_3
3. Chien, C.C., Jenkins, W.F., Gerstoft, P., Zumberge, M., Mellors, R.: Automatic classification with an autoencoder of seismic signals on a distributed acoustic sensing cable. Comput. Geotech. **155**, 105223 (2023)
4. Gamboa-Chacón, S., Meneses, E., Chaves, E.J.: Analysis of earthquake detection using deep learning: evaluating reliability and uncertainty in prediction methods. Comput. Geosci. **197**, 105877 (2025). https://doi.org/10.1016/j.cageo.2025.105877
5. Joyce, J.M.: Kullback-Leibler Divergence, pp. 720–722. Springer, Heidelberg (2011). https://doi.org/10.1007/978-3-642-04898-2_327
6. van der Laat, L., Baldares, R.J.L., Chaves, E.J., Meneses, E.: OKSP: a novel deep learning automatic event detection pipeline for seismic monitoring in Costa Rica (2021). https://arxiv.org/abs/2109.02723
7. Moulavi, D., Jaskowiak, P.A., Campello, R.J.G.B., Zimek, A., Sander, J.: Density-based clustering validation. In: Proceedings of the 2014 SIAM International Conference on Data Mining, pp. 839–847. SIAM (2014)
8. Mousavi, S.M., Ellsworth, W.L., Zhu, W., Chuang, L.Y., Beroza, G.C.: Earthquake transformer—an attentive deep-learning model for simultaneous earthquake detection and phase picking. Nat. Commun. **11**(1), 1–12 (2020)
9. Münchmeyer, J.: Pyocto: a high-throughput seismic phase associator. arXiv preprint arXiv:2310.11157 (2023)
10. Murtagh, F.: Multilayer perceptrons for classification and regression. Neurocomputing, **2**, 183–197 (1991). https://api.semanticscholar.org/CorpusID:7949799
11. OVSICORI: Universidad nacional de costa rica - station maps (2025). https://www.ovsicori.una.ac.cr/index.php/sismologia/mapas-de-estaciones
12. Ross, Z.E., Yue, Y., Meier, M.A., Hauksson, E., Heaton, T.H.: PhaseLink: a deep learning approach to seismic phase association. J. Geophys. Res. Solid Earth **124**(1), 856–869 (2019)

13. scikit-learn Developers: sklearn.metrics.davies_bouldin_score. scikit-learn (2025). aPI reference, version 0.20
14. Shahapure, K.R., Nicholas, C.: Cluster quality analysis using silhouette score. In: 2020 IEEE 7th International Conference on Data Science and Advanced Analytics (DSAA), pp. 747–748 (2020). https://doi.org/10.1109/DSAA49011.2020.00096
15. Wirth, R., Hipp, J.: Crisp-dm: towards a standard process model for data mining (2000). https://api.semanticscholar.org/CorpusID:1211505
16. Xie, J., Girshick, R., Farhadi, A.: Unsupervised deep embedding for clustering analysis (2016). https://arxiv.org/abs/1511.06335
17. Zhu, W., McBrearty, I.W., Mousavi, S.M., Ellsworth, W.L., Beroza, G.C.: Earthquake phase association using a Bayesian gaussian mixture model. J. Geophys. Research: Solid Earth **127**(5), e2021JB023249 (2022)

Mathematical Interpretations of the Acoustic Complexity Indices: Relationship with the Concept of Total Variation and Their Numerical Approximations

Fernando Martínez-Tabares[1(✉)] and Mauricio Orozco-Alzate[1,2]

[1] Universidad Nacional de Colombia - Sede Manizales, Grupo de Control y Procesamiento Digital de Señales, Km 7 vía al Magdalena, 170003 Manizales, Caldas, Colombia
fmartinezt@unal.edu.co

[2] Universidad Nacional de Colombia - Sede Manizales, Departamento de Informática y Computación, Km 7 vía al Magdalena, 170003 Manizales, Caldas, Colombia

Abstract. Acoustic Complexity Indices (ACIs) estimate sonic information variability along time and across frequencies. Even though these indices were proposed with an application-oriented motivation, nothing —to the best of our knowledge— has been discussed so far regarding their mathematical interpretations. Here we distinguish between punctual and total ACIs, as well as between their discrete and continuous versions. For the discrete punctual ACIs, we present two numerical approximations using backward and central differences, respectively. Based on that, we define the discrete total ACI and propose a geometric interpretation. Afterwards, we present an analytic interpretation that allows to generalize to the continuous case and show the relationship between total variation (TV) and the continuous total ACI; moreover, we also show that the later is numerically approximated by the discrete total ACI and that it can be computed by a proposed procedure based on the estimation of the TV. Finally, differences between the numerical approximations of the punctual ACIs are illustrated using a chirp signal and the equivalences between the approximations of the continuous total ACI are shown for a real-world acoustic recording.

Keywords: Acoustic complexity index · Ecoacoustics · Numerical derivatives · Soundscapes · Total variation

1 Introduction

Robust and large-capacity audio recorders for acoustic ecosystem monitoring are nowadays available in the market at relatively low prices, allowing scientists to

D. Chaves et al. (Eds.): CIARP 2025, LNCS 16528, pp. 292–307, 2026.
https://doi.org/10.1007/978-3-032-23161-1_21

easily record nature sounds during long periods of time and under rush environmental conditions. Those technological advancements, along with Audio Signal Processing techniques and expert knowledge from Biology and Ecology, have boosted the fields of Bioacoustics and Ecoacoustics.

Ecoacoustics [5], in particular, consists in the study of soundscapes; that is, it refers to the characterization and analysis of environmental sounds in natural landscapes. Such a study is typically based on extracting features from the recorded audio signals, either generic features for audio content analysis [19] or specific ones designed for answering ecological research questions [1]. These features, in the context of Ecoacoustics, have been termed as acoustic indices. Most of them are computed from a preprocessed version of the soundscape audio recordings, typically from a time-frequency representation of the signal such as the Short-Time Fourier Transform (STFT).

Many specially designed acoustic features are currently used by the academic community in Ecoacoustics [2,6,9,20,24]. The Acoustic Complexity Indices (ACIs) —according to the authors of [6]— are among the most commonly used ones; for instance in studies, to name a few recent ones, for assessing marine biodiversity [3], estimating the effects of wind turbines [18] and climate change [13] on bird communities, distinguishing frog species [8] and planning urban development with lower environmental impacts [7,22]. Moreover, the study in [23] empirically showed (in an unsupervised way) that the ACIs are very competitive to characterize soundscapes, either when used individually or as part of a set of representative acoustic features.

The ACIs for an audio signal are computed from a matrix (often known as sonic matrix) containing the squared magnitude of its STFT; in other words, they are computed from the spectrogram. The ACIs can be intuitively understood as summations of normalized differences of the signal energy, either across frequencies or during time intervals. Such normalized differences can be called punctual ACIs and, in consequence, their summations along an axis of the spectrogram (either frequency or time) can be referred as total ACIs. The punctual ACIs, in their classic definitions given in [6], are straightforwardly related to the numerical approximation of the derivative of some continuous function.

Recently, Farina [4] provided a theoretical foundation for the ACI; nonetheless, further mathematical properties of the ACIs —to the best of our knowledge— have not been studied yet, particularly neither in terms of their alternative numerical approximations, nor about their geometric and analytic interpretations. Moreover, their close formulations to similar concepts that are used in mathematical analysis has not been pointed out yet. Therefore, this paper (which is part of the thesis in [11]) is aimed at presenting equivalent definitions for the punctual ACIs using forward, backward and central differences. These definitions, in turn, allow us to propose a continuous punctual ACI; based on that, we also define two versions of the total ACI, namely: the discrete total ACI and the continuous total ACI. In analogy with the geometric interpretation for the derivative of a function, we provide a geometric interpretation for the punctual ACIs; besides, an analytic interpretation for the discrete and continuous total ACIs, which we relate to the concept of total variation of a function, is given.

The structure of this paper is the following one: the classical definitions of the punctual ACIs and two alternative numerical approximations for them are presented in Sect. 2. Then, our analyses and results are presented in Sect. 3, corresponding to novel geometric and analytic interpretations of the ACIs, and computational examples for both synthetic and real-world audio signals; Sect. 4 presents conclusions, considerations and opportunities for future work.

2 Methods

2.1 The Acoustic Complexity Indices (ACIs)

The ACIs were originally proposed in [16] as a way to quantify the biotic singing activity in a soundscape recording. These acoustic indices are formally defined as follows. Let $x(t)$ be an audio signal, once a spectral analysis of the discrete-time signal x is carried out (typically a Short Time Fourier Transform (STFT) [17]), a so-called sonic matrix $X_{(m+1)\times(n+1)}$ is obtained. Rows in this matrix represent $m+1$ frequencies, whereas columns correspond to $n+1$ time instants. Both n and m depend of the sample rate and the length of the window, which are chosen such that the Nyquist theorem is guaranteed. Each entry X_{ij} in the sonic matrix contains the squared magnitude of the Fourier coefficient (in the case of the STFT) at the time instant t_j and for the frequency ω_i.

Two different ACIs can be defined depending on the approach: one for a given frequency (ω_i) across n time instants (denoted as $X_{i:}$), and another for a fixed time instant (t_j) across m frequencies (denoted as $X_{:j}$). Both are defined as follows:

$$\overrightarrow{ACI}(X_{i:}) = \sum_{j} \frac{|X_{ij} - X_{i(j+1)}|}{X_{ij} + X_{i(j+1)}} \quad (1) \qquad \overrightarrow{ACI}(X_{:j}) = \sum_{i} \frac{|X_{ij} - X_{(i+1)j}|}{X_{ij} + X_{(i+1)j}} \quad (2)$$

The first ACI definition is presented in Eq. (1), which calculates the variation across time instants for a specific frequency. On the other hand, Eq. (2) shows the ACI definition that considers the variation across frequencies for a fixed time instant.

For the sake of notation convenience, we will denote a particular term in the sums in Eqs. (1) and (2) as $\overrightarrow{ACI}_{X_{i:}}(t_j)$ and $\overrightarrow{ACI}_{X_{:j}}(\omega_i)$, respectively; that is,

$$\overrightarrow{ACI}_{X_{i:}}(t_j) = \frac{|X_{ij} - X_{i(j+1)}|}{X_{ij} + X_{i(j+1)}} \quad (3) \qquad \overrightarrow{ACI}_{X_{:j}}(\omega_i) = \frac{|X_{ij} - X_{(i+1)j}|}{X_{ij} + X_{(i+1)j}} \quad (4)$$

Notice that we have used an arrow on top of ACI in Eqs. (1) to (4) because the differences, in the classical formulation of the ACIs, are computed either between the i-th and the $(i+1)$-th terms or between the j-th and the $(j+1)$-th terms; that is, by using forward differences. Therefore, expressions in Eqs. (3)

and (4) can be called *punctual forward ACIs*. Similarly, we may refer to Eqs. (1) and (2) as *total forward ACIs*.

Forward ACI for a Particular Function. Let us now restrict ourselves to either a single row ($X_{:i}$) or column ($X_{:j}$) in the sonic matrix X and consider it as the discrete version of a strictly positive function f defined on $[a, b]$ and evaluated in an uniform partition of length N

$$\mathcal{P} = \{x_k\}_{k=0}^{N}, \quad x_k = a + kh, \quad h = \frac{b-a}{N} \tag{5}$$

of the interval $[a, b]$ (time or frequency). According to that, the punctual and total ACIs can be written in general for a function $f > 0$ and a partition $\mathcal{P}$ like this (with $f_k = f(x_k)$ for $k = 0, 1, \cdots, N$):

- *Punctual forward ACI:*

$$\overrightarrow{ACI}_f(x_k) = \frac{|f_k - f_{k+1}|}{f_k + f_{k+1}} \tag{6}$$

- *Total forward ACI:*

$$\overrightarrow{ACI}(f) = \sum_{k=0}^{N-1} \overrightarrow{ACI}_f(x_k) \tag{7}$$

2.2 Numerical Approximations of the ACI

In order to introduce alternative formulations of the ACI, let us remind the following properties of the Big-O notation for the numerical approximation of functions [12].

Big-O Notation. If the function $p(h)$ approximates to the function $f(h)$ and there is a constant $M > 0$, such that,

$$\frac{|f(h) \quad p(h)|}{|h^n|} < M \text{ , with } h \to 0$$

we say that $p(h)$ approximates to $f(h)$ with approximation order $O(h^n)$ and we write

$$f(h) = p(h) + O(h^n)$$

If $f(h) = p(h) + O(h^m)$, $g(h) = q(h) + O(h^n)$ then we can say that: $p(h) + q(h)$, $p(h) \cdot q(h)$ and $\frac{p(h)}{q(h)}$ approximates to $f(h) + g(h)$, $f(h) \cdot g(h)$ and $\frac{f(h)}{g(h)}$ respectively, with approximation order $O(h^r)$, where $r = \min\{m, n\}$.

By the Taylor's theorem for the partition defined in Eq. (5), we obtain the following approximations for $f(x)$, its derivative and its integral:

Approximations for $f(x)$ and its Derivative $f'(x)$:

– *Backward formulas*: If $f \in C^2[a, b]$

$$f(x_k) = \frac{f_{k-1} + f_k}{2} + O(h), \quad f'(x_k) = \frac{f_k - f_{k-1}}{h} + O(h) \tag{8}$$

- *Forward formulas*: If $f \in C^2[a,b]$

$$f(x_k) = \frac{f_k + f_{k+1}}{2} + O(h), \quad f'(x_k) = \frac{f_{k+1} - f_k}{h} + O(h) \tag{9}$$

- *Centered formulas*: If $f \in C^3[a,b]$

$$f(x_k) = \frac{f_{k-1} + f_{k+1}}{2} + O(h^2), \quad f'(x_k) = \frac{f_{k+1} - f_{k-1}}{2h} + O(h^2) \tag{10}$$

Trapezoidal Rule Approximation for the Integral of $f(x)$: The integral of $f(x)$ can be approximated as:

$$\int_a^b f(x)dx = \frac{h}{2}(f(a) + f(b)) + h \sum_{k=1}^{N-1} f(x_k) + O(h^2) \tag{11}$$

Redefinition of the ACI. Consider a strictly positive function f with at least three continuous derivatives, as required by the centered formula; see Eq. (10). Besides, let us consider the uniform partition $\mathcal{P}$ of the $[a,b]$ interval as in Eq. (5). Under these considerations and according to the above-mentioned formulas for backward and central approximations of the derivative —refer again to Eqs. (8) and (10), respectively— the punctual ACI defined in Eq. (6) can be reformulated as:

- *Punctual backward ACI:*

$$\overleftarrow{ACI}_f(x_k) = \frac{|f_{k-1} - f_k|}{f_{k-1} + f_k} \tag{12}$$

- *Punctual central ACI:*

$$\overleftrightarrow{ACI}_f(x_k) = \frac{|f_{k-1} - f_{k+1}|}{f_{k-1} + f_{k+1}} \tag{13}$$

Then we can now define the *Discrete total ACI*, by considering the forward, central and backward ACIs, as follows:

$$\overleftrightarrow{ACI}(f) = \overrightarrow{ACI}_f(a) + \sum_{k=1}^{N-1} \overleftrightarrow{ACI}_f(x_k) + \overleftarrow{ACI}_f(b) \tag{14}$$

3 Analysis and Results

3.1 Geometric Interpretation of the ACI

Consider again the function f and the partition $\mathcal{P}$ as it was described in Sect. 2.1. Now, for each $x_k \in \{a = x_0, x_1, \cdots, x_N = b\}$, let us consider a circumference centered at $C = (c_k, m_k)$ and with radius r_k, as shown in Fig. 1. Since we need to define the center and radius in different ways depending on whether x_k is one of the extreme points of the interval or not, we need the following auxiliar indicatrix:

$$\Delta_a = \begin{cases} 0 & x_k = a \\ 1 & x_k \neq a \end{cases}, \quad \Delta_b = \begin{cases} 1 & x_k \neq b \\ 0 & x_k = b \end{cases} \tag{15}$$

Then, the center of the circumference has coordinates

$$c_k = x_k + \frac{\Delta_b - \Delta_a}{2}h, \quad m_k = \frac{f(x_k - \Delta_a h) + f(x_k + \Delta_a h)}{2} \tag{16}$$

and its radius is

$$r_k = \frac{|f(x_k + \Delta_b h) - f(x_k - \Delta_a h)|}{2} \tag{17}$$

Then, we build a right triangle $\triangle ABC$ which, as illustrated in Fig. 1, is given by

$$\triangle ABC : A = (c_k, 0)\,,\ B = (c_k + r_k, m_k) \,\text{and}\, C = (c_k, m_k)$$

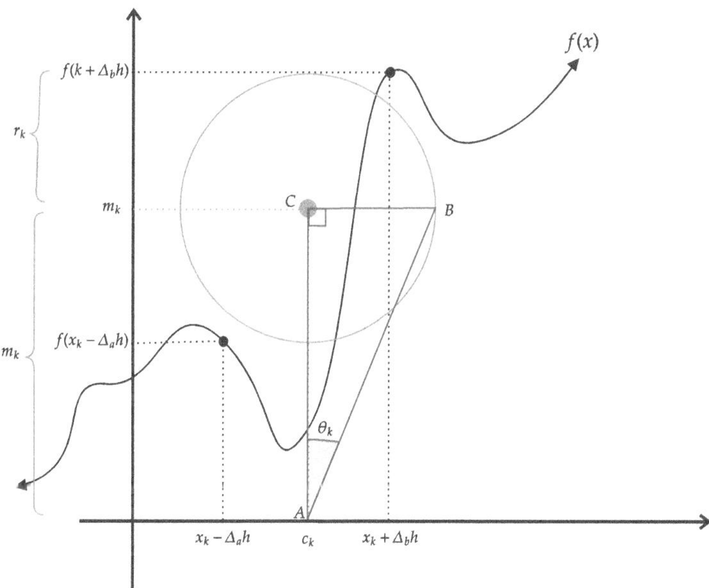

Fig. 1. Circumference and angle θ_k associated to the punctual ACI in x_k.

In these triangle, let us consider the angle $\theta_k = B\hat{A}C$ and relate it with the punctual ACIs according to the following proposition.

Proposition 1. *Let $f \in C^3[a,b]$ and $\mathcal{P}$ an uniform partition of $[a,b]$; then, in the triangle $\triangle ABC$*

$$|\tan(\theta_k)| = \begin{cases} \overrightarrow{ACI}_f(x_0) & k = 0 \\ \overleftrightarrow{ACI}_f(x_k) & 0 < k < N \\ \overleftarrow{ACI}_f(x_N) & k = N \end{cases}$$

Proof. In all cases in the right triangle $\triangle ABC$, it holds that

$$|\tan\theta_k| = \frac{BC}{AC} = \frac{r_k}{m_k}$$

Then, replacing for each case of k in Eqs. (15), (16) and (17) we obtain:

- *For $x_0 = a$,*

$$\frac{r_0}{m_0} = \frac{|f_1 - f_0|}{f_1 + f_0} = \overrightarrow{ACI}_f(a)$$

- *For $a < x_k < b$,*

$$\frac{r_k}{m_k} = \frac{|f_{k+1} - f_{k-1}|}{f_{k+1} + f_{k-1}} = \overleftrightarrow{ACI}_f(x_k)$$

- *For $x_N = b$,*

$$\frac{r_N}{m_N} = \frac{|f_N - f_{N-1}|}{f_N + f_{N-1}} = \overleftarrow{ACI}_f(b)$$

Let us consider the following scenario as illustrated in Fig. 2: two changes of the same amount but occurring at instants where the values of f are significantly different. Notice in that scenario that $r_a = r_b$, $\alpha_a = \alpha_b$, then $\tan(\alpha_a) = \tan(\alpha_b)$ which represent the usual approximations of the derivative at x_a and x_b, respectively. However, $\theta_a > \theta_b$ and, therefore, $\tan(\theta_a) > \tan(\theta_b)$; in other words, since $\tan(\theta)$ increases for $\theta \in (0, \frac{\pi}{2})$, the punctual ACI value is larger for x_a than for x_b since changes with the same derivative are penalized differently depending on the scale they occur: those in lower scales are divided by smaller values.

Remember that penalizing changes on different scales, in order to make them comparable, corresponds to the notion of a logarithmic scale, which is precisely what the $\overrightarrow{ACI}_f(x_0)$, $\overleftrightarrow{ACI}_f(x_k)$ and $\overleftarrow{ACI}_f(x_N)$ achieves numerically. Furthermore, since the auditory perception operates on a logarithmic scale [14], we can say that the ACI inherently takes into account this perceptual property of the human ear.

3.2 Analytic Interpretation of the ACIs

As a generalization of the results from the previous section, let us now define the continuous versions of the ACIs as follows:

Definition 1. *Consider $f \in C^3[a, b]$, $f > 0$, we define:*

i) $ACI_f(x)$ is called the continuous punctual ACI, *which is given by*

$$ACI_f(x) = \left|\frac{d}{dx}\ln(f(x))\right| \tag{18}$$

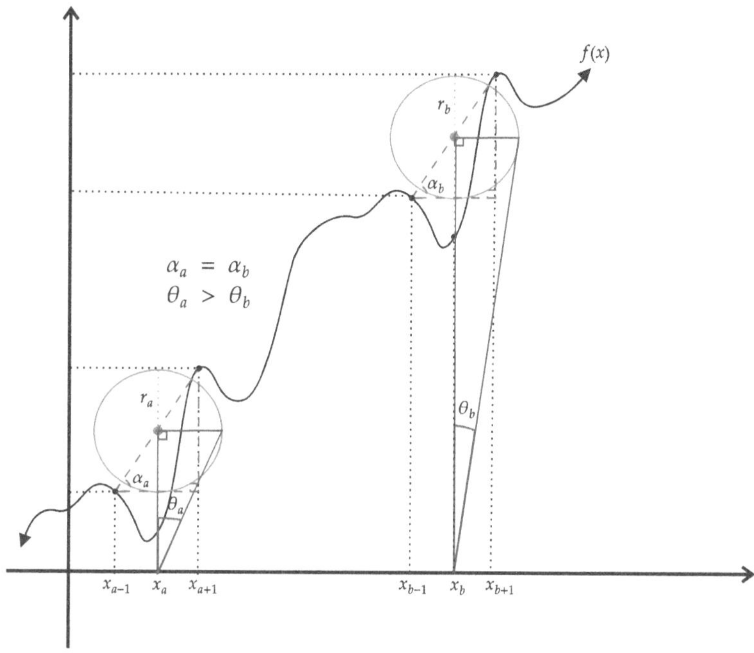

Fig. 2. A scenario where the punctual ACI values differ for points having the same derivative but at different scales of the values of f.

ii) $ACI(f)$ is called the continuous total ACI, *given by*

$$ACI(f) = \int_a^b ACI_f(x)dx \tag{19}$$

Since $ACI_f(x) = \left|\frac{d}{dx}\ln(f(x))\right| = \frac{|f'(x)|}{f(x)}$, we use this fact in the following proposition for notational convenience.

Proposition 2. *Let $f \in C^3[a,b]$, $f > 0$ and an uniform partition $\mathcal{P}$ of $[a,b]$, then*

$$\frac{|f'(x_k)|}{f(x_k)} = \begin{cases} \frac{2}{h}\overrightarrow{ACI}_f(x_0) + O(h), & x_0 = a \\ \frac{1}{h}\overleftrightarrow{ACI}_f(x_k) + O(h^2), & a < x_k < b \\ \frac{2}{h}\overleftarrow{ACI}_f(x_N) + O(h), & x_N = b \end{cases}$$

Proof. *Using the properties of the big-O notation that were introduced in Sect. 2.2, the definitions of the discrete punctual ACIs presented in Eqs. (6), (12) and (13), as well as the three approximations for $f'(x)$ that were given in Eqs. (8), (9) and (10), we have*

- *For $x_0 = a$, remember that $f(a) = f(x_0) = f_0$*

$$\frac{|f'(a)|}{f(a)} = \frac{\left|\frac{f_1-f_0}{h}\right| + O(h)}{\frac{f_0+f_1}{2} + O(h)} = \frac{2}{h}\frac{|f_1 - f_0|}{f_1 + f_0} + O(h) = \frac{2}{h}\overrightarrow{ACI}_f(a) + O(h)$$

- *For $a < x_k < b$,*

$$\frac{|f'(x_k)|}{f(x_k)} = \frac{\left|\frac{f_{k+1}-f_{k-1}}{2h}\right| + O(h^2)}{\frac{f_{k-1}+f_{k+1}}{2} + O(h^2)} = \frac{1}{h}\frac{|f_{k+1} - f_{k-1}|}{f_{k+1} + f_{k-1}} + O(h^2)$$
$$= \frac{1}{h}\overleftrightarrow{ACI}_f(x_k) + O(h^2)$$

- *For $x_N = b$, remember that $f(b) = f(x_N) = f_N$*

$$\frac{|f'(b)|}{f(b)} = \frac{\left|\frac{f_N-f_{N-1}}{h}\right| + O(h)}{\frac{f_{N-1}+f_N}{2} + O(h)} = \frac{2}{h}\frac{|f_N - f_{N-1}|}{f_N + f_{N-1}} + O(h) = \frac{2}{h}\overleftarrow{ACI}_f(b) + O(h)$$

ACI and Total Variation. The study of functions of bounded variation is a well-known topic in Mathematical Analysis. Below, we provide a number of definitions about it, along with a novel theorem that links the concepts of ACI and total variation.

Let f be a real-valued function, defined on $[a, b] \subset \mathbb{R}$, and $\mathcal{P} = \{x_0, x_1, \cdots, x_n\}$ a uniform partition of $[a, b]$ as in Eq. (5), we denote:

- The variation of f for the partition $\mathcal{P}$

$$V_{\mathcal{P}}(f) = \sum_{k=0}^{n-1} |f(k) - f(k+1)|$$

- The Total Variation (TV) of f over the interval $[a, b]$

$$TV(f) = \sup_{\mathcal{P}} V_{\mathcal{P}}(f) \tag{20}$$

  the supremum is taken over the set of all partitions of the interval $[a, b]$.
- We denote as $BV[a, b]$ to the set of the functions of bounded variation, that is

$$BV[a, b] = \{f : TV(f) < \infty\}$$

For continously differentiable functions, the TV defined in Eq. (20) can be computed by using the following theorem, which is proven in [10]:

Theorem 1. *If f is continously differentiable on $[a, b]$ then $f \in BV[a, b]$ and*

$$TV(f) = \int_a^b |f'(x)|\, dx$$

From the definitions and theorem above, we can now establish the following theorem, which is the main contribution of this paper.

Theorem 2. *Let $f \in C^3[a,b]$, $f > 0$, then*

i) $ACI(f) = TV(\ln(f))$
ii) For a uniform partition $\mathcal{P}$, we have

$$ACI(f) = \overleftrightarrow{ACI}(f) + O(h^2)$$

Proof. *i) From Eqs. (18) and (19), we obtain*

$$ACI(f) = \int_a^b ACI_f(x)dx = \int_a^b \left| \frac{d}{dx} [\ln f(x)] \right| dx$$

and using Theorem 1, we conclude that

$$ACI(f) = TV(\ln(f))$$

ii) From the formulation of the continuous total ACI defined in Eq. (19) and by the trapezoidal rule given in Eq. (11), we have

$$ACI(f) = \frac{h}{2}(ACI_f(a) + ACI_f(b)) + h \sum_{k=1}^{N-1} ACI_f(x_k) + O(h^2)$$

Now, for the punctual case, consider the definition of the continuous punctual ACI given in Eq. (18) and the fact $\left|\frac{d}{dx}\ln(f(x))\right| = \frac{|f'(x)|}{f(x)}$, *then*

$$ACI(f) = \frac{h}{2}\left(\frac{|f'(x_0)|}{f(x_0)} + \frac{|f'(x_N)|}{f(x_N)}\right) + h \sum_{k=1}^{N-1} \frac{|f'(x_k)|}{f(x_k)} + O(h^2)$$

By Proposition 2, the equation above can be written as in Eq.(21).

$$\begin{aligned} ACI(f) =& \frac{h}{2}\left(\frac{2}{h}\overrightarrow{ACI}_f(x_0) + O(h) + \frac{2}{h}\overleftarrow{ACI}_f(x_N) + O(h)\right) + \\ & h \sum_{k=1}^{N-1} \left[\frac{1}{h}\overleftrightarrow{ACI}_f(x_k) + O(h^2)\right] + O(h^2) \end{aligned} \tag{21}$$

Taking into account the properties of the big-O notation, and the facts

$$hO(h) = O(h^2), \quad hO(h^2) = O(h^3)$$

if $h = \frac{b-a}{N}$, *then*

$$\sum_{k=1}^{N} O(h^3) = NO(h^3) = O(h^2)$$

Thus, we can conclude that

$$\int_a^b ACI_f(x)dx = \overrightarrow{ACI}_f(x_0) + \sum_{k=1}^{N-1} \overleftrightarrow{ACI}_f(x_k) + \overleftarrow{ACI}_f(x_N) + O(h^2)$$

Finally, by the definition of the discrete total ACI in Eq. (14), we obtain

$$ACI(f) = \overleftrightarrow{ACI}(f) + O(h^2)$$

Estimating Total ACI Through Total Variation Approach. From Theorems 1 and 2, we propose here a new way to compute a numerical approximation of the total ACI (TV-based ACI), as described in the following procedure:

Procedure 1
Step 1. *Calculate the natural logarithm of* f_k *for* $x_k \in \mathcal{P}$.
Step 2. *Approximate* $\left|\frac{d}{dx}\left[\ln(f_k)\right]\right|$ *using the forward difference formula for* $k = 0$, *the central differences formula for* $1 \leq k \leq N-1$, *and the backward difference formula for* $k = N$.
Step 3. *Approximate* $\int_a^b \left|\frac{d}{dx}\left[\ln(f)\right]\right| dx$ *using the trapezoidal rule.*

3.3 Computational Examples

In this section, two illustrative examples are presented. The first one uses the spectrum of a synthetic chirp signal and is aimed to show —under controlled conditions— the differences when calculating the punctual ACI via three methods: the forward (classic) formulation, its central redefinition, and an approximation for the continuous case; see Eq. (6), Eq. (13) and Procedure 1 until the second step, respectively. The second example is based on a real-world audio signal and is intended to demonstrate —in a realistic scenario— the behavior of the total ACI when computing it also through three methods, namely: the classic formulation from Eq. (7), the so-called discrete total version defined in Eq. (14), and the TV-based approximation given in Procedure 1.

Punctual ACIs for the Spectrum of a Chirp Signal. We considered a linear chirp signal with the parameters described in Table 1, mixed with additive white noise having zero mean and a standard deviation of 0.001.

Table 1. Chirp signal parameters

Parameter	Value
Sampling frequency	$f_s = 1024$ Hz
Initial frequency	$f_0 = 50$ Hz
Final frequency	$f_1 = 350$ Hz
Time interval	$[0, 2\pi]$ seconds

For this synthetic example, we calculated the spectrum of both the original signal and the signal with noise, which are shown in the first row of Fig. 3 —on the left and the right sides, respectively. Afterwards, the three above-mentioned

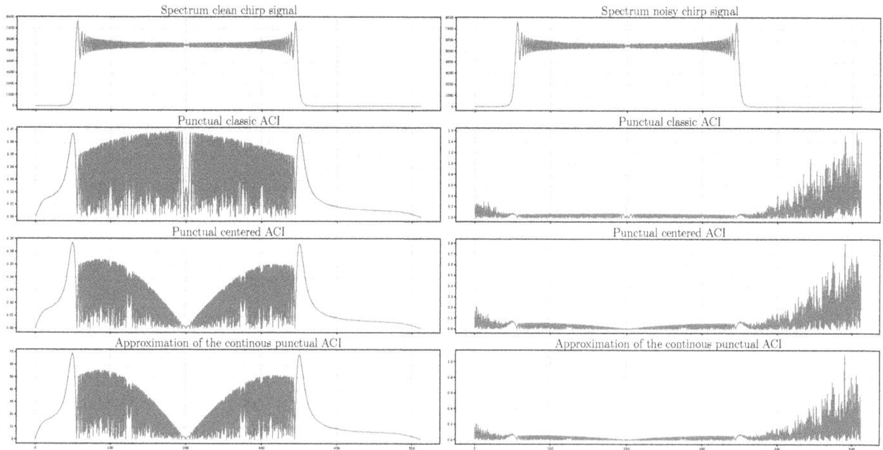

Fig. 3. Approximations of the countinuous punctual ACIs for the spectrum of a clean chirp signal (left) and a noisy chirp signal (right).

methods for computing the punctual ACI were applied for both versions of the chirp signal, whose corresponding results are shown in the second, third and fourth rows in Fig. 3.

These results highlight the ability of both the central ACI and the continuous punctual ACI approximation to effectively capture the dynamic nature of the signal spectrum, even in the presence of noise. Even though these methods excel at revealing spectral variability in clean signals (see plots on the left in Fig. 3), it is evident that noise significantly impacts the estimatios of the punctual ACIs, particularly for the classic and centered methods (see plots in the second and third rows on the right side of Fig. 3). In spite that further analyses might be needed (also to explain the differences in the scales between the classic and centered estimations of the punctual ACI), the continuous approximation method seem to exhibit greater robustness to noise compared to the classic and centered methods. However, all methods tend to underestimate the amplitude of the peaks in the presence of noise. These findings suggest that, in order to obtain a more accurate estimation of the spectrum in noisy conditions, it is necessary to combine ACI calculations with pre-filtering techniques.

Total ACIs for a Real-World Audio Signal. To illustrate the application of the three methods for computing the total ACI, we analyzed a 1-minute soundscape recorded near Colegio San Luis Gonzaga in Manizales, Colombia [15]. This recording contains various acoustic events, including anthropogenic, biophonic, and technophonic sounds. A short-time Fourier transform (STFT) was applied to this signal, using a window length of 1024 samples and an overlap of 768 samples. The resulting spectrogram (sonic matrix) is a matrix of dimensions 513×6598, as shown in Fig. 4.

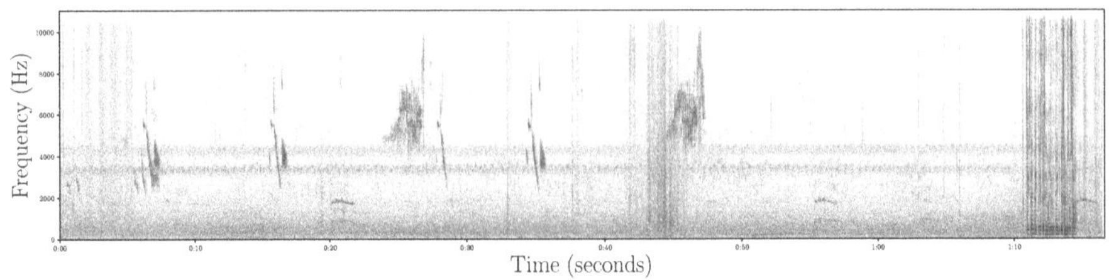

Fig. 4. The spectrogram of the real-world acoustic signal.

Since the negative effect of the noise was noticed in the case of the chirp signal, for this example we tried to mitigate the impact of the background noise by preprocessing the recording with the specialized bioacoustic filter proposed in [21]. This filter identifies and attenuates background noise, assuming that it remains constant throughout the recording. Subsequently, the three numerical approximations of the total ACI (classical, Discrete total, and TV-based) were computed on the filtered spectrogram, both in the time and frequency domains. The corresponding results are presented in Fig. 5 —on the left for the time domain and, on the right, for the frequency domain.

Figure 5 reveals a remarkable similarity between the three total ACI approximations, confirming their equivalence. However, the TV-based approximation exhibits a slightly better robustness for capturing fine details of the spectrum, particularly in the low-frequency range where spectral variability is higher.

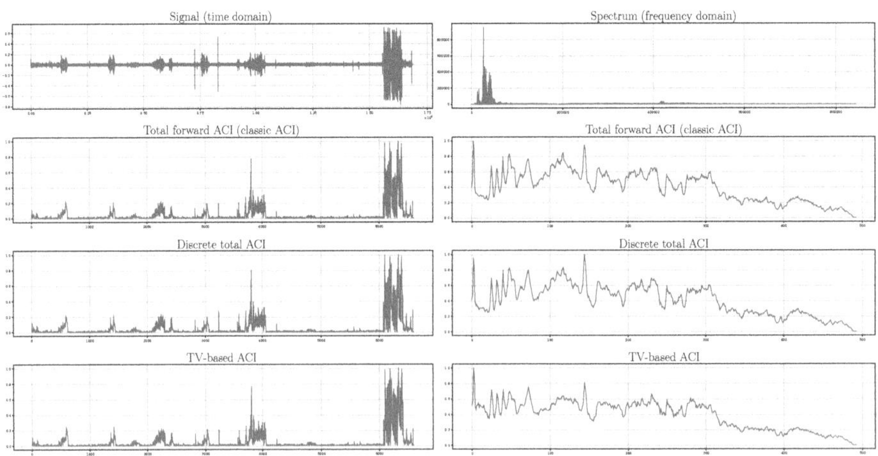

Fig. 5. Approximations of the continuous total ACIs for the spectrogram of the real-world acoustic signal along the time domain (left) and across the frequency domain (right).

4 Conclusion

Based on numerical approximations of the derivative of a function, alternative definitions of the punctual ACI were presented. Those definitions let us to present a geometric interpretation for the punctual version of the ACI, an analytic interpretation for the continous versions of it and to propose novel definitions of the total ACI in both discrete and continous versions.

The geometric interpretation allowed to visually understand that the punctual ACI is calculated using an approximation of the derivative of the function on a logarithmic scale. On the other hand, the analytic interpretation revealed the relationship between the total ACI and the concept of total variation, opening a whole field of study if the mathematical theory of bounded variation functions is applied to the fields of Bioacoustics and Ecoacoustics. Indeed, we proposed a procedure to compute the ACI via the estimation of the total variation.

It should be taken into account that the ACI, as presented in this paper, corresponds to its simplest version; however, it was originally proposed as a whole family of indices. In fact, other versions of the ACI can also be calculated using more points (clumping) that may also be geometrically and analytically studied by using approximations of higher order derivatives.

Acknowledgments. The authors acknowledge support from Universidad Nacional de Colombia - Sede Manizales and from Minciencias - Colombia [Excellence Doctoral Studies scholarship].

Data Availibility Statement. The audio file data for the illustrative example in Sect. 3.3 was obtained from the public Xeno-canto repository [15].

Disclosure of Interests. The authors have no competing interests to declare that are relevant to the content of this article.

References

1. Alcocer, I., Lima, H., Sugai, L.S.M., Llusia, D.: Acoustic indices as proxies for biodiversity: a meta-analysis. Biol. Rev. **97**(6), 2209–2236 (2022). https://doi.org/10.1111/brv.12890
2. Boelman, N.T., Asner, G.P., Hart, P.J., Martin, R.E.: Multi-trophic invasion resistance in Hawaii: bioacoustics, field surveys, and airborne remote sensing. Ecol. Appl. **17**(8), 2137–2144 (2007). https://doi.org/10.1890/07-0004.1
3. Davies, B.F.R., Attrill, M.J., Holmes, L., Rees, A., Witt, M.J., Sheehan, E.V.: Acoustic Complexity Index to assess benthic biodiversity of a partially protected area in the southwest of the UK. Ecol. Ind. **111**, 106019 (2020). https://doi.org/10.1016/j.ecolind.2019.106019
4. Farina, A.: The acoustic complexity index (ACI): theoretical foundations, applied perspectives and semantics. Oikos **2025**(1), e10760 (2025). https://doi.org/10.1111/oik.10760
5. Farina, A., Gage, S.H.: Ecoacoustics: a new science. In: Ecoacoustics: The Ecological Role of Sounds, chap. 1, pp. 1–11. John Wiley and Sons (2017). https://doi.org/10.1002/9781119230724.ch1

6. Farina, A., Li, P.: Methods in Ecoacoustics: The Acoustic Complexity Indices, Frontiers in Ecoacoustics, vol. 1. Springer, Cham (2021). https://doi.org/10.1007/978-3-030-82177-7
7. Holgate, B., Maggini, R., Fuller, S.: Mapping ecoacoustic hot spots and moments of biodiversity to inform conservation and urban planning. Ecol. Ind. **126**, 107627 (2021). https://doi.org/10.1016/j.ecolind.2021.107627
8. Indraswari, K., Bower, D.S., Tucker, D., Schwarzkopf, L., Towsey, M., Roe, P.: Assessing the value of acoustic indices to distinguish species and quantify activity: a case study using frogs. Freshw. Biol. **65**(1), 142–152 (2020). https://doi.org/10.1111/fwb.13222
9. Kasten, E.P., Gage, S.H., Fox, J., Joo, W.: The remote environmental assessment laboratory's acoustic library: an archive for studying soundscape ecology. Ecol. Inform. **12**, 50–67 (2012). https://doi.org/10.1016/j.ecoinf.2012.08.001
10. Lind, M.: Functions of bounded variation. C-level thesis, Karlstad University, Division for Mathematics, Karlstad, Sweden (2006). https://www.diva-portal.org/smash/get/diva2:5850/FULLTEXT01.pdf
11. Martínez-Tabares, F.: Factorización no negativa de matrices e índices acústicos para el análisis perceptual y la caracterización de paisajes sonoros. Ph.D. thesis, Universidad Nacional de Colombia - Sede Manizales, Manizales, Colombia (2025). Doctorado en Ingeniería - Automática
12. Mathews, J.H., Fink, K.K.: Numerical Methods Using Matlab (4th Edition). Pearson (2004)
13. McGrann, M.C., Wagner, B., Klauer, M., Kaphan, K., Meyer, E., Furnas, B.J.: Using an acoustic complexity index to help monitor climate change effects on avian diversity. Ecol. Ind. **142**, 109271 (2022). https://doi.org/10.1016/j.ecolind.2022.109271
14. Müller, M.: Fundamentals of Music Processing - Audio, Analysis, Algorithms, Applications. Springer, Cham (2015). https://doi.org/10.1007/978-3-319-21945-5
15. Nazarit, Y.: XC805861 (2023). https://xeno-canto.org/805861
16. Pieretti, N., Farina, A., Morri, D.: A new methodology to infer the singing activity of an avian community: the acoustic complexity index (ACI). Ecol. Ind. **11**(3), 868–873 (2011). https://doi.org/10.1016/j.ecolind.2010.11.005
17. Plonka, G., Potts, D., Steidl, G., Tasche, M.: Numerical Fourier Analysis. Applied and Numerical Harmonic Analysis, Birkhäuser, Cham (2018). https://doi.org/10.1007/978-3-030-04306-3
18. Raynor, E.J., Whalen, C.E., Brown, M.B., Powell, L.A.: Grassland bird community and acoustic complexity appear unaffected by proximity to a wind energy facility in the Nebraska Sandhills. Condor **119**(3), 484–496 (2017). https://doi.org/10.1650/condor-16-164.1
19. Serizel, R., Bisot, V., Essid, S., Richard, G.: Acoustic Features for Environmental Sound Analysis, pp. 71–101. Springer, Cham (2018). https://doi.org/10.1007/978-3-319-63450-0_4
20. Sueur, J., Pavoine, S., Hamerlynck, O., Duvail, S.: Rapid acoustic survey for biodiversity appraisal. PLoS ONE **3**(12), e4065 (2008). https://doi.org/10.1371/journal.pone.0004065
21. Towsey, M.W.: Noise removal from wave-forms and spectrograms derived from natural recordings of the environment. Technical report, Queensland University of Technology (2013). https://eprints.qut.edu.au/216373/
22. Tsaligopoulos, A., Karapostoli, A., Radicchi, A., Economou, C., Kyvelou, S., Matsinos, Y.G.: Ecological connectivity of urban quiet areas: the case of Mytilene.

Greece. Cities Health **5**(1–2), 20–32 (2021). https://doi.org/10.1080/23748834.2019.1599093
23. Uribe-Hurtado, A.L., Martínez-Tabares, F., Orozco-Alzate, M.: Unsupervised selection of acoustic indices: an experimental comparison for characterizing unlabeled audio recordings from sub-Andean forest soundscapes. Appl. Acoust. **228**, 110363 (2025). https://doi.org/10.1016/j.apacoust.2024.110363
24. Villanueva-Rivera, L.J., Pijanowski, B.C., Doucette, J., Pekin, B.: A primer of acoustic analysis for landscape ecologists. Landscape Ecol. **26**(9), 1233–1246 (2011). https://doi.org/10.1007/s10980-011-9636-9

Enhanced 3D Pipeline Installation Reconstruction and Modeling for Industrial Inspection

Ioannis Karantaidis, Alexandros Zamioudis(✉), and Ioannis Pitas

School of Informatics, Aristotle University of Thessaloniki, University Campus, 54124 Thessaloniki, Central Macedonia, Greece
azamiou@csd.auth.gr

Abstract. Although industrial installation inspection has been widely discussed in the literature, 3D shape reconstruction and modeling of industrial pipeline installation still has many poorly researched topics. Pipeline installation can be estimated from sensed data, which are either 3D point clouds obtained by LiDAR or multiview RGB images or videos. UAV imaging is typically used to acquire the sensed data. Existing 3D pipeline modeling algorithms from a 3D point cloud or RGB image data is error prone, as they either completely lack outlier removal or they utilize trivial methods to this end. This can lead to suboptimal 3D pipe modeling in real world scenarios, where cluttered industrial environments introduce outliers. In this paper, we propose the **Enhanced Robust Cylinder Fitting (ERCFit)**, a novel industrial 3D pipeline reconstruction algorithm. ERCFit exhibits enhanced outlier removal capabilities, improving the state-of-the-art Robust Cylinder Fitting algorithm by incorporating the Local Outlier Factor (LOF) algorithm into Robust-PCA. Consequently, our proposed algorithm can use local neighbourhood density information for enhanced outlier removal. We further propose a comprehensive 3D pipeline installation reconstruction framework, which makes use of the ERCFit algorithm to fully model and reconstruct all pipelines present in such a 3D industrial scene. Extensive experiments show that our proposed 3D pipeline modeling algorithm outperforms the SOTA pipeline modeling methods. Our 3D pipeline installation reconstruction framework successfully reconstructs all straight pipelines present in a real world scene of a typical chemical plant.

Keywords: 3D reconstruction · Pipeline · RobustPCA · Local Outlier Factor · 3D point cloud · Minimum Covariance Determinant

1 Introduction

Unmanned Aerial Vehicles (UAVs) have become increasingly vital in industrial inspection tasks, particularly in critical infrastructures such as petrochemical refineries [5,18]. Their versatility, flexibility, and ability to efficiently capture

D. Chaves et al. (Eds.): CIARP 2025, LNCS 16528, pp. 308–322, 2026.
https://doi.org/10.1007/978-3-032-23161-1_22

high-resolution data have led to widespread adoption across numerous industries. In the context of industrial infrastructure inspection, UAVs offer a non-invasive method to monitor extensive pipeline networks, even in challenging environments.

Pipelines, often exhibiting cylindrical geometry, are among the most frequently encountered structures in these settings. Their condition is crucial for the smooth operation of industrial processes, making damage detection a primary focus in pipeline inspection [11,13]. Over time, factors such as sudden pressure increases, corrosion, traffic stress, and ground movement significantly elevate the risk of pipeline damages and leaks [4]. Such damages can cause substantial financial losses and environmental harm. UAVs provide a safe, efficient, and cost-effective solution for detecting faults without halting normal operations—a particularly useful advantage for online flares where traditional methods require facility shutdowns for visual inspections. Typical pipeline damages are illustrated in Fig. 1.

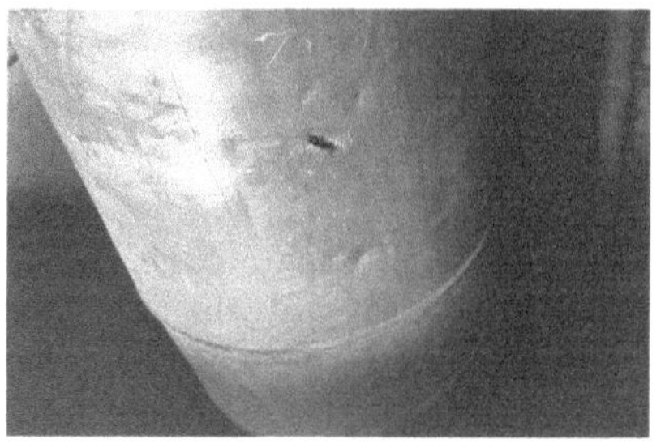

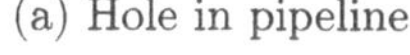

(a) Hole in pipeline

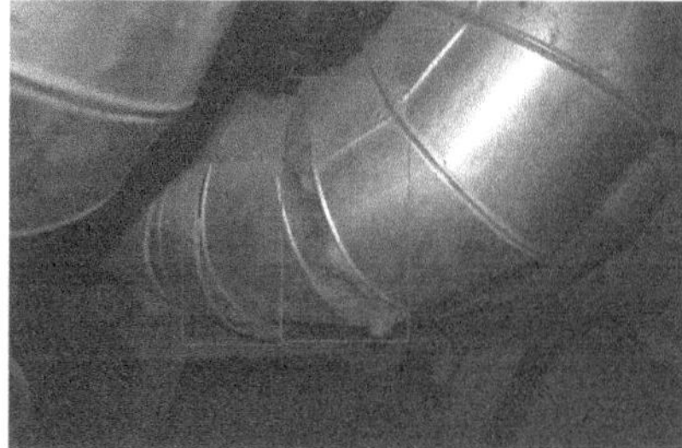

(b) Open insulation

Fig. 1. Examples of pipeline damages captured by drone from Greek facilities in PDI [11].

Traditional inspection techniques, such as ground-based visual checks, are time-consuming and hazardous, requiring downtime and physical access to the structure. In contrast, UAVs provide a more agile and safer solution, especially for monitoring pipelines located in remote or hazardous areas where manual inspection is not feasible [4]. The typical drone inspection process involves three key stages: first, using drones to gather high-resolution imagery of the target infrastructure; second, reconstructing the 3D point cloud of the scene using computer vision and photogrammetric algorithms, and applying machine learning algorithms to analyze faults and assess damage propagation; and third, testing and validating the approach in real-world conditions.

While 2D damage detection can be achieved by analyzing images captured by drones using deep learning algorithms [11], human inspectors require precise 3D damage location information to perform effective repairs. Thus, reconstructing the 3D scene and geometrically modeling the pipeline installations are key components of the pipeline inspection process.

Given the cylindrical nature of pipelines, fitting 3D cylinders to the sensed data is a logical approach for effective modeling. Each cylinder is characterized by six geometrical parameters [12]: four for modeling the cylinder's axis (the spine) and two for the cylinder's radius and length. The 3D point cloud representing the pipeline can be derived either from LiDAR data or from RGB images processed through Structure from Motion (SfM) algorithms.

Despite their advantages, UAV-based inspections also present challenges. The sheer size and complexity of industrial installations require detailed data collection and advanced algorithms for efficient data processing and interpretation. Converting the 2D images captured by drones into usable 3D information is of significant difficulty. Moreover, pipelines are often surrounded by various other structures, including support beams, valves, and joints. These additional elements can interfere with data acquisition, introducing noise into the point cloud. Without robust outlier detection methods, erroneous data points may lead to incorrect modeling, making subsequent damage analysis unreliable [12].

To address these challenges, research in robust geometric fitting has evolved significantly, focusing on methods that can estimate cylinder parameters accurately while minimizing the impact of noisy data. Cylinder fitting methods range from simple least-squares approaches to advanced algorithms integrating outlier detection mechanisms, such as the RANSAC algorithm [9], to enhance robustness.

This paper extends the method of Robust Cylinder Fitting (RCF) [12] to enhance pipeline modeling by removing outliers found in real-world industrial environments. It also establishes a reliable framework for reconstructing entire 3D pipeline installations. To improve the robustness of cylinder parameter estimation, the Local Outlier Factor (LOF) algorithm is integrated into the RCF method, yielding a new approach termed Enhanced Robust Cylinder Fitting (ERCFit). The RCF algorithm serves as the baseline for evaluating the performance of ERCFit, which is assessed using synthetic data representing cylinders of various predefined lengths, orientations, radii, and positions. The performance of ERCFit is compared against both the baseline RCF method and a neural network approach. Furthermore, multiple-cylinder modeling of the pipeline installation is conducted using a 3D point cloud derived from SfM applied to RGB images of the entire installation. The connected-component labeling algorithm [8] is utilized to isolate individual pipes from the global 3D point cloud, allowing ERCFit to be applied for accurate modeling of each pipe installation. This sequential process ensures precise identification and modeling of the pipeline components.

The paper is organized as follows: Sect. 2 provides a comprehensive review of state-of-the-art techniques in 3D pipeline reconstruction. Section 3 introduces essential background on single-cylinder geometrical modeling and cylinder parameter regression using a neural network. Section 4 details the robust outlier removal workflow. Section 5 presents a framework for segmenting a point cloud into separate components for individual modeling. Section 6 discusses the experimental results, highlighting the effectiveness of the proposed approach. Section 7 summarizes the key findings and conclusions of the research.

2 Related Work

3D cylinder modeling for fitting refers to the process of estimating the parameters of a cylinder from a given point cloud, such as position, radius, length and orientation. Recent advances in pipeline inspection methods have demonstrated the potential for multi-view imaging techniques to enhance 3D pipeline shape reconstruction accuracy. For example, the method described in [20] involves moving long pipelines through a 3D shape reconstruction system with multiple cameras. However, this approach introduces the need to move the pipeline, which can lead to alignment errors and complicates the process for extremely long pipelines.

Pre-processing of raw 3D point clouds to obtain labeled cylindrical data is proposed in [12]. The insufficiency of self-sufficient extraction of multiple cylinders in complex environments like industrial plants, is highlighted in the same paper. This limitation hinters handling intricate pipeline scenes without prior data preparation.

Neural network approaches have been explored for fitting and recognizing simple geometric primitives from point clouds, using PointNet as the backbone architecture [15]. While this method shows promising results in automating 3D pipeline modeling, it cannot estimate cylinder length, which is a crucial 3D cylinder model parameter for real-world applications, such as industrial inspection and robotics. Moreover, its modeling accuracy falls short, when compared to traditional non-neural network methods.

RANSAC (Random Sample Consensus) [9] has long been a foundational method for model fitting in computer vision and 3D reconstruction tasks, including cylinder fitting in point clouds. Its main advantage lies in its robustness to noise and outliers, making it highly effective in real-world datasets where point clouds often contain a significant amount of extraneous data. By randomly sampling subsets of points, RANSAC iteratively searches for the best-fit model, discarding points that do not align with the detected cylinder. However, RANSAC has a key limitation: it can be computationally expensive, especially when dealing with large point clouds. Moreover, its accuracy is heavily dependent on the choice of parameters, such as the threshold for inlier detection, and the method may struggle with complex structures where cylindrical features are partially obscured or fragmented.

Recent advancements in deep learning have introduced methods like DeepFit [6], which leverage neural networks to directly learn geometric primitives from point clouds. DeepFit is capable of handling noisy and incomplete data more effectively than traditional methods, as it generalizes well across different types of datasets. Its primary strength is its ability to learn local geometric features that correspond to shapes like cylinders, making it particularly useful in environments where the point cloud data is sparse or contains occlusions. However, like many deep learning approaches, DeepFit requires a large amount of training data to achieve high accuracy. Furthermore, the training process itself can be computationally expensive, and the method may not generalize well to unseen data if the training set does not adequately represent the variety of shapes present in the test data.

DiffusionNet [17] offers a novel approach to point cloud analysis by utilizing diffusion-based representations of surface geometry. This method has the potential to provide smooth and robust cylinder fitting, even in the presence of noise or incomplete data, as it leverages the diffusion process to capture smooth transitions in surface geometry. DiffusionNet is particularly effective in preserving geometric structures, which makes it well-suited for identifying and fitting cylindrical shapes that may be obscured by noise. However, one of the main disadvantages of DiffusionNet is its complexity and computational overhead, particularly when applied to large point clouds. The method may also be sensitive to the parameters governing the diffusion process, requiring careful tuning to achieve optimal results.

The limitations of the aforementioned approaches for pipeline modeling and 3D reconstruction of pipeline scenes are the primary motivation for this work. Subsequently, we briefly review a state-of-the-art statistical approach, Robust Cylinder Fitting (RCFit) [12], and a DNN-based 3D cylinder parameter regression method. The primary objective of this paper is to enhance the robustness and accuracy of the RCFit algorithm. It employs a robust version of Principal Component Analysis (PCA), and introduces a 3D reconstruction framework for modeling a 3D scene containing multiple pipelines.

3 Background

3.1 Robust PCA

Principal Component Analysis (PCA) identifies directions that maximize data variability by calculating the covariance matrix Σ and applying eigen analysis to determine the Principal Components (PCs), in the form of eigenvalues and eigenvectors. However, PCA is highly sensitive to outliers, which can significantly affect its accuracy. ROBPCA algorithm [10], has been developed to improve PCA robustness by effectively removing outliers, before identifying the Robust Principal Components (RPCs). ROBPCA combines the Minimum Covariance Determinant (MCD) algorithm with the detection of multivariate outliers. In this study, we exclusively utilize the MCD method, as the low dimensionality of the point cloud data enables its effective application without significant challenges. Consequently, we refer to this approach as Robust PCA. When Robust PCA is applied to cylinder modeling from a 3D point cloud of its surface, the first eigenvector i.e. Principal Component PCA1 is selected to represent the cylinder axis, as it captures maximum data variability. The 3D point cloud is then projected onto PCA1 for determining the cylinder length by calculating the difference between the maximum and minimum projection values along the cylinder axis. Accurate cylinder axis estimation relies on the assumption that the cylinder's length exceeds its radius, thereby ensuring the proper orientation of the principal directions.

By projecting the data onto the second and third principal components (PC2 and PC3), the cylinder cross-section (circle) is identified. The Hyper algorithm [3] is then employed to accurately compute the circle radius and center. It offers several advantages, including efficient circle fitting to arcs, precise data point sampling along circle circumference, superior performance for modeling circles with small radius, and fast computation compared to traditional geometric fitting methods [12]. Additionally, Hyper algorithm results are invariant under 3D data point rotation and translation. Moreover, results can be improved by applying Least-trimmed Squares (LTS) regression [12]. LTS employs only a 3D data point subset of n observations for circle fitting to reduce the influence of outliers. The subset cardinality n is chosen to have the minimum sum of sorted square residuals. It is empirically set to $n = \lfloor 0.5N \rfloor$, where N is the number of the whole 3D data point.

The Hyper algorithm computes the center of the circular arc in the x and y coordinates, yet the z coordinate must be estimated to uniquely determine the unique position of the cylinder along the z axis. The z coordinate corresponds to the location along the estimated orientation axis and is positioned at the midpoint of the cylinder's length.

3.2 DNN Cylinder Parameter Regression

Recently, DNN's have been employed to regress parameters for various primitive surfaces, including planes, spheres, cylinders, cones, and tori [15]. In the DNN context, each 3D surface is considered to be one 3D shape class, whose parameters should be regressed. This approach typically involves two main steps: shape classification and parameter regression [15]. The PointNet architecture [14] serves as the foundational network, initially extracting features from the input 3D point cloud. Subsequently, a classification network performs 3D surface classification to the primitive shape class. A specific regression DNN is assigned to each 3D shape class. These regression networks are tailored with unique final layers comprising MLP (Multi-Layer Perceptron) heads to produce the output vector (3D shape parameters). The cylinder regression network is used to estimate the cylinder parameters, including orientation, center, and radius.

4 Enhanced Robust Cylinder Fitting (ERCFit)

The proposed Enhanced Robust Cylinder Fitting (ERCFit) algorithm integrates the Local Outlier Factor (LOF) algorithm [7] to improve RCFit robustness, by effectively removing outliers from the 3D point cloud of a chemical plant pipework, resulting in more accurate cylinder fitting to the straight pipelines.

The proposed ERCFit workflow can be seen Fig. 2. In the original RCFit algorithm [12], the 3D point cloud is projected onto the first principal component (PC1) to compute the cylinder length. However, some outliers persist, even after

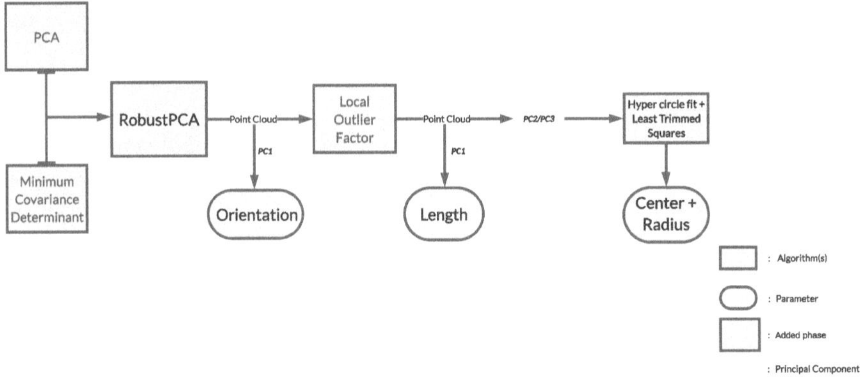

Fig. 2. ERCFit algorithm workflow.

applying Minimum Covariance Determinant (MCD), because MCD need prior knowledge of the outlier percentage, which is critical for maximizing outlier exclusion. Despite effectively removing most outliers, the presence of even a single outlier can significantly impact the accuracy of the computed cylinder length.

This limitation can be seen in Fig. 3. The blue points represent those that belong to the cylinder, while the red x markers represent outliers excluded using the Minimum Covariance Determinant method. Even though most of the outliers have been removed, the circled blue points that will be projected onto the x axis will overestimate the length of the cylinder, severely reducing the accuracy of the length parameter.

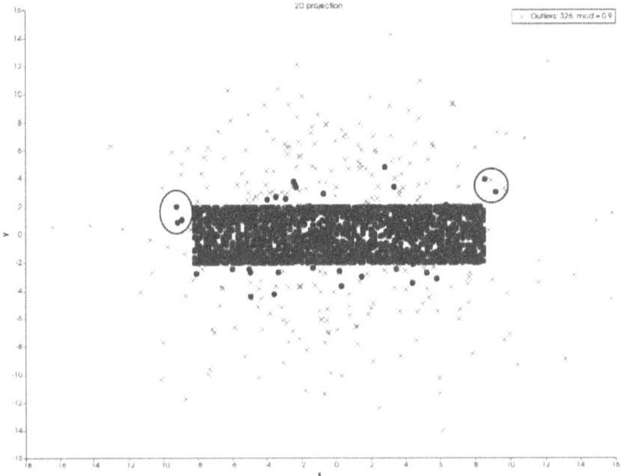

Fig. 3. Cylinder with scattered outliers following the Guassian distribution.

To mitigate this issue, the Local Outlier Factor (LOF) [7] is applied to the 3D points projected onto PC1 and PC2, rather than only on PC1, thereby revealing the density of the data points more clearly. Only after LOF has successfully removed the outliers, the 2D points (projected onto PCA1 and PCA2) are then projected onto PC1 for the estimation of cylinder length. LOF evaluates the outlierness of each data point based on its deviation from the density of its nearest neighbors within a predefined threshold k. This approach ensures that points not adequately filtered by MCD are excluded as outliers based on their local neighborhood density.

In Fig. 4, the effectiveness of the Enhanced Robust Cylinder Fitting (ERCFit) method is demonstrated. Initially, the Minimum Covariance Determinant (MCD) was used to exclude outliers, represented by red x markers, while the blue points represent the inliers that belong to the cylinder. Now, using the ERCFit method, the outliers (previously marked as blue dots) are now successfully removed, as indicated by the black dots in the figure.

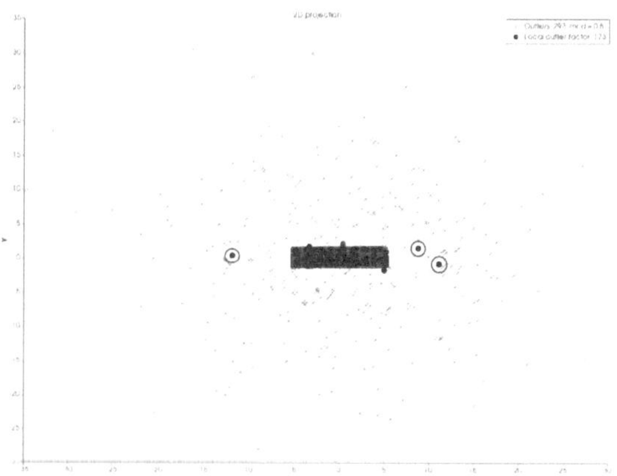

Fig. 4. Outlier removal - Local Outlier factor.

Mathematically, the LOF of a point x_i is defined as:

$$\mathcal{L}(\mathbf{x_i}) = \frac{\sum_{x_j \in N_k(x_i)} r(x_i, x_j)}{|N_k(x_i)|} \tag{1}$$

where:

- $N_k(x_i)$ represents the cardinality of the k nearest neighbor set of x_i within a predefined threshold,
- $r(x_i, x_j)$ denotes the reachability distance between x_i and x_j, defined as the maximum of the distance between x_i and x_j.

$\mathcal{L}(\mathbf{x_i})$ range is [0, 1]. Values greater than 1 indicates that the point x_i is an outlier, as it deviates significantly from its neighbors in terms of density.

Conversely, a $\mathcal{L}(\mathbf{x_i})$ value close to 1 suggests that x_i is similar to its neighbors in terms of density and is less likely to be an outlier.

By integrating LOF into the robust cylinder fitting process, outliers that escape detection by MCD due to uncertain outlier percentages are effectively identified and removed. This enhancement improves the robustness of computing cylinder parameters, such as length and center of a cylinder, by ensuring that outliers with anomalous relationships to their local neighborhoods do not influence the calculations. Thus, the combination of MCD and LOF in the proposed novel ERCFit algorithm, provides a more comprehensive and reliable approach to outlier detection in the context of 3D point cloud analysis. This is particularly useful 3D in environments prone to noise and clutter as is the case of industrial pipeline installations.

5 3D Pipe Scene Reconstruction System

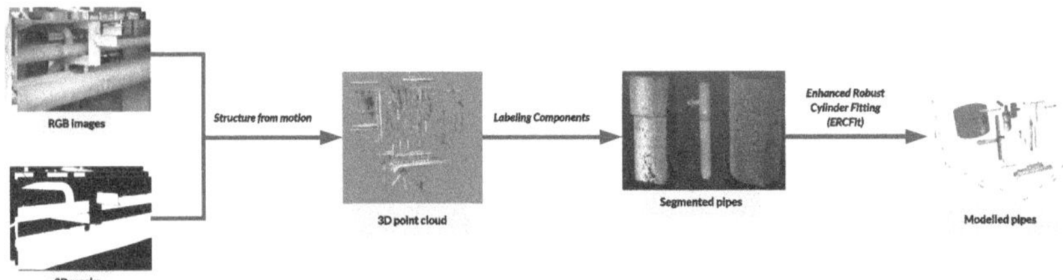

Fig. 5. Reconstruction system workflow.

3D pipeline installations can be imaged using LiDAR and/or RGB cameras. Both sensors can be airborne using UAV imaging or ground-based. LiDAR provides a global 3D point cloud of the pipeline installation surfaces. RGB cameras provide multi-view RGB images that can feed a Structure from Motion (SfM) algorithm [19] to obtain the 3D surface point cloud.

The RGB images of pipelines can be segmented using DNN-based methods [13]. Segmentation masks corresponding to the class pipeline contain image pixels that are projections of the 3D pipelines. In the 3D pipeline reconstruction process, only image pixels of pipelines can feed the SfM algorithm. This approach greatly reduces 3D point cloud and clutter during the process of generating a 3D point cloud of the scene. However, due to the use of DNN-based segmentation algorithm errors, some background noise may remain in the 3D point cloud.

The 3D pipe scene reconstruction system consists of two main components: a) the Connected-component labeling algorithm [8] and b) the ERCFit algorithm, as presented in Fig. 5. Alternatively, a DNN cylinder modeling algorithm can replace the ERCFit algorithm.

The global 3D point cloud contains numerous cylindrical elements, such as pipes and storage tanks. For further analysis, the point cloud is segmented into individual pipe segments using a Connected-component labeling algorithm, which identifies and isolates different connected components within the point cloud. The labeling process assigns a unique label to each 3D connected component by analyzing the spatial 3D point connectivity. 3D connected components having few possible sparse 3D points are discarded as noise.

Finally, the ERCFit algorithm is applied iteratively to fit each segmented cylinder in order to produce the final 3D pipeline scene.

6 Experimental Results

The main goal of the experiments is to test the robustness of the proposed methodology and its competitors, compare the performance of classical and DNN-based 3D cylinder modeling approaches, and validate the 3D pipe installation reconstruction system of a chemical plant data set.

To test the robustness of the proposed method, a dataset containing 3D surface point clouds of pipes was created containing 100 cylinders, each having 1000 3D surface points; when needed, an arbitrary amount outliers were added following the Gaussian distribution with $\sigma = 5m$. The ERCFit algorithm was compared with the RCFit and the PointNet-based DNN.

A key concept of the minimum covariance determinant algorithm, is the determination of the 3D data point percentage ("support of the estimate") to be included in the core subset for estimating the point cloud mean and covariance [16]. This parameter ensures that the algorithm can effectively manage the trade-off between excluding noise and capturing the true underlying distribution of the data. Experiments were conducted to determine the optimal percentage (10%, 30%, 50%, 70%) while injecting Gaussian noise at data point percentages ranging from 10% to 80%. The effectiveness of each setting was assessed using the Area Under Curve (AUC) metric for every experiment and parameter combination. This approach facilitated the identification of the most suitable hyperparameter value for future following experiments.

Figure 6 displays the Root Mean Square Error (RMSE) values for four cylinder parameters (radius, center, length, orientation) across varying numbers of outliers and different Support of Estimate (SoE) values. Each subplot illustrates how RMSE changes with increasing outliers, with multiple lines representing different SoE values. To determine the optimal SoE value (λ^*), the mean RMSE across all parameters i is computed and evaluated for each SoE value λ:

$$\lambda^* = \arg\min_{\lambda} \frac{1}{4} \sum_{i=1}^{4} \text{RMSE}_{i\lambda} \tag{2}$$

Here, $\text{RMSE}_{i\lambda}$ denotes the RMSE value for parameter i at SoE value λ. Across all evaluations, $\lambda^* = 0.5$ consistently yielded the lowest mean RMSE for all parameters. This finding indicates that using SoE with a value of 0.5 optimally reduces overall RMSE values in estimating cylinder parameters.

After establishing the optimal hyperparameter value for the MCD algorithm, the proposed ERCFit was compared to the baseline (RCFit) one. To evaluate the

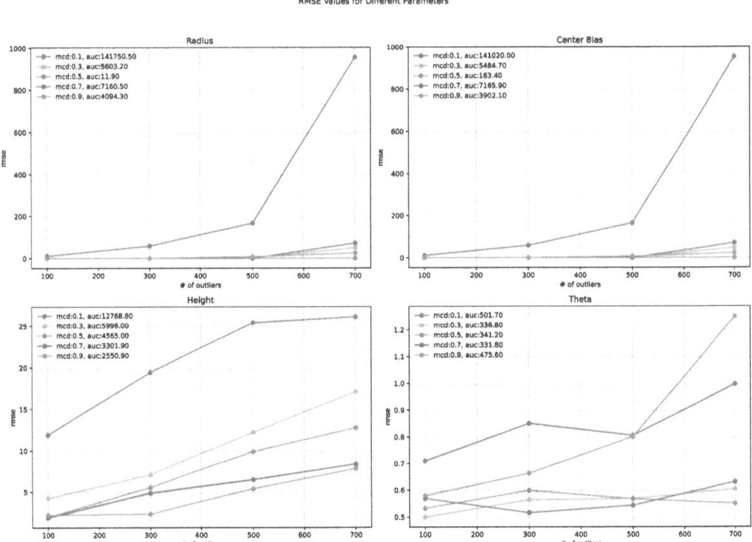

Fig. 6. RMSE values for different parameters (Radius, Center Bias, Length, Orientation) across varying numbers of outliers and support mcd values.

robustness of the two approaches, 20% of the cylinder points were added as outliers following the Gaussian distribution with $\sigma = 5m$. The experimental results are presented in Table 1. It was observed that incorporating the Local Outlier Factor (LOF) algorithm significantly lowered the RMSE for the center and length parameters. Specifically, the RMSE for the center parameter decreased from 0.254 to 0.21, and the RMSE for the length parameter was drastically reduced from 2.313 to 0.141. The Radius and Orientation parameters maintained consistent RMSE values of 0.007 and 0.646, respectively, across both methods. These findings indicate that integrating LOF into the methodology effectively reduces the RMSE for critical parameters, notably the length, thereby enhancing the overall accuracy of the cylinder fitting process.

Additionally, to compare the PointNet-based DNN method with ERCFit, 10,000 cylinders with varying radii, positions, lengths, and orientations were generated and provided with known parameters. These cylinders were used to train the neural network from scratch, with a subset of 100 cylinders serving as the test set. However, PointNet-based DNN does not predict the length parameter, so the comparison focused solely on the radii, positions, and orientations. Moreover, the comparison was conducted without introducing any outliers into the data, since the PointNet-based DNN does not include a mechanism for outlier removal.

The RMSE values for the radius, center, and orientation parameters of both algorithms are presented in Table 2. The ERCFit method is much better than

Table 1. Comparison of ERCFit and RCFit cylinder parameter estimation accuracy

Parameter	Robust Cylinder Fit (RMSE) [12]	ERCFit (RMSE)	Improvement (%)
Radius (m)	0.007	0.007	0%
Center (m)	0.254	0.21	17.32%
Length (m)	2.313	0.141	93.90%
Orientation (deg)	0.646	0.646	0%

Table 2. Comparison of ERCFit and PointNet cylinder parameter estimation accuracy.

Parameter	PointNet-based DNN (RMSE) [15]	ERCFit (RMSE)	Improvement (%)
Radius (m)	0.394	0.004	98.98%
Center (m)	0.281	0.228	18.86%
Length (m)	-	0.087	-
Orientation (deg)	1.731	0.511	70.49%

the PointNet-based DNN approach across multiple parameters. Specifically, the ERCFit method achieves a remarkably low RMSE of 0.004 for Radius, compared to 0.394 for the PointNet-based DNN. For the Center parameter, ERCFit records an RMSE of 0.228, which is lower than the PointNet-based DNN RMSE of 0.281. Additionally, ERCFit demonstrates superior performance in estimating Orientation, with an RMSE of 0.511, substantially better than the PointNet-based DNN RMSE of 1.731. The length parameter is only computed by ERCFit, with a very low RMSE of 0.087, substantiating its robustness.

6.1 Chemical Plant 3D Pipeline Installation Reconstruction

The performance of the proposed 3D pipe scene reconstruction system is validated by applying the system to a real world scenario of an chemical plant pipeline installation. Initially, the Structure From Motion (SFM) algorithm was employed to generate the comprehensive 3D global point cloud of the pipeline installation using 169 UAV images, processed using Agisoft Metashape software [1]. Next, the 3D point cloud was segmented into individual components using the Connected-component labeling algorithm using the CloudCompare software [2]. The 3D point cloud produced by the SfM and Connected-component labeling algorithms are presented in Fig. 7.

Each labeled 3D point component was modeled using the ERCFit method, to estimate its radius, center, length, and orientation. ERCFit ensures that the

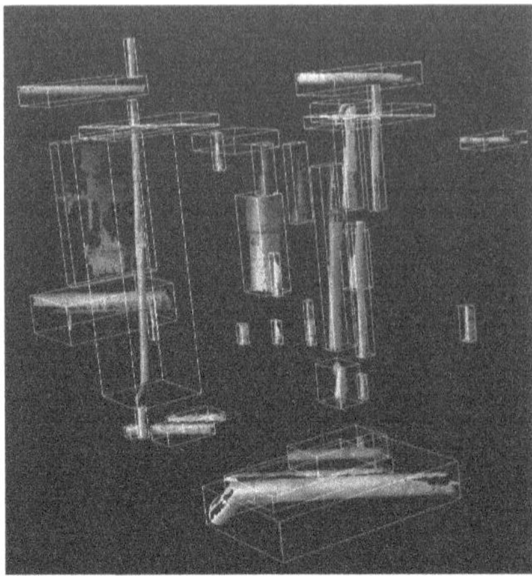

Fig. 7. Segmented and filtered 3D point cloud after applying the connected-component labeling algorithm.

calculated parameters are not significantly affected by noise and outliers present in the point cloud data.

To visually validate the computed parameters, only the straight pipes were geometrically modeled and plotted onto the point cloud, as presented in Fig. 8. Since ERCFit is not designed to process non-pipe elements or pipes with elbows, these components were manually excluded from the model. All of the accurately reconstructed 3D point clouds corresponding to individual pipes were successfully geometrically modeled, resulting in a total of 38 pipes. This modeling process confirmed the effectiveness of the proposed method and provided a clear representation of the spatial arrangement of the pipes within the plant. The robust cylinder fit enabled an accurate reconstruction of the 3D model of the straight pipelines.

The integration of the baseline method with the Local Outlier Factor (LOF) algorithm proved to be particularly effective in reducing RMSE values, especially for the length parameter, as demonstrated in the comparative analysis. The resulting 3D model showcases the potential of the proposed approach to deliver precise and reliable pipe modeling in complex industrial environments.

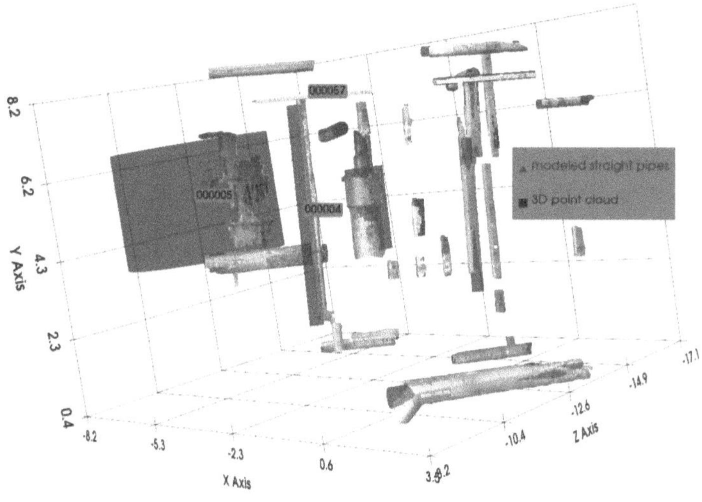

Fig. 8. 3D point cloud along with straight modeled pipes.

7 Conclusion

This paper introduces Enhanced Robust Cylinder Fitting (ERCFit), a novel comprehensive methodology for modeling pipes in industrial settings. By integrating Robust Principal Component Analysis (RobustPCA) with the Local Outlier Factor (LOF) algorithm, ERCFit significantly mitigates the impact of outliers in cylinder modeling, thereby enhancing the accuracy of parameter estimation.

Our experiments demonstrated that ERCFit outperforms the RCFit [12] and PointNet methods in terms of RMSE across various cylinder parameters, including its radius, center, length, and orientation. The optimal value for the Minimum Covariance Determinant (MCD) parameter was calculated, ensuring robust performance across different outlier scenarios.

The application of our method to a real-world 3D point cloud from an chemical plant highlighted its practical utility. Accurate reconstruction of the pipe network within the point cloud underscores ERCFit robustness and precision, making it a valuable tool for industrial facility inspection, digital twin modeling and maintenance.

Acknowledgments. This work has received funding from the European Union's Horizon research and innovation program under grant agreement number 101070604 (SIMAR).

References

1. Agisoft photoscan standard (version 2.0.3) (2023). https://www.agisoft.com/

2. Cloudcompare (version 2.13) [GPL software] (2024). http://www.cloudcompare.org
3. Al-Sharadqah, A., Chernov, N.: Error analysis for circle fitting algorithms. Electron. J. Stat. **3**, 886–911 (2009)
4. Aljuaid, K.G., Albuoderman, M.A., AlAhmadi, E.A., Iqbal, J.: Comparative review of pipelines monitoring and leakage detection techniques. In: 2020 2nd International Conference on Computer and Information Sciences (ICCIS), pp. 1–6 (2020)
5. Asadzadeh, S., de Oliveira, W.J., de Souza Filho, C.R.: UAV-based remote sensing for the petroleum industry and environmental monitoring: state-of-the-art and perspectives. J. Petrol. Sci. Eng. **208**, 109633 (2022)
6. Ben-Shabat, Y., Lindenbaum, M., Fischer, A.: Deepfit: 3D surface fitting via neural network weighted least squares. In: Proceedings of the IEEE Conference on Computer Vision and Pattern Recognition (CVPR), pp. 5119–5128. IEEE (2018)
7. Breunig, M.M., Kriegel, H.P., Ng, R.T., Sander, J.: LoF: identifying density-based local outliers. In: Proceedings of the 2000 ACM SIGMOD International Conference on Management of Data, pp. 93–104 (2000)
8. Di Stefano, L., Bulgarelli, A.: A simple and efficient connected components labeling algorithm. In: Proceedings 10th International Conference on Image Analysis and Processing, pp. 322–327 (1999)
9. Fischler, M.A., Bolles, R.C.: Random sample consensus: a paradigm for model fitting with applications to image analysis and automated cartography. Commun. ACM **24**(6), 381–395 (1981)
10. Hubert, M., Rousseeuw, P.J., Vanden Branden, K.: Robpca: a new approach to robust principal component analysis. Technometrics **47**(1), 64–79 (2005)
11. Mentesidis, P., Papaioannidis, C., Pitas, I.: Advancing industrial inspection: a dataset for automated damage detection in insulated pipes. In: IEEE International Conference on Acoustics, Speech, and Signal Processing Workshop (ICASSP Workshop) (2024)
12. Nurunnabi, A., Sadahiro, Y., Lindenbergh, R.: Robust cylinder fitting in three-dimensional point cloud data. Int. Arch. Photogrammetry Remote Sens. Spatial Inf. Sci. **XLII-1/W1**, 63–70 (2017)
13. Psarras, D., Papaioannidis, C., Mygdalis, V., Pitas, I.: A unified DNN-based system for industrial pipeline segmentation. In: 2024 IEEE International Conference on Acoustics, Speech and Signal Processing (ICASSP), ICASSP 2024, pp. 7785–7789 (2024)
14. Qi, C.R., Su, H., Mo, K., Guibas, L.J.: Pointnet: deep learning on point sets for 3D classification and segmentation (2017)
15. Romanengo, C., et al.: Shrec 2022: fitting and recognition of simple geometric primitives on point clouds. Comput. Graph. **107**, 32–49 (2022)
16. Rousseeuw, P.J., Driessen, K.V.: A fast algorithm for the minimum covariance determinant estimator. Technometrics **41**(3), 212–223 (1999)
17. Sharp, N., Attaiki, S., Crane, K., Ovsjanikov, M.: Diffusionnet: discretization agnostic learning on surfaces. ACM Trans. Graph. **39**(6), 1–16 (2020)
18. Wanasinghe, T.R., Gosine, R.G., De Silva, O., Mann, G.K.I., James, L.A., Warrian, P.: Unmanned aerial systems for the oil and gas industry: overview, applications, and challenges. IEEE Access **8**, 166980–166997 (2020)
19. Westoby, M.J., Brasington, J., Glasser, N.F., Hambrey, M.J., Reynolds, J.M.: 'Structure-from-motion' photogrammetry: a low-cost, effective tool for geoscience applications. Geomorphology **179**, 300–314 (2012)
20. Zhang, T., Liu, J., Liu, S., Tang, C., Jin, P.: A 3D reconstruction method for pipeline inspection based on multi-vision. Measurement **98**, 35–48 (2017)

Non-linear Model Comparison for Cordoba Steer Beef Cattle Growth Curve

Ronnal Ortiz-Cuadros[1](✉), Lorena Inés Mestra-Vargas[2], Diego Andres Medina-Herrera[2], Olga Lucia Mayorga-Mogollón[1], Andrea Milena Sierra-Alarcón[1,3], Vincent Pamugas Reyes[3], Lorena Aguayo-Ulloa[2], and Satoshi Ohkura[3]

[1] Corporación Colombiana de Investigación Agropecuaria – Agrosavia, Centro de Investigación Tibaitatá, Km. 14, Vía Mosquera - Bogotá, Mosquera, Cundinamarca, Colombia
rortiz@agrosavia.co

[2] Corporación Colombiana de Investigación Agropecuaria – Agrosavia, Centro de Investigación Turipaná, Km. 13 Vía Montería - Cereté, Cereté, Córdoba, Colombia

[3] Graduate School of Bioagricultural Sciences, Nagoya University, Nagoya, Japan

Abstract. This study compared the performance of five nonlinear growth models (Gompertz, Richards, von Bertalanffy, Logistic, and Michaelis-Menten) for describing cattle growth patterns across five sub-regions of Córdoba, Colombia. Body weight data from 731 calves across 16 beef farms were analyzed from birth to weaning (0–240 days). Three analytical approaches were employed: individual-level fitting, regional grouping, and regional averaging. Model selection used a voting method based on AIC, BIC, MSE, MSEP, MAE, and R^2 criteria.

Birth weight averaged 32.3±9.9 kg, consistent with crossbred tropical cattle. Individual-level analysis favored Gompertz models (66% selection), followed by von Bertalanffy (35%) and Logistic (2%). Regional and average analyses showed different model preferences, with von Bertalanffy dominating regional analysis and Gompertz preferred in average analysis. Asymptotic weights ranged from 428-563 kg across analytical approaches. Maximum growth rates for Gompertz models (700-900 g·day¯1) were more biologically plausible than von Bertalanffy estimates (<300 g·day¯1). Inflection points occurred at 200-250 days for Gompertz models, aligning with expected puberty timing.

Regional variations were substantial, with San Jorge showing superior growth performance and Costanera exhibiting constraints (204 kg asymptotic weight in regional analysis). Data aggregation methodology significantly influenced model selection outcomes, suggesting that individual-level modeling suits genetic evaluation while population-level approaches better serve management recommendations. Enhanced growth monitoring protocols and consideration of regional environmental factors are essential for implementing growth models in tropical cattle production systems.

Keywords: Cattle modeling · calves growing · tropical livestock

D. Chaves et al. (Eds.): CIARP 2025, LNCS 16528, pp. 323–333, 2026.
https://doi.org/10.1007/978-3-032-23161-1_23

1 Background

Growth models serve as mathematical representations of the growth-age relationship across different biological areas, such as plant, forestry and animal sciences [1]. In animal sciences, growth models have been described for different species, including birds, dogs, cats, cattle, sheep, rabbits [2], buffaloes [3], and other animals [4]. These models typically provide good approximations of biological features such as daily weight gain and mature weight. They are valuable tools for estimating daily feed requirements, slaughter weights, and the timing of specific biological events, including puberty, weaning, and first mating [5, 6]. Additionally, they help evaluate the effects of various conditions such as management practices, environmental factors, and other variables on growth performance. Consequently, these models serve as essential tools for supporting control and optimization of farm management decisions [4, 7].

Among the various approaches to growth modeling, sigmoidal functions have proven particularly effective. An appropriate growth function can summarize information from observed data over time using a few biologically meaningful parameters such as the inflection point, initial weight, final weight, and maximum growth rate [8]. Several sigmoidal models are commonly used to describe the growth patterns, including the Gompertz (GZ), Richards (RS), von Bertalanffy (vB), Logistic (LC), and Michaelis-Menten (MM) models [4, 9–12]. The inflection point deserves special attention because it represents the maximum daily growth rate, a crucial indicator for assessing the economic impact of the livestock operations [5]. The inflection point deserves special attention because it represents the maximum daily growth rate, a crucial indicator for assessing the economic impact of the livestock operations. The inflection point deserves special attention because it represents the maximum daily growth rate, a crucial indicator for assessing the economic impact of the livestock operations [5, 13, 14]. In this context, selecting the most appropriate mathematical expression presents a significant challenge. Various goodness-of-fit criteria can be applied, including Akaike Information Criteria – AIC, Bayesian Information Criteria – BIC, Mean Square Error - MSE, Mean Square Error Prediction – MSEP, and Mean Absolute Error [4, 10, 13].

Despite the availability of these analytical tools, several methodological challenges persist in growth curve analysis. Model fitting is typically performed on datasets containing observations throughout the full growth trajectory of cattle, encompassing specific life stages such as birth, weaning, puberty, fattening, and slaughter [4, 5]. However, the impact of incomplete data at critical stages of the curve, particularly at birth and maturity, remains poorly documented [15–17]. Furthermore, successful model fitting requires a thorough understanding of growth patterns to provide appropriate initial parameter values necessary for achieving convergence of the mathematical models [4, 5, 13].

Livestock farming contributes 6% of employment in Colombia [18], mainly based on forage grazing, which represents the most economical option in terms of dry matter supply for feed [19], occupying 31.59 million hectares with a cattle population of 29.3 million animals and 7.8% of them concentrated in Córdoba [20]. This department has 1,673 hectares of forests with strong influence of various pressures such as climate change, fragmentation, human settlements, and agricultural conversion due to the fertility of its soils [21], aspects that affect the forage availability and growing performance of animals [22].

This study aimed to compare the performance of five nonlinear growth models (GZ, RS, vB, LC, and MM) in describing calf growth patterns across five sub-regions of Córdoba, Colombia, using multiple goodness-of-fit criteria to determine the most suitable model for each regional context.

2 Materials and Methods

2.1 Dataset Collection

Body weight data were collected from sixteen beef farms located across five subregions, in the Córdoba department, Colombia. The information was obtained through direct communication with farmers and livestock managers who granted access to herd management notebooks and software used to monitor production and reproduction on each farm. Body weight was measured using cattle scales available at each farm. It was recorded from birth until weaning, specifically at 0, 30, 60, 120 and 240 days of age.

2.2 Preprocessing and Data Analysis

Two steps of cleaning were performed: first, the null or empty records for ID, day and weight were deleted, and second, the calves with fewer than three growth records were excluded from the fitting analysis. An overall summary of days and weight was performed to evaluate the fitting of the non-linear growth models GZ, LC, RS, MM, and vB reported in Appendix 1.

The analysis considered three stages: first, an individual fitting models comparison (Individual) was performed for the 731 calves. In the second stage, the individual data were grouped by sub-region and the fitting comparison was performed (Regional). Finally, the sub-regional grouping was maintained, but the comparison used the average values for the calves in each group (Average). Additionally, as the observed data were collected until weaning age, a random reference value of the asymptotic growth was added to enhance the performance of the growth models, assuming a normal distribution with $463 \perp 42$ kg; these parameters correspond to the mean and standard deviation of the Cordoba slaughter weight reported for the last quarter of 2024 [23].

The fit and accuracy of the growth models were compared using the voting method to select the best model based on the winner for each criterion, using the lowest values for AIC, BIC, MSE, MSEP, MAE, and the highest value of R^2. The analysis was conducted using dplyr [24], tidyr [25], and nls2 [26] libraries in R.

3 Results and Discussion

A total of 7,399 records for 1,057 calves were collected, with the first cleaning process dropping 2,914 records for null or empty values on identification, weight, and day of measurement, maintaining a total of 1,034 calves. Additionally, the calves with fewer than three weight measurements were deleted. The final dataset maintained 3,832 records for 731 calves from the five sub-regions of Córdoba. The birth weight for all sub-regions was 32.3 ± 9.9 kg comparable with values reported for Brown swiss dairy cattle [12] who found a mean weight of 38.8 ± 0.59 kg and lower than 50.8 kg for *Bos taurus* x *Bos indicus* cross breeds [27]. However, the minimum observed value is slightly lower than the expected value around 20 kg for *Bos indicus* breeds, supporting the crossbreed nature and tropical adaptation of the study population. The mean weight between birth and 300 days of age for San Jorge was 19.5, 20.0, 38.0, and 43.1 kg higher than Sinú medio, Costanera, Alto Sinú, and Sabana, respectively (Table 1).

Table 1. Overall summary of weight records for each sub-region.

Sub-region	Records	Minimum	Mean	Maximum	Sd	CV
Alto Sinú	296	24	78.4	210	48.8	62.3
Costanera	1728	15	96.4	277	54.3	56.3
Sabana	77	19	73.3	140	33.8	46.1
San Jorge	319	23	116.4	273	69.7	59.9
Sinú medio	1412	18	96.9	280	57.1	58.9

The frequency of the best-fitted curve for Individual, Regional, and Average analyses is shown in Table 2. For individual modeling, GZ was selected for 66% of calves, followed by vB (35%) and LC (2%), while vB and GZ were chosen in the Regional and Average analyses. The predominance of GZ in individual fitting aligns with findings from [5], who recommended GZ models for holstein friesian heifers from birth to first mating. However, this contrasts with [3], who found Brody models superior for Friesian crossbred and buffalo calves during fatting periods, suggesting that optimal model selection depends on both the growth phase studied and data aggregation level.

Table 2. Model frequency adjusted on Individual, Regional, and Regional average curves.

Model	Individual	Regional	Average
Von Bertalanffy – vB	255	3	2
Gompertz – GZ	460	2	3
Logistic – LC	15	0	0
Without convergency	1	0	0

The summary of the Individual parameterization for each sub-region is presented in Table 3. For all fitted curves, the asymptotic weight was estimated to be around 480 kg, which falls within ranges reported for similar production systems. [5] reported comparable values between 354 to 369 kg for Holstein heifers, while [3] found estimates of 450 and 420 kg for Friesian crossbred and buffalo, respectively. The consistency across these studies suggest that our parameter estimates realistic growth for crossbred cattle under tropical conditions. MGR for vB showed the lowest values with estimations below 300 g·day^{-1}, while the same parameter for GZ curves was more plausible for the biological expression of this type of animal within the influence area of this work (700 to 900 g·day^{-1}). These MGR values are more conservative than reported [27] for *Bos taurus* x *Bos indicus* crosses in Mexican humid tropics (1,150 g.day^{-1}), reflecting the extensive production system constraints characteristics of Colombian cattle operations. The vB models consistently produced estimates that, while mathematically valid, fall below expected biological ranges for cattle.

The highest asymptotic weight was estimated using the LC curve in Sabana (525 ± 32.5 kg) with an MGR of 600 g·day^{-1} at 586 days of age and an estimated weight of 263 kg at the inflection point. Conversely, the lowest value was found in Alto Sinú using the same curve model, which reached the inflection point earliest at 187 days with an estimated weight of 218 kg but showed an MGR (1,700 g·day^{-1}) higher than biologically expected values for grazing animals of the predominant breed in this sub-region. The inflection point for the GZ curve was determined between 200 and 250 days with weights near 170 kg, which aligns closely with biological expectations for puberty onset in tropical cattle. [5] reported similar patterns, with GZ estimating puberty at 193 days (130.4 kg), supporting the biological validity of our parameter estimates. The LC curve showed widespread estimation for these parameters from 187 to 470 days with weights between 218 and 244 kg, whereas vB had the longest time to reach this stage with age over 300 days and weight around 300 kg.

The fitting of non-linear growth models with the average observed weight to avoid the effect of the individual variation is a common practice [4], therefore the studies with the fitting comparisons and the parametrization for each animal are not reported. However, the individual results of this work are comparable with [3] for frisian crossbred calves with asymptotic values between 450 to 574 kg with GZ, LC and vB models, while MGR for some sub-regions with LC and vB are different with [27] whom reported daily gain between 680 to 1,310 g.day^{-1}.

Table 3. Average parameters for Individual calf curve models adjusted by region

Sub-region	Model	N	b_0	b_1	b_2	Inflex t	Inflex w	MGR
Alto Sinú	Bertalanffy	8	458.7 ± 51.1	0.002 ± 0.001	−19.3 ± 5.2	474.6 ± 103.3	305.8 ± 34	0.2 ± 0.03
Alto Sinú	Gompertz	50	473.6 ± 36.3	0.99 ± 0.085	0.005 ± 0.001	207.2 ± 41.4	174.2 ± 13.3	0.9 ± 0.2
Alto Sinú	Logistic	4	436.3 ± 43	17.6 ± 1.8	0.015 ± 0.002	187.7 ± 28	218.1 ± 21.5	1.7 ± 0.22
Costanera	Bertalanffy	85	456.5 ± 49.7	0.003 ± 0.001	−36.3 ± 9.9	341.5 ± 90.3	304.4 ± 33.1	0.3 ± 0.06
Costanera	Gompertz	243	478.4 ± 46	0.9 ± 0.102	0.004 ± 0.001	229 ± 64.3	176 ± 16.9	0.7 ± 0.22
Costanera	Logistic	7	477 ± 54.4	15.1 ± 3.4	0.011 ± 0.005	319.6 ± 161.7	238.5 ± 27.2	1.2 ± 0.54
Sabana	Bertalanffy	1	471.8	0.003	−16.9	417.4	314.5	0.2
Sabana	Gompertz	14	479.5 ± 39.6	0.87 ± 0.11	0.004 ± 0.002	257.1 ± 109.8	176.4 ± 14.6	0.7 ± 0.29
Sabana	Logistic	2	525 ± 32.5	12.4 ± 1.1	0.004 ± 0.001	586.7 ± 96.4	262.5 ± 16.2	0.6 ± 0.04
San Jorge	Bertalanffy	32	496.7 ± 75.1	0.002 ± 0.001	−33.5 ± 11.5	554.6 ± 164.6	331.2 ± 50	0.2 ± 0.03
San Jorge	Gompertz	33	479.1 ± 49.1	0.8 ± 0.13	0.004 ± 0.001	229.3 ± 73	176.3 ± 18.1	0.7 ± 0.19
Sinú medio	Bertalanffy	129	483.7 ± 64.5	0.002 ± 0.001	−27.7 ± 11.5	542.3 ± 174.1	322.5 ± 43	0.2 ± 0.04
Sinú medio	Gompertz	126	483.8 ± 44.3	0.91 ± 0.1	0.004 ± 0.001	254.2 ± 65.6	178 ± 16.3	0.7 ± 0.11
Sinú medio	Logistic	2	488.7 ± 27.4	16.9 ± 3.5	0.007 ± 0.004	473.1 ± 224.4	244.4 ± 13.7	0.8 ± 0.4

MGR is the maximum growth rate (gr.day^{-1}). b_0 is the asymptotic value for weight (kg) , b_1 and b_2 are shape parameters.

The comparative analysis between Regional and Average approaches revealed distinct patterns in growth curve fitting and model selection, representing a novel finding in cattle modeling literature. The Regional analysis showed considerable heterogeneity in both model selection and parameter estimates. While this grouping was dominated by the vB model in Costanera, San Jorge and Sinú medio with asymptotic weights of 205, 442 and 563 kg, respectively, GZ was selected for Alto Sinú and Sabana with 474 and 498 kg (Fig. 1 Left). In contrast, the Average analysis provided more consistent and biologically plausible results across regions with more uniform growth patterns that better reflect population-level expectations for cattle development in the region. The selected models were vB for Alto Sinú (429 kg) and San Jorge (465 kg); GZ for Costanera (513 kg), Sabana (519 kg), and Sinú medio (492 kg) (Fig. 1 Right).

The shift in model preference from individual to average data demonstrated that data aggregation methodology significantly influences model selection outcomes, suggesting that individual variation, when preserved in regional analysis, may enhance different mathematical expressions than population level trends revealed through averaging. On the other hand, GZ is a model that provide good fits for averaged growth trajectories, likely due to their more flexible sigmoid shape that better describe population patterns.

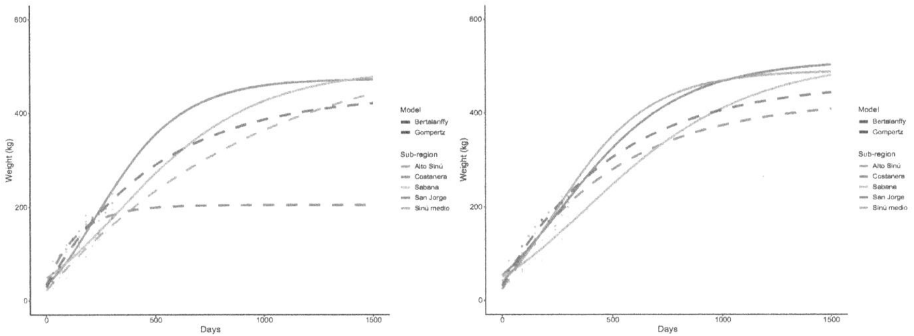

Fig. 1. Growth curves for cattle calves across five sub-regions of Córdoba, Colombia. (Left) Regional analysis: Models fitted using all individual data grouped by sub-region. (Right) Average analysis: Models fitted using mean weight values for each sub-region. Solid line indicates Gompertz models, dash line indicates von Bertalanffy models. Models were selected using a voting method based on AIC, BIC, MSE, MSEP, MAE, and R2 criteria.

The detailed parameterizations for the Regional and Average analyses are presented in Tables 4 and 5, respectively. The asymptotic weight had similar values to the Individual estimation, between 441 and 562 kg and 428 and 518 kg for Regional and Average analyses, respectively. However, this parameter for Costanera (204 kg) was considerably lower in the Regional analysis compared with the b_0 estimated in the other groupings. The considerable variation in asymptotic weights across sub-regions highlights the importance of environmental factors in cattle growth, with Costanera´s notably low asymptotic weight suggesting constraints that may limit productive potential in that sub-region.

The lowest age to the inflection point in the Regional analysis was 138 days with an MGR of 265 g·day^{-1} in Costanera, whereas the highest growth rate was estimated with the GZ curve for Alto Sinú with 766 g·day^{-1}, where the age and weight at the inflection point were estimated at 219 days and 174 kg, respectively. The results in the Average analysis showed more consistent values across all sub-regions, with age and weight at the inflection point between 230 and 526 days and 180 and 310 kg, respectively. The MGR was estimated from 163 to 717 g·day^{-1} for Alto Sinú and Sinú medio, respectively. The more consistent parameter estimates obtained through average analysis suggest that population level predictions may provide more stable targets for production planning, with practical implications for extension services and technical recommendations.

Considering the performance of both curves for Regional and Average analyses, GZ had more biologically consistent parameterization compared to vB. The latter showed a trend of late maturation process in the sub-regions where it was selected, with the longest time and lowest growth rates below 300 g·day$^-$1 at the inflection point, whereas GZ reached this growth stage around nine months with weights near 200 kg, values that are more realistic compared to the observed values for steer cattle in Córdoba. This finding supports the recommendations of [5] for using GZ in prediction applications, particularly for management decisions related to breeding timing and nutritional planning.

Table 4. Curve models parametrization with all calf growing data grouped by region.

Sub-region	Model	b_0	b_1	b_2	Inflex t	Inflex w	MGR
Alto Sinú	Gompertz	473.9	0.962	0.004	219	174.4	0.766
Costanera	Bertalanffy	204.5	0.007	−25.4	138	136.3	0.265
Sabana	Gompertz	498.1	0.843	0.003	309	183.3	0.500
San Jorge	Bertalanffy	441.8	0.002	−34	489	294.5	0.179
Sinú medio	Bertalanffy	562.8	0.001	−39.3	746	375.2	0.152

MGR is the maximum growth rate (gr.day^{-1}). b_0 is the asymptotic value for weight (kg) , b_1 and b_2 are shape parameters.

Table 5. Curve models parametrization with average calf weight data by region.

Sub-region	Model	b_0	b_1	b_2	Inflex t	Inflex w	MGR
Alto Sinú	Bertalanffy	428.5	0.002	−28.8	526	285.7	0.163
Costanera	Gompertz	512.9	0.812	0.003	253	188.7	0.605
Sabana	Gompertz	518.7	0.846	0.002	362	190.8	0.445
San Jorge	Bertalanffy	465.3	0.002	−34.9	526	310.2	0.175
Sinú medio	Gompertz	491.7	0.913	0.004	230	180.9	0.717

MGR is the maximum growth rate (gr.day^{-1}). b_0 is the asymptotic value for weight (kg) , b_1 and b_2 are shape parameters.

The regional variations observed in this study reflect the environmental heterogeneity characteristic of Córdoba cattle production systems. San Jorge´s superior growth performance, achieving 310 kg inflection weight, exceeds the weaning weight reported by [22] for Romosinuano cattle (182 kg at 240 days) in the same region, but with a larger period of growth (above of 500 days). This difference likely reflects the crossbred nature of our study population compared to pure Romosinuano cattle. The methodological insight that data aggregation influences model selection extends beyond statistical considerations to practical applications as the fact of individual level modeling may be more appropriate for genetic evaluation programs, where individual performance variations is crucial, while population level approaches may better serve management recommendations and production system optimization.

4 Final Remarks

This study demonstrates the sensitivity of non-linear growth curve model parameterization to methodological approaches in data handling and the temporal distribution of growth records. The findings reveal that data aggregation methodology significantly influences model selection outcomes, with individual-level analysis favoring different mathematical expressions than population-level approaches. This methodological insight extends beyond statistical considerations to practical applications, suggesting that enhanced methods for monitoring and registering growth on cattle farms could support improved fitting practices and provide more accurate tools for management decisions in livestock operations.

However, several methodological limitations must be acknowledged. The incorporation of reference weight values to enhance model performance, while biologically justified, introduced variability in the baseline parameters used for model fitting. The random assignment of slaughter weights as a reference for mature growth, although based on regional slaughter data (463 $\pm$ 42 kg), means that individual, regional, and average analyses did not share identical reference points for evaluating model fit. This limitation potentially affected the comparability of model performance across different analytical approaches.

Future research should prioritize the evaluation of data imputation effects when observed data are unavailable at critical growth phases, particularly around inflection points where growth rate changes are most pronounced. The development of standardized growth monitoring protocols that ensure adequate data coverage throughout the growth trajectory would enhance the accuracy and reliability of growth curve applications in cattle production systems.

In conclusion, while non-linear growth models provide valuable tools for cattle production management, their implementation requires careful consideration of data collection protocols, analytical approaches, and regional environmental factors. The methodology used for data aggregation should align with the intended application: individual-level modeling for genetic evaluation and breeding decisions, and population-level approaches for management recommendations and production system optimization.

Acknowledgments. The authors wish to express their appreciation for the financial support provided by Science and Technology Research Partnership for Sustainable Development (SATREPS) project research through JST/JICA (Grant No. JPMJSA2106 which was selected in 2021.

Disclosure of Interests. The authors have no competing interests to declare that are relevant to the content of this article.

Appendix

Appendix 1. Mathematical expressions and parameterizations of growth curve models that were fitted.

Equation	Model	Inflexion t	Inflexion w	MGR
$w = b_0 * exp^{(-exp^{(b_1-(b_2*t))})}$	Gompertz [2, 14]	$\frac{b_1}{b_2}$	$\frac{b_0}{exp}$	$\frac{b_2*b_0}{exp}$
$w = \frac{b_0}{1+b_1*exp^{-b_2*t}}$	Logistic [5, 14]	$\frac{\log b_1}{b_2}$	$\frac{b_0}{2}$	$\frac{b_0*b_2}{4}$
$w = \frac{b_0}{(1+exp^{b1(-b2*t)})^{\frac{1}{b3}}}$	Richards [1, 14, 28]	$\frac{b_1+\log b_3}{b_2}$	$b_0 * \left(\frac{b_3}{b_3+1}\right)^{\frac{1}{b_3}}$	$\frac{b_0*b_2*b_3*(b_3-1)^{\frac{1}{b_3}}}{(b_3+1)^{\frac{b_3+1}{b_3}}}$
$w = \frac{b_2 {b_3}^{b_1}+b_0 t^{b_1}}{{b_3}^{b_1}+t^{b_1}}$	Michaelis-Menten [8, 14][a]	$b_3 * \left(\frac{b_1-1}{b_1+1}\right)^{\frac{1}{b_1}}$	$\frac{b_2*b_3^{b_1}+b_0*Inflexiont^{b_1}}{b_3^{b_1}+Inflexiont^{b_1}}$	$\frac{b_0*b_1*{b_3}^{b_1}*\frac{b_1-1}{(b_1+1)^{\frac{b_1-1}{b_1}}}}{2^{2*b_1}*{b_3}^{b_1}}$
$w = b_0 * (1-exp^{-b_1*(t-b_2)})$	Bertalanffy [2, 5, 14]	$b_2 + \frac{\log 3}{b_1}$	$b_0 + \frac{2}{3}$	$\frac{b_0*b_1}{3*\sqrt{3}}$

Inflexion t, and inflexion w correspond to days, and weight at inflexion point of the curve, respectively. MGR is the maximum growth rate. b_0, b_1, b_2, and b_3 are shape parameters. [a] The parametrization exists with $b_1 > 1$.

References

1. Teleken, J.T., Galvão, A.C., Robazza, W.S.: Use of modified Richards model to predict isothermal and non-isothermal microbial growth. Braz. J. Microbiol. **49**, 614–20 (2018)
2. Vázquez, J.A., Lorenzo, J.M., Fuciños, P., Franco, D.: Evaluation of non-linear equations to model different animal growths with mono and bisigmoid profiles. J. Theor. Biol. **314**, 95–105 (2012)
3. Salem, M.M.I., Kassem El-Hedainy, D.: Comparison of Non-Linear Growth Models to Describe The Growth Curves in Fattening Friesian Crossbred and Buffalo Male Calves [Internet]. https://www.researchgate.net/publication/321146255
4. Teleken, J.T., Galvão, A.C., Robazza, W.D.S.: Comparing non-linear mathematical models to describe growth of different animals. Acta Sci. **39**, 73 (2017). http://periodicos.uem.br/ojs/index.php/ActaSciAnimSci/article/view/31366

5. Budimulyati, S.L., Noor, R.R., Saefuddin, A., Talib, C.: Comparison on accuracy of Logistic, Gompertz and von Bertalanffy models in predicting growth of new born calf until first mating of Holstein Friesian heifers. J. Indones. Trop. Anim. Agric. **37**, 151–160 (2012)
6. Monteiro, M.S., Carnevale, R.F., Muro, B.B.D., Mezzina, A.L.B., Carnino, B.B., Poor, A.P., et al.: The Role of Nutrition Across Production Stages to Improve Sow Longevity. Multidisciplinary Digital Publishing Institute (MDPI), Animals (2025)
7. ICAR-International Committee for Animal Recording. ICAR Guidelines - Section 3: ICAR Guidelines for Beef Cattle Production Recording. The Global Standard for Livestock Data [Internet] (2018). [cited 2025 Jun 28]. https://www.icar.org/Guidelines/03-Beef-Cattle-Recording.pdf
8. López, S., France, J., Gerrits, W.J., Dhanoa, M.S., Humphries, D.J., Dijkstra, J.: A generalized Michaelis-Menten equation for the analysis of growth. J. Anim. Sci. (2000). https://academic.oup.com/jas/article-abstract/78/7/1816/4625797
9. Magotra, A., Bangar, Y.C., Yadav, A.S.: Growth curve modeling and genetic analysis of growth curve traits in Beetal goat. Small Ruminant Res. **195**, 106300 (2021). https://linkinghub.elsevier.com/retrieve/pii/S0921448820302492
10. Balafrej, M., Fagroud, M., Sraïri, M.T.: Nonlinear modeling of growth and study of a new characterization point applied to Timahdit breed sheep under uncertain paternity. Trop Anim. Health Prod. **55** (2023)
11. Ernandes, J., et al.: Evaluation of non-linear models for growth curve in Brazilian tropical goats. https://doi.org/10.1007/s11250-021-02598-2
12. Bayram, B., Yanar, M., Akbulut, Ö.: Relationships among Richards growth curve parameters, reproductive and milk production traits in brown Swiss cattle sag Feisfcfoh. J. Appl. Anim. Res. **26**, 29–32 (2004)
13. Gbangboche, A.B., Glele-Kakai, R., Salifou, S., Albuquerque, L.G., Leroy, P.L.: Comparison of non-linear growth models to describe the growth curve in West African Dwarf sheep. Animal **2**, 1003–1012 (2008)
14. Goshu, A.T.: Derivation of inflection points of nonlinear regression curves - implications to statistics. Am. J. Theor. Appl. Stat. **2**, 268 (2013)
15. Hossein-Zadeh, N.G., Golshani, M.: Comparison of non-linear models to describe growth of iranian guilan sheep. Revista Colombiana de Ciencias Pecuarias. **29**, 199–209 (2016)
16. Bathaei, S.S., Leroy, P.L.: Growth and mature weight of Mehraban Iranian fat-tailed sheep. Small Ruminant Research (1996)
17. Domínguez-Viveros, J., Canul-Santos, E., Rodríguez-Almeida, F.A., Burrola-Barraza, M.E., Ortega-Gutiérrez, J.Á., Castillo-Rangel, F.: Defining growth curves with nonlinear models in seven sheep breeds in Mexico. Rev Mex Cienc Pecu. **10**, 664–675 (2019)
18. FEDEGAN. Cifras de referencia del sector ganadero colombiano. Fedegan **55** (2018). www.fedegan.org.co
19. Moreno, L.A., Rueda, C., Andreade, G.: Biodiversidad 2017. Estado y tendencias de la biodiversidad continental de Colombia [Internet]. Instituto de Investigación de Recursos Biológicos Alexander von Humboldt. Bogotá D.C, Colombia (2018). http://repository.humboldt.org.co/handle/20.500.11761/34619
20. Fedegan. Balance_Perspectivas_ganaderia_colombiana_2021_2022_. Balance Perspectivas Ganaderas 1–29 (2022)
21. Ruiz Vega, R., Saab, H.P.: Diversidad florística del bosque seco tropical en las subregiones bajo y medio Sinú, Córdoba. Colombia. Rev Biol Trop. **68**, 167–179 (2020)
22. Ossa, G., Suárez Tronco, M., Pérez, J.: Efecto del medio y la herencia sobre el peso al destete de terneros de la raza Romosinuano. RevMVZ Cordoba. **10**, 673–83 (2005). http://www.scielo.org.co/scielo.php?script=sci_arttext&pid=S0122-02682005000200009&lng=en&tlng=es

23. Departamento Administrativo Nacional de Estadística (DANE). Boletín técnico. Encuesta de Sacrificio de Ganado (ESAG) I trimestre de 2025 (2025). https://www.dane.gov.co/files/operaciones/ESAG/bol-ESAG-Itrim2025.pdf
24. Wickham, H., François, R., Henry, L., Müller, K., Vaughan, D.: dplyr: A Grammar of Data Manipulation. R package version 1.1.4 (2025). https://dplyr.tidyverse.org
25. Wickham, H., Vaughan, D., Girlich, M.: tidyr: Tidy Messy Data. R package version 1.3.1 (2024). https://tidyr.tidyverse.org
26. Grothendieck, G., R Core Team (nls). nls2: Non-Linear Regression with Brute Force. R package version 0.3–4 (2024). https://github.com/ggrothendieck/nls2
27. Córdova, A., Rodríguez, G., Córdova, M., Córdova, C., José, P.: Ganancia diaria y peso al destete en terneros de cruces Bos taurus con Bos indicus en trópico húmedo. RevMVZ Cordoba **10**, 589–92 (2005). http://www.scielo.org.co/scielo.php?script=sci_arttext&pid=S0122-02682005000100009&lng=en&tlng=es
28. Wang, X., Liu, S., Huang, Y.: A study on the rapid parameter estimation and the grey prediction in Richards model. J. Syst. Sci. Inf. **4**, 223–234 (2016)

Relevance Metric for Bird Sound Detection

Gabriela Paola Sereniski[1], Juliano Henrique Foleis[1], Gabriel Lima Medina Rosa[2], and Diego Bertolini[1](✉)

[1] Department of Computing, Federal University of Technology – Paraná, Campo Mourão, PR, Brazil
diegobertolini@utfpr.edu.br

[2] Department of Animal and Vegetable Biology, State University of Londrina, Londrina, Paraná, Brazil

Abstract. This work proposes a relevance metric strategy for automated bird vocalization detection, tested on a newly compiled dataset of annotated recordings collected from multiple rainforest fragments. Our approach combines a shallow CNN architecture with a relevance metric to better handle diverse real-world recording conditions. Our experiments were conducted using a private dataset comprising 849 samples with bird vocalizations, totaling 15,867 regions of interest (ROIs) that were labeled by experts. The marked ROIs correspond to the calls or songs of 161 different bird species. We compare our method against a traditional energy-based baseline, which achieves F1-scores of 0.42 and 0.44 when using Per-Channel (PCEN) Energy Normalization and Spectral Subtraction, respectively. By contrast, our best CNN variant, leveraging PCEN spectrograms, obtains an F1-score of 0.66. While the approach maintains consistently high precision, recall remains variable in more challenging scenarios. Overall, the results highlight that modest increases in network depth and appropriate spectrogram processing can substantially improve detection performance in complex soundscapes.

Keywords: bird song detection · machine learning · relevance metric

1 Introduction

Birds are important indicators of ecosystem health because their presence, abundance, and behavior reflect changes in environmental conditions [7]. They can be found in nearly every habitat and play fundamental roles in ecosystem functions such as pest control, seed dispersal, and pollination [21]. As a result, monitoring bird populations can provide an understanding of biodiversity status and habitat quality, which can be used as a tool for ecological research and facilitate the development of conservation strategies.

Ornithologists have relied on manual audio inspection to annotate recordings. This process is time-consuming and labor-intensive, especially when dealing with massive datasets. To address this challenge, a pre-filtering step that discards segments or entire recordings without bird activity can significantly reduce both

D. Chaves et al. (Eds.): CIARP 2025, LNCS 16528, pp. 334–348, 2026.
https://doi.org/10.1007/978-3-032-23161-1_24

manual annotation effort and computational overhead [9]. By focusing only on the audio portions containing avian sounds, specialists save time and memory resources, allowing subsequent analysis to be applied more efficiently [16].

Deep learning models have shown considerable success in the automatic detection and classification of bird vocalizations [10]. In particular, CNN-based approaches can often handle noisy environments and overlapping calls more effectively than traditional methods [22]. However, as CNN architectures grow deeper, they can become computationally demanding, leading to longer training times and the need for larger labeled datasets. This becomes particularly challenging when bird monitoring organizations have limited computing and human resources available.

In this work, we propose the use of a relevance metric [6] to highlight the presence of bird vocalizations in a given audio track. By indicating how confidently a segment is assigned to a particular class, the metric helps clarify which parts of the audio can be considered regions of interest (ROIs). To evaluate the metric, we developed a comparatively shallow CNN model to handle spectrogram-based feature extraction and classification without the need for extensive computational resources. We compare the methodology to a simple energy based threshold baseline.

The rest of this paper is organized as follows: Sect. 2 describes the database employed. The state-of-the-art is presented in Sect. 3. The relevance metric approach is detailed in Sect. 4. Section 5 and Sect. 6 describe the experimental settings and results obtained by the proposed relevance metric methodology and baseline. Finally, Sect. 7 concludes the paper and presents future works.

2 Database

During four campaigns between October 2023 and June 2024, recorders were installed at 54 distinct points within rainforest fragments to monitor their soundscapes. Audiomoth recorders [5] were used, configured to record environmental sounds at a rate of six minutes per hour (one minute every 10 min), resulting in 144 one-minute recordings per day. Battery autonomy allowed data collection for over a month, and the files were built in an uncompressed WAV format using a 16-bit depth and a 48kHz sampling rate. This data collection method is called Passive Acoustic Monitoring (PAM) and is one of the most advanced methods for ecological monitoring currently available [18,19].

In order to identify the occurrence of acoustically active species, a subset of 1.207 recordings were thoroughly inspected, and a total of 15.867 ROIs were identified across 849 recordings. The remaining tracks had no species detected. In this context, ROIs refer to manually annotated segments within the audio spectrograms, precisely delimiting the temporal (start and end times) and frequency boundaries of bird vocalizations as identified by expert ornithologists. Although only a portion of the inspected recordings have bird vocalizations, the remaining soundscapes were used as negative examples for the proposed methodology.

The species of the bird was identified for each ROI, as well as the level of identification certainty, sociobiological type of signal emission (e.g., song or call), intraspecific social context of signals (e.g. single individual calling, individuals calling in duets, or multiple individuals in chorus) and, whenever possible, the period (time in seconds) between same-individual consecutive calls, resulting in a sequence of signals from which we can calculate call rate. A total of 161 unique bird species were cataloged. For this segmentation process, the segmentation app available in the R package monitoraSom [17] was used, which was specifically designed for the PAM workflow.

We highlight that the complexity of constructing the segmentation database stems from its high dependency on contextual information. Specialists must be able to discriminate vocalizations from multiple species singing simultaneously and disambiguate bird species identities based on biological, behavioral, and ecological contexts. Specialists properly trained for this task are considerably rare.

Figure 1 depicts a one-minute excerpt of a soundscape in which 11 different bird species vocalize across a broad spectral band. Some species produce rapid chirps lasting well under a second while others sustain notes for several seconds. Many calls overlap in both time and frequency, often two or three species sing simultaneously. This dense, multi-species overlap and the wide variability in call rate and duration underscore the complexity of the presented dataset.

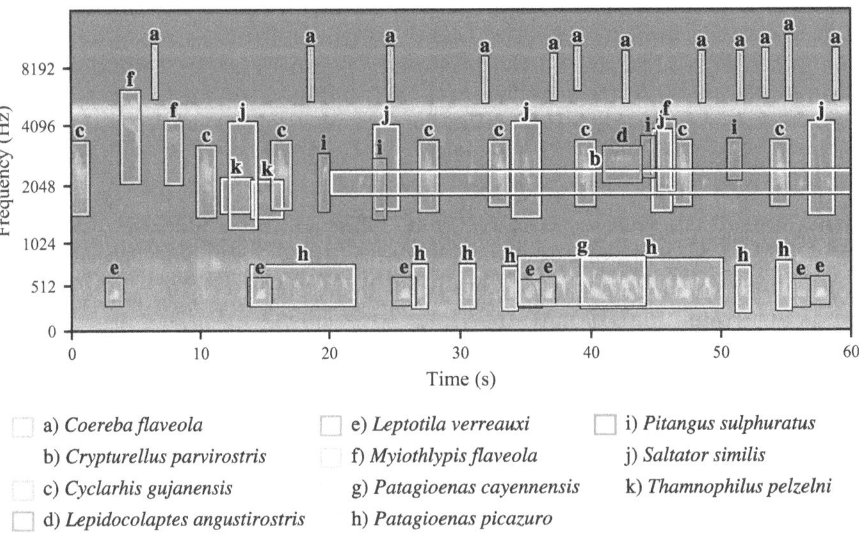

Fig. 1. Example of a manually annotated soundscape.

Overall, the annotations span a wide acoustic range, with the majority of vocalizations concentrated between 1.5–4.5 kHz, with only 0.32% extended above 16 kHz. It is worth noting that a total of 57.1% of the annotated ROIs present time-wise overlaps with other bird vocalizations. These overlaps compose 26% of the total annotated vocalization time.

3 Related Works

The development of bird sound recognition systems has been greatly promoted by the popularization of acoustic monitoring technology [12]. Automatic bird species identification has leveraged various machine learning and neural network architectures [14], designed specifically to address the complexities of bird sound segmentation and classification [4].

A logistic regression-based acoustic occupancy model using sound frequency percentiles for bird sound detection was presented in [20]. Their logistic regression approach demonstrated an overall accuracy of 75% with a 16% false negative rate, illustrating a computationally efficient alternative to CNN-based methods suitable for large-scale acoustic monitoring.

In [2], the authors presented TweetyNet, a neural network that automated the segmentation and annotation of birdsong directly from spectrograms, without a segmentation step. TweetyNet demonstrated robust performance across multiple datasets, achieving an average syllable error rate of 0.9% and 5.17%, and a frame error rate of 1.56% and 2.91%, for Bengalese finches and canaries, respectively. This was attained with only 8–11 min of manually annotated training data per individual, demonstrating the network's potential to scale up the annotation of large datasets.

In the approach described in [4], the authors explore explainable artificial intelligence (XAI) methods for interpreting CNNs in bird sound classification. They employed a simplified network architecture tailored for this task. Input audio signals were first processed using Mel-frequency cepstral coefficients (MFCCs). The extracted features were then passed through a CNN which achieved a classification accuracy of 92%. The integration of XAI techniques facilitated model interpretability, with SHAP demonstrating slightly superior stability and interpretability compared to LIME.

In [15], the use of an encoder-decoder CNN architecture was proposed for simultaneous segmentation and classification of bird song spectrograms. The network provides pixel-wise predictions that identify bird syllables and classify their species. Evaluation on real-world datasets, showed that, in the segmentation task, the method obtained a 98.8% true positive rate, and in the classification task, a Hamming loss of 9.24 was achieved.

In [8], a feature extraction and bird sound recognition model is proposed based on neural network architecture search. Using the Differentiable Architecture Search (DARTS) algorithm, the model automatically selects the optimal neural network structure and the optimal number of Mel filters. The resulting model achieved an average classification accuracy of 92.44% on a dataset containing 264 species. The self-optimization improved classification accuracy and reduced model complexity.

4 Relevance Metric

We propose the use of an importance measure, namely the relevance metric. This metric measures class separability by combining the output of N one-versus-all

$E_i, i \in \{1, 2...N\}$ classifiers, which we call experts. To apply the relevance metric, every audio is described as a finite sequence of segments, and each segment is classified independently from others in the same track. An expert is trained for each target class. All segments corresponding to a class are considered positive examples, and all segments from other classes are considered negative examples.

In the prediction process, an expert E_i emits an estimate of probability $P_i(k)$ that segment k belongs to class i. If only one expert predicts a high value for $P_i(k)$, k is associated with a low prediction uncertainty, so k is considered a relevant segment for the track classification. Conversely, if multiple experts predict similarly high probabilities $P_i(k)$ for segment k, it indicates uncertainty, as no single class clearly dominates. Likewise, if all probabilities are low, it also suggests uncertainty because the segment likely does not belong confidently to any known class. This uncertainty can be measured by Shannon's Entropy [3]:

$$H(k) = -\sum_{i=1}^{N} P_i(k) \log_N P_i(k) \quad | \quad \sum_{i=1}^{N} P_i(k) = 1 \tag{1}$$

Shannon's entropy approaches 1 when all probabilities are equal, which means the result's uncertainty is maximum. Conversely, when one event has a much higher probability than others, it approaches 0, indicating a low uncertainty. This interpretation is used to estimate the relevance $R(k)$ of segment k in regards to classes $1, ..., N$ as the complement of the entropy:

$$R(k) = 1 - H(k) \tag{2}$$

Equation 2 can generate misleadingly high results when $P_i(k)$ is low for all i, due to a normalization step to guarantee that P is a probability mass function. In this case, the importance of k is low, even if $R(k)$ is high. We solve this by weighting $R(k)$ by the value of $P_i(k)$:

$$R_i(k) = R(k) P_i(k) \tag{3}$$

This means that high $R_i(k)$ values indicate segments that have a high chance of belonging to class i. Any classifier algorithm can be used to calculate the relevance metric as long as it can output probabilities. This allows the use of experts based on different classifier algorithms.

4.1 Relevance Based Classifier

The specialists are CNN models developed specifically to deal with audio classification problems. The network is called *SpecNet*. The input files are transformed into spectrograms with 128 frequency bins and a length of t. The frequency bins are divided into L bands, each processed independently by a one-dimensional convolution layer. For each band j, the convolutional layer learns a set of f_j distinct convolutional filters. The convolution is not padded, hence the output has fewer samples ($t' < t$).

The outputs are then concatenated into a single vector of size $F = \sum_{j=1}^{L} f_j$. These features are aggregated over consecutive segments with average pooling, resulting in even fewer segments ($t'' < t'$). The features are then passed on to a fully connected layer with HN neurons, and followed by a softmax classifier that outputs the probability that a segment belongs to the expert's class or not. The estimates are combined by summing the results over all segments, resulting in a 1×2 vector. Lastly, the outputs are normalized so they sum up to 1.

The *SpecNet* expert networks were parameterized as follows: the input is divided into $L = 4$ bands, and each one learns $f_j = 42$ filters. A kernel of size 4 is used in the one-dimensional convolution layers. The average pooling window was of length 10, with 50% overlap. The number of neurons in the first fully connected layer was $HN = 30$. The expert network architecture can be visualized in Fig. 2.

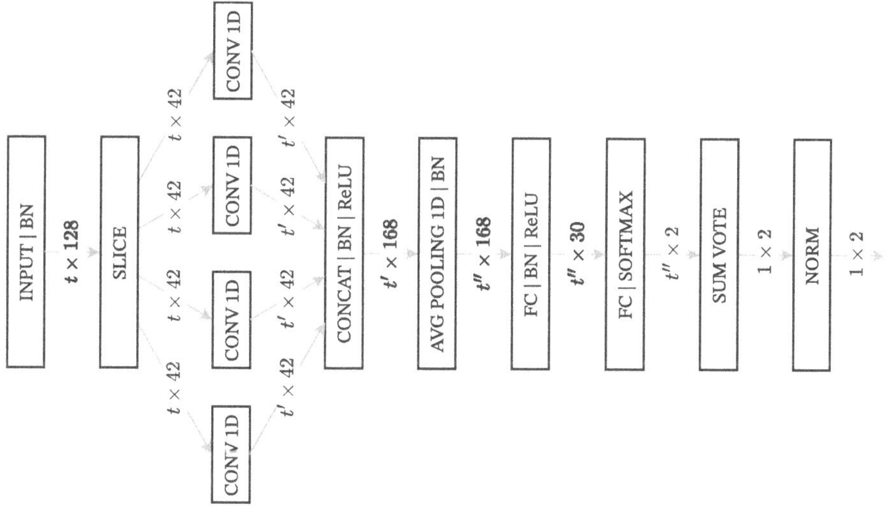

BN - Batch Normalization; CONV 1D: One-dimensional convolution layer; CONCAT - Feature concatenation; ReLU - Activation layer; AVG POOLING - One-dimensional average pooling; FC - Fully connected layer; SOFTMAX - Softmax classifier; SUM VOTE - Estimate combination; NORM - Output Normalization.

Fig. 2. *SpecNet* Architecture.

The Adam optimizer [11] was used to train the networks, and the Søresen Dice coefficient served as the loss function. Early-stopping with a patience of 100 epochs was employed to avoid overfitting, considering a minimal improvement of 0.05 over the validation loss. Each expert was trained 3 times to reduce the impact of random weight initialization. When stopped, the weights from the best epoch were reloaded into the model, but this time, the outputs of the Softmax classifier were not aggregated by sum, which means the classifier made

t'' predictions for each input spectrogram. The training was done in batches of 32 spectrograms, and the maximum number of epochs was set to 500.

Slight model variations were introduced to evaluate the impact of increasing the network's depth: The model *SpecNetC* has a second convolutional layer for every frequency band, and *SpecNetD* has a second fully connected layer. *SpecNetCD* is the combination of both. It is worth noting that all models are especially shallow, ranging from 28.128 to 57.846 trainable parameters (*SpecNet* and *SpecNetCD* architectures, respectively). Each model outputs the probabilities for both the negative and positive classes. The positive class probabilities are isolated, summed, and normalized. These results are used to calculate R_i, according to Eq. 3. Being c the class one wishes to highlight, R_c smoothed by applying a zero-phase filter via a bidirectional majority vote. Concretely, we apply a sliding window of size $s = 3$ (thresholded at 0.5) first in the forward direction and then again in reverse. The results are merged so a frame is marked positive if it was flagged in either pass.

5 Experimental Settings

Before describing the settings of the *SpecNet* and relevance metric approach, we establish a straightforward benchmark to evaluate the task with a common audio event detection technique. We implement a classic energy-thresholding pipeline as a minimal baseline: simply put, any time–frequency region whose spectral energy exceeds a learned percentile is flagged as a vocalization. With that baseline in place, we then turn to our relevance-based *SpecNet* models.

5.1 Energy Thresholding Baseline

As a baseline, we implemented an energy-based segmentation method. This approach processes the audio signal through a series of transformations to identify time segments with significant energy variations while attenuating background noise. The underlying assumption is that the target signals, in this case, bird songs and calls, exhibit higher energy levels than other acoustic components.

The audio is resampled to 44.1 kHz and converted to a mono signal. A Short-Time Fourier Transform (STFT) is computed using a window size of 2048 samples and a hop length of 512 samples. A Mel spectrogram with 128 bins is generated, retaining frequencies between 100 Hz and 15 kHz. Per-Channel Energy Normalization (PCEN) [13] is applied to compensate for amplitude variations and enhance transient events. Additionally, we evaluated an alternative approach without PCEN, employing spectral subtraction (SSub) via mean removal per frequency band.

Spectral energy is computed as the sum of squared magnitudes across frequency bins. A low-pass filter with a window size of 10 is applied six times to smooth short-term fluctuations. Segmentation is performed using an adaptive threshold set at the 80th percentile of the energy distribution, classifying frames exceeding this threshold as containing vocalizations. A binary opening operation with a window size of 10 frames is applied, followed by a binary closing operation with a window size of 60 frames. These morphological operations aim to fill gaps within detected segments, improving the temporal coherence of the segmentation. The parameters presented were optimized using Optuna [1] via 5-fold cross-validation. Aside from slight variations, the parameter settings across the five folds were the same.

5.2 Relevance Application

As for the training and validation process of the *SpecNet* models, given the ROI timestamps of each valid soundscape, we first split the tracks with Tukey windowing to generate an audio file containing the concatenation of its non-interesting regions. These files are grouped with soundscapes that contain no bird identification to compose the nonROI class. The complete soundscapes with identified bird sounds compose the ROI class.

The splitting process generates windowing artifacts. To ensure that the model would not learn to separate the classes based on the presence of these artifacts, we introduced random Tukey window effects on the ROI class and non-identified soundscapes. This means that both expert sets will contain environmental sounds and windowing artifacts, but only the ROI set will contain bird sounds.

The audio files of both classes are resampled to 32 kHz and codified as Mel spectrograms with 128 bins. For this, an STFT with a window size of 1024 samples and a hop length of 512 samples was computed. We also evaluated the model with log Mel spectrograms and with PCEN applied to Mel spectrograms.

As a data augmentation technique, for each ROI spectrogram, a random ROI timestamp is used to select a frame equivalent to 25 to 30 s of the spectrogram. If the random frame size is smaller than 30 s, zero padding is added to the right side of the spectrogram. As for the nonROI class, a random frame with a size of the same interval is selected, with zero padding added if necessary. This random selection is applied in each epoch, making the models learn and validate with different snippets of the same spectrogram, reducing the impact of the limited dataset size.

For the testing set, spectrograms generated from the original complete soundscapes were used, without the addition of windowing artifacts. The models were evaluated using 6-fold cross-validation. A dedicated 15% portion of the training set was used to assess loss values for early stopping purposes. The train, test, and loss evaluation splits were done according to the base soundscape file, which means that segments from the same track are guaranteed to be in the same set. In an attempt to balance the species that would appear in each set, iterative stratification was employed. The data flow and transformations applied in each step are depicted in Fig. 3.

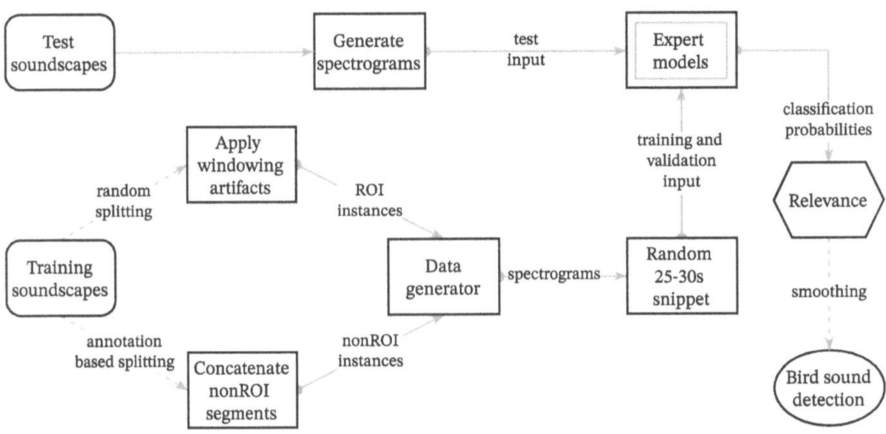

Fig. 3. Overview of the *SpecNet* training, validation and classification process.

6 Results and Discussion

Table 1 shows the baseline results with PCEN and Spectral Subtraction (SSub) noise reduction techniques. These results are for the entire dataset, with the best parameters obtained via cross-validation, as stated in Sect. 5.

The results show that both noise-attenuation algorithms performed very similarly. Although, on average, SSub outperformed PCEN in both Precision and F1-Score, the differences are not statistically significant (Student's T-test, $p > 0.05$).

Table 1. Classification metrics across all recordings (Baseline)

	Precision	Recall	F1-Score
	Baseline		
PCEN	0.58 ± 0.31	0.41 ± 0.18	0.42 ± 0.16
SSub	0.66 ± 0.31	0.41 ± 0.21	0.44 ± 0.17

The metrics are calculated for each recording since there is a prediction for each audio frame. The high standard deviations shown in Table 1 may indicate that both algorithms cannot deal with the differing noise levels among recordings. To verify this, Fig. 4(a) shows the distributions of F1-score, precision, and recall among all recordings for both PCEN and SSub, respectively.

In comparing the two noise suppression techniques, both the spectral subtraction and PCEN baselines exhibit high precision but limited recall. The spectral subtraction baseline shows a median precision of 0.75 (third quartile 0.95) and a median recall of 0.35 (first quartile 0.27), resulting in an F1-score with a median of 0.45 (IQR 0.36–0.55). Similarly, the PCEN baseline achieves a median precision of 0.61 (third quartile 0.87) and a median recall of 0.38 (IQR 0.31–0.48), leading to a median F1-score of 0.46 (third quartile 0.53). In both cases, precision is consistently higher than recall, indicating that while false positives are relatively infrequent, many true bird vocalizations are missed.

Table 2 summarizes the performance of various *SpecNet* architecture variations using different spectrogram representations. Overall, introducing an additional convolutional layer consistently improves recall and F1-score compared to the base *SpecNet* model, with only modest impacts on precision. Similarly, incorporating an extra dense layer provides smaller but consistent gains, indicating the benefit of deeper networks for capturing more complex, hierarchical audio features.

Table 2. Performance of *SpecNet* architecture variants and spectrogram representations. Classification metrics across all recordings.

	Precision	Recall	F1-Score
SpecNet			
Mel	0.77 ± 0.22	0.59 ± 0.34	0.59 ± 0.28
logMel	0.76 ± 0.22	0.63 ± 0.33	0.61 ± 0.27
PCEN	0.73 ± 0.23	0.68 ± 0.31	0.63 ± 0.26
Conv			
Mel	0.79 ± 0.21	0.60 ± 0.33	0.60 ± 0.28
logMel	0.73 ± 0.22	0.71 ± 0.31	0.65 ± 0.26
PCEN	0.71 ± 0.22	0.70 ± 0.30	0.65 ± 0.25
Dense			
logMel	0.74 ± 0.23	0.65 ± 0.33	0.61 ± 0.27
PCEN	0.72 ± 0.23	0.69 ± 0.30	0.64 ± 0.25
ConvDense			
logMel	0.73 ± 0.22	0.70 ± 0.31	0.65 ± 0.26
PCEN	0.71 ± 0.22	0.72 ± 0.29	0.66 ± 0.24

The best overall performance, particularly in recall and F1-score, is observed when combining both convolutional and dense layers (*ConvDense* variants). Specifically, the PCEN spectrogram in the *ConvDense* model yields the highest average recall (0.72) and the best F1-score (0.66), as illustrated in the distribution plots (Fig. 4(e)). This indicates that moderate expansion of the network architecture, coupled with per channel energy normalization or log-scaling, notably enhances the model's capability to effectively detect relevant features in audio segmentation tasks.

In terms of spectrogram representations, PCEN consistently outperforms Mel and logMel variants, especially regarding recall, highlighting the effectiveness of per-frequency normalization in enhancing sensitivity without significantly sacrificing precision. The distribution plots (Fig. 4(b)–(e)) illustrate how the

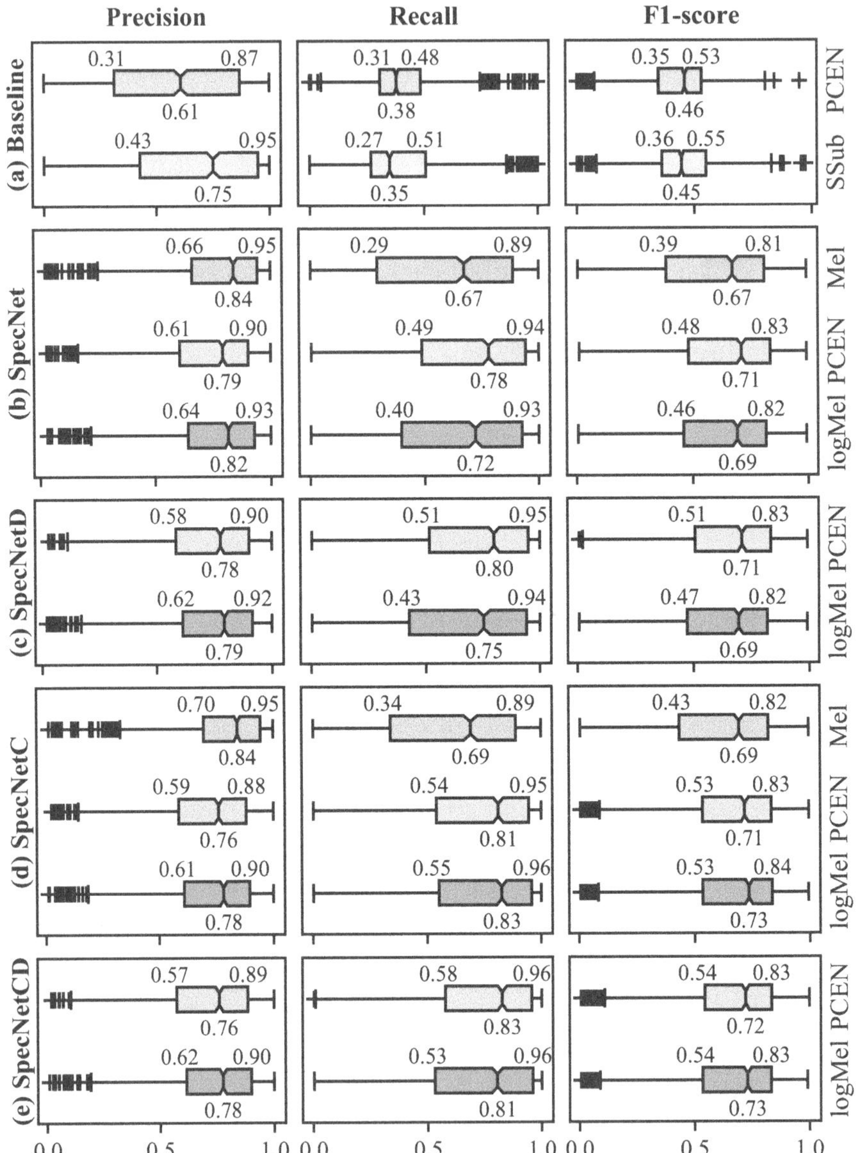

Fig. 4. Distribution of F1-Score, Precision, and Recall among all recordings.

PCEN and log Mel spectrograms reduce variability and improve performance consistency across recordings compared to the standard Mel spectrograms. For instance, the *SpecNet* model with Mel spectrograms exhibits a large spread in F1-scores, ranging widely between 0.42 and 0.82 (first to third quartiles). In

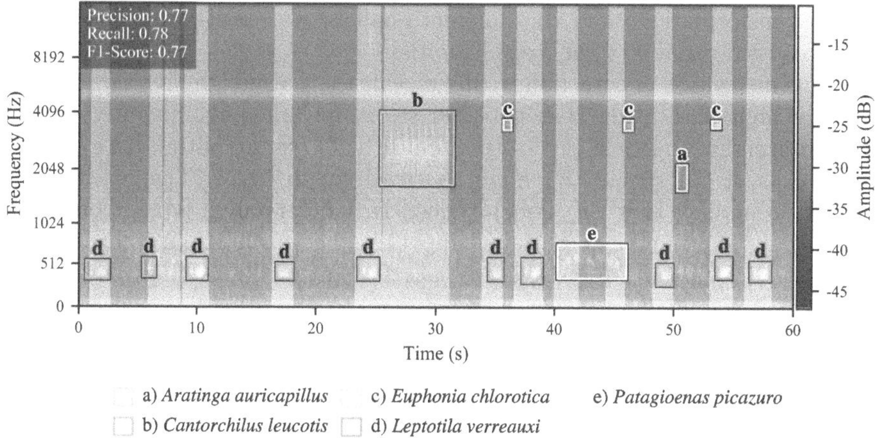

Fig. 5. *SpecNetCD* detections for a log Mel spectrogram. Whiter shades represent bird sound detections.

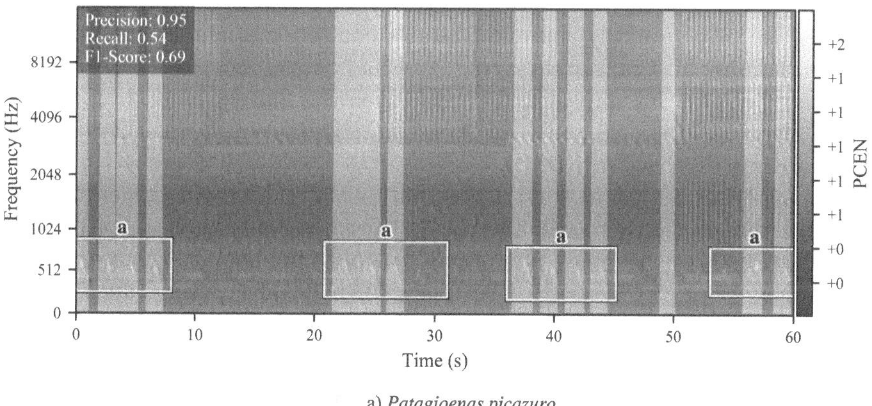

Fig. 6. *SpecNetCD* detections for a PCEN spectrogram. Whiter shades represent bird sound detections.

contrast, using log Mel or PCEN reduces this variability and notably raises the lower quartile values, ensuring more robust results even on challenging samples.

Precision, on the other hand, generally remains higher and more tightly distributed relative to recall. This pattern implies that, in most recordings, the model avoids many false positives, but recall can still be low in certain conditions (as reflected in the low first quartile values for recall). Consequently, while the upper quartiles often demonstrate that the model can achieve solid performance in favorable recordings, a subset of audio clips with more background noise or lower amplitude bird vocalizations pulls down the first quartile. These distributions highlight both the strengths of the deeper or more sophisticated transformations and the inherent challenges of real-world audio data, where acoustic variability can be substantial.

Visual inspection of detection results reveals that, although precision remains generally high, indicating that detected segments are predominantly correct, recall often suffers, particularly for vocalizations of lower amplitude, as evidenced by the calls of the species *Euphonia cloritica* (c) in Fig. 5. The recall is also affected by the fact that the model frequently detects only a portion of a ROI, failing to highlight lower amplitude snippets of the vocalization, as exemplified in Fig. 6 or in the call of *Patogioenas picazuro* (e) of Fig. 5. High background noise can often impair precision, as it happens at the same frequency as other bird sounds.

7 Conclusion

In this study, we proposed a relevance metric-based strategy to improve the automated detection of bird vocalizations in complex acoustic environments. Our methodology was evaluated against a traditional energy-based segmentation baseline using a newly compiled dataset composed of real-world recordings with overlapping bird calls and varying levels of background noise. The proposed CNN-based approach, coupled with PCEN spectrogram representation, demonstrated clear performance gains, achieving an F1-score improvement from 0.44 (baseline) to 0.66 (*SpecNetCD*).

Our experiments led to several observations. Precision was generally high, but recall exhibited larger variability, suggesting that low-amplitude or partially overlapping calls often remain undetected. Notably, enhancing spectrogram representations using log Mel or PCEN significantly improved overall model sensitivity and robustness to noise, while moderate architectural expansions further enhanced the network's capacity to identify challenging acoustic patterns.

Based on these findings, our current methods demonstrate encouraging performance. Future research will explore the integration of this relevance metric directly into species classification workflows and investigate more sophisticated CNN architectures, such as two-dimensional convolutions, to better handle overlapping and low-amplitude vocalizations commonly encountered in passive acoustic monitoring tasks.

Acknowledgment. This study was supported by Fundação Araucária (NAPI Biodiversidade: Serviços Ecossistemicos - Grant 156/2023). We also thank the Federal University of Technology – Paraná (UTFPR) for its support in carrying out this research.

References

1. Akiba, T., Sano, S., Yanase, T., Ohta, T., Koyama, M.: Optuna: a next-generation hyperparameter optimization framework. In: The 25th ACM SIGKDD International Conference on Knowledge Discovery & Data Mining, pp. 2623–2631 (2019)
2. Cohen, Y., Nicholson, D.A., Sanchioni, A., Mallaber, E.K., Skidanova, V., Gardner, T.J.: Automated annotation of birdsong with a neural network that segments spectrograms. eLife **11**, e63853 (2022)
3. Cover, T.M., Thomas, J.A.: Elements of Information Theory (Wiley Series in Telecommunications and Signal Processing). Wiley-Interscience (2006)
4. Das, N., Padhy, N., Dey, N., Paul, H., Chowdhury, S.: Exploring explainable ai methods for bird sound-based species recognition systems. Multimed. Tools Appl. **83**(24), 64223–64253 (2024)
5. Devices, O.A.: AudioMoth: A Low-Cost, Full-Spectrum Acoustic Logger. https://www.openacousticdevices.info/audiomoth
6. Foleiss, J.H., Tavares, T.F.: Segment relevance estimation for audio analysis and weakly-labelled classification (2019)
7. Gregory, R., Noble, D., Field, R., Marchant, J., Raven, M., Gibbons, D.: Using birds as indicators of biodiversity. Ornis Hungarica **12**, 11–24 (2003)
8. Jiang, Y., Zhang, L.L., Yang, F.: Bird sound feature extraction and recognition model design based on neural network architecture search (2024), preprint (Version 1) available at Research Square
9. Joly, A., et al.: Overview of LifeCLEF 2019: identification of Amazonian plants, South & North American birds, and niche prediction. In: Crestani, F., et al. (eds.) CLEF 2019. LNCS, vol. 11696, pp. 387–401. Springer, Cham (2019). https://doi.org/10.1007/978-3-030-28577-7_29
10. Kahl, S., Wood, C.M., Eibl, M., Klinck, H.: Birdnet: a deep learning solution for avian diversity monitoring. Eco. Inform. **61**, 101236 (2021)
11. Kingma, D.P., Ba, J.: Adam: a method for stochastic optimization (2017)
12. Liu, D., Xiao, H., Chen, K.: Research progress in bird sounds recognition based on acoustic monitoring technology: a systematic review. Appl. Acoust. **228**, 110285 (2025)
13. Lostanlen, V., Salamon, J., Farnsworth, A., Kelling, S., Bello, J.P.: Robust sound event detection in bioacoustic sensor networks. PLoS ONE **14**(10), e0214168 (2019)
14. Maithripala, A., Arachchi, S.M., Karunanayaka, K., Perera, R., Pallewatta, P.: A review of automated bird sound recognition and analysis in the new ai era. In: 2024 8th SLAAI International Conference on Artificial Intelligence (SLAAI-ICAI), pp. 1–6 (2024)
15. Narasimhan, R., Fern, X.Z., Raich, R.: Simultaneous segmentation and classification of bird song using CNN. In: 2017 IEEE International Conference on Acoustics, Speech and Signal Processing (ICASSP), pp. 146–150 (2017)
16. Rognan, C., Szewczak, J., Morrison, M.: Autonomous recording of great gray owls in the Sierra Nevada. Northwest. Nat. **93**, 138–144 (2012)
17. Rosa, G.L.M., Zurano, J.P., Torres, I.M.D., Simoes, C.R., dos Anjos, L., de Araujo, C.B.: monitoraSom: a toolbox for automated sound detection applied to Bioacoustics and Passive Acoustic Monitoring (2025), r package version 0.3.1
18. Ross, S.R.P.J., et al.: Passive acoustic monitoring provides a fresh perspective on fundamental ecological questions. Funct. Ecol. **37**(4), 959–975 (2023)
19. Sugai, L.S.M., Silva, T.S.F., Ribeiro, José Wagner, J., Llusia, D.: Terrestrial passive acoustic monitoring: Review and perspectives. BioScience **69**(1), 15–25 (2018)

20. Tseng, Y.C., Eskelson, B.N.I., Martin, K., LeMay, V.: Automatic bird sound detection: logistic regression based acoustic occupancy model. Bioacoustics **30**(3), 324–340 (2021)
21. Vallecillo, S., Maes, J., Polce, C., Lavalle, C.: A habitat quality indicator for common birds in Europe based on species distribution models. Ecol. Ind. **69**, 488–499 (2016)
22. Venkatesh, S., Moffat, D., Miranda, E.R.: You only hear once: a YOLO-like algorithm for audio segmentation and sound event detection. Appl. Sci. **12**(7) (2022)

Source Code Minimaps at Scale: Feature-Oriented Analysis of 100 Projects

Munif Gebara Jr.[1(✉)], Igor S. Wiese[2], Hua-Liang Wei[3], and Yandre M. G. Costa[1,4]

[1] Graduate Program in Computer Science, State University of Maringá (UEM), Maringá, Brazil
pg55752@uem.br
[2] Federal University of Technology (UTFPR), Campo Mourão, Brazil
[3] School of Electrical and Electronic Engineering and Centre of Machine Intelligence (CMI), The University of Sheffield, Sheffield, UK
[4] Department of Informatics, State University of Maringá (UEM), Maringá, Brazil

Abstract. In this work, we present a large-scale study of *source code minimaps* that spans 685,906 images from 100 open-source projects across 10 programming languages. The corpus is standardized to 128×128 pixels, anonymized, and accompanied by a file-level catalog that enables three complementary objectives: *type* (63 classes), *project* (99), and *author* (97). We evaluate convolutional backbones under a unified protocol (batch size 64, maximum 100 epochs, early stopping with patience = 5), comparing from-scratch and ImageNet-pretrained variants. The best validation results are obtained with EfficientNet-B0 (pretrained), reaching accuracies of 0.9333 (*type*), 0.8604 (*project*), and 0.8705 (*author*); ROC-AUC (macro/micro) remains near or above 0.995 across tasks, and optimizer choice (Adam vs. SGD) has negligible impact at the adopted precision. To interpret the models, we compute Integrated Gradients and aggregate attributions over the validation set, revealing that predictions for *type* rely on structural bands (indentation, delimiters), *project* emphasizes repository conventions and boilerplate, and *author* exhibits more diffuse stylistic cues. These findings establish a reproducible benchmark for learning from visual abstractions of code under severe class imbalance and motivate future work on imbalance-aware objectives, dataset scale and class imbalance; we also provide post-hoc attributions with Integrated Gradients. Code and dataset are publicly available at https://github.com and https://zenodo.org.

Keywords: Transfer learning · Class imbalance · Integrated Gradients · Software visualization

1 Introduction

Software systems are becoming increasingly complex, spanning diverse domains and relying on heterogeneous technologies. Understanding and analyzing these

D. Chaves et al. (Eds.): CIARP 2025, LNCS 16528, pp. 349–363, 2026.
https://doi.org/10.1007/978-3-032-23161-1_25

systems at scale requires novel approaches that go beyond traditional static analysis or rule-based tools. Recent studies have shown the potential of source code minimaps, which transform code into compact visual abstractions that capture structural and syntactic patterns. These visual signatures enable the application of machine learning (ML) and computer vision techniques to source code, supporting tasks such as stereotype identification, architectural analysis, and anomaly detection. However, despite promising results, most existing efforts remain limited to small datasets or shallow models, offering little interpretability regarding the internal decision-making processes of classifiers. [3]

Software visualization has long been a key strategy to support program comprehension and software evolution. Novais et al. [10] proposed the SourceMiner Evolution tool, which employs coordinated visualizations such as Treemaps and Polymetric Views to reveal structural and evolutionary aspects of software systems. Similarly, other visualization approaches such as SArF Map [9] explored feature-oriented clustering and user feedback visualization, reinforcing the role of visual abstractions in enhancing the understanding of large-scale systems. While effective, these methods typically require predefined rules, manual curation, or domain-specific mappings, which limit their scalability and adaptability. Our work builds on this foundation by employing source code minimaps as generic visual signatures that can be automatically processed by machine learning models.

In parallel, several studies investigated the transformation of non-visual data into visual domains for machine learning applications. Costa et al. [4,5] demonstrated how spectrograms of audio signals, combined with texture descriptors such as Local Binary Patterns (LBP), can achieve competitive results in music genre classification. This methodology inspired the application of texture-based techniques to software minimaps, as introduced in our previous work [3]. However, prior approaches were limited to a relatively small dataset and classical ML models, providing little insight into model decision-making. More recent advances in Explainable Artificial Intelligence (XAI) [1,13] highlight the importance of interpretability, particularly when deep learning models are applied to high-dimensional data. In this context, our study expands on earlier contributions by introducing a large dataset of 100 projects in 10 languages, evaluating deep learning architectures, and applying XAI methods such as Integrated Gradients to interpret how visual code patterns influence classification outcomes.

These developments motivate a next step that unifies scale and interpretability. We therefore extend minimap-based learning with a substantially larger corpus and modern deep networks, and we integrate XAI to audit model behavior, clarifying *which* visual cues are leveraged and *why* predictions differ across objectives.

To address these limitations, we introduce a large-scale dataset of source code minimaps comprising 100 projects across 10 programming languages, including Java, Python, JavaScript, C, C#, C++, PHP, Go, Ruby, and Kotlin. This dataset enables comprehensive evaluation of machine learning methods under realistic and diverse conditions. We investigate state-of-the-art deep learning models trained directly on minimaps and employ Explainable Artificial Intelligence (XAI) techniques, such as Integrated Gradients, to analyze which visual patterns contribute most to classification decisions. This combination of dataset expansion, deep learning, and interpretability provides new insights into how visual representations of source code can support feature detection and software comprehension at scale.

To provide an overview of the workflow adopted in this study, Fig. 1 illustrates the main stages of our approach. The process begins with the selection of programming languages and repositories, followed by the generation of minimaps and associated metadata. From this, relevant attributes are chosen for classification tasks, leading to the training phase. Finally, post-hoc explainability (XAI) methods are applied to interpret the results.

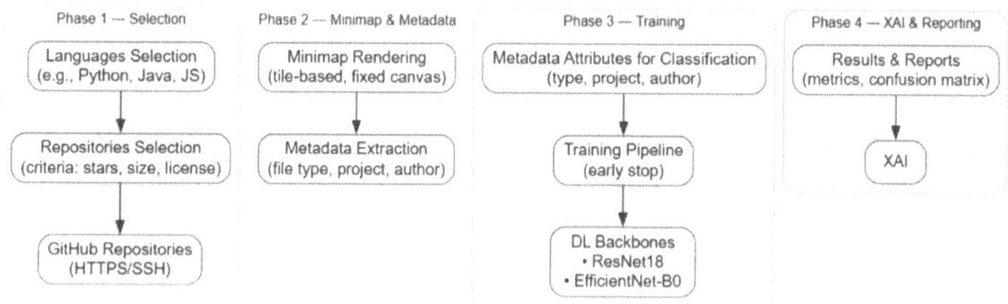

Fig. 1. Overview of the workflow adopted in this study.

The process begins with *repository selection*. A corpus of 100 public repositories across 10 programming languages is considered, prioritizing active maintenance, conventional tooling, and clear licensing. Language coverage is balanced as far as possible. Forks and mirrors are excluded to prevent duplication, and for each repository, an identifier and basic metadata (hosting platform, number of stars, default branch) are recorded.

Next comes *file filtering and annotation.* Source files are retained according to language-appropriate extensions, whereas generated, vendored, and build artifacts (e.g., `node_modules/`, `dist/`, `build/`, and VCS directories) are removed. Binary content is excluded via MIME inspection, and near-duplicates are eliminated using content hashes. Each remaining file receives three labels corresponding to the supervised objectives: a *type* label representing a coarse file role, a *project* label corresponding to the repository identifier, and an *author* label derived from version-control blame using last-committer attribution. File paths are stored as salted hashes to avoid leaking sensitive names.

Anonymization focuses on what is replaced versus what is preserved. Identifiers, string literals, and comments are substituted by placeholders, while whitespace, indentation, line counts, punctuation, and block delimiters are retained. The transformation is deterministic and language-aware so that structural and layout cues remain available to the downstream vision models even though lexical content is suppressed.

For *minimap generation*, files are rendered as images with a fixed resolution of **128×128** pixels by default. **Grayscale** rendering is the standard setting, with RGB available as an option when explicitly required by an experiment. Short files are padded to the target canvas; long files are uniformly downsampled to preserve global layout. Pixel intensities are normalized to the $[0, 1]$ range. Variable resolutions are supported for ablation studies but are not used in the main benchmark.

The *catalog and dataset splits* record, for each image, the repository identifier, a file identifier (SHA-256), a hashed relative path, the programming language, and the labels for the three objectives (*type*, *project*, *author*). Splits are stratified per objective. To prevent leakage, the *project* task enforces repository disjointness across train, validation, and test, and the *author* task avoids placing the same author in both training and test partitions.

Training and evaluation follow a standardized protocol. Images are resized to 128×128; random horizontal flips are applied with probability 0.5, while vertical flips are disabled to respect layout semantics. The default backbone is **ResNet-18**; **EfficientNet-B0** and **ViT-B/16** are available as alternatives. Optimization uses Adam with learning rate 10^{-3} and weight decay 10^{-4}; mini-batches contain 256 images and training runs for up to 100 epochs with early stopping (patience 10) on validation macro-F1. Class imbalance is handled with weighted cross-entropy using inverse-frequency weights, and imbalance is reported via IR_{global} and IR_{avg}. Three random seeds are executed and results are reported as mean $\pm$ standard deviation. The computational environment (GPU/CPU model, OS, and library versions) is described in a dedicated subsection for reproducibility.

Finally, *explainability* is integrated through Integrated Gradients configured with a black baseline and 50 steps, using the same input normalization as training. Attributions are rescaled to $[0, 1]$ and overlaid on the input image to highlight class-relevant regions. Parameters are kept constant across the three objectives to ensure qualitative comparability.

The remainder of this paper is organized as follows: Sect. 3 presents the dataset; Sect. 4 describes the training protocol and results; Sect. 5 introduces the explainability analyses; and Sect. 6 provides conclusions and directions for future work.

To facilitate reproducibility and reuse, we release both the source code and the full dataset. The training and evaluation pipeline, including configuration files and scripts for generating minimaps and reports, is available at

https://anonymous.4open.science/r/minimap-learner-B626/README.md.

The curated dataset (catalog and images) can be accessed and cited via Zenodo at https://zenodo.org/records/16929672. These resources enable end-to-end replication of our experiments and support further analyses.

The central contribution is a scalable, framework-agnostic methodology whose primary advantage is its reproducibility across **ten** distinct programming languages, indicating robustness beyond any single ecosystem. A corpus of 685906 minimaps drawn from 100 projects is assembled, together with a catalog and executable scripts that can be rerun on new repositories without language-specific tailoring, supporting three supervised objectives—*type*, *project*, and *author*. The anonymization stage removes lexical content while preserving structural and visual cues essential for learning; training and evaluation follow a fixed protocol with explicit imbalance reporting (IR_{global}, IR_{avg}); and Integrated Gradients is incorporated into the benchmark to expose class-relevant evidence in the images. Taken together, these choices render the approach robust and genuinely language-agnostic, facilitating faithful replication across heterogeneous codebases and toolchains.

Experiments were conducted on a workstation equipped with a single NVIDIA RTX 3070 (8 GB) GPU, a 6-core CPU, and 64 GB of RAM, running Ubuntu 24.04 LTS. The software stack comprised Python 3.11, PyTorch 2.x with CUDA 12.x, TorchVision, and Captum for explainability. Random seeds were fixed across runs, and early stopping criteria were identical for all model families to limit variance due to training dynamics.

2 Related Work

Software visualization has long supported program comprehension, especially for reasoning about feature evolution and maintenance. Novais et al. [11] demonstrated this with SourceMiner Evolution (SME), which offers multiple interactive and coordinated views to analyze how feature code changes across versions. These visual metaphors—including Treemaps, Polymetric views, and dependency graphs—highlight how structure and evolution become more tractable when exposed visually. In contrast to interactive, preconfigured views, the present approach leverages automated learning directly from source code minimaps rendered at fixed resolution, aiming to infer software stereotypes without manual mappings or hand-crafted dashboards. The result is a pipeline designed to be framework-agnostic and reproducible across languages and repositories.

Architectural specificities also shape how design problems surface in practice. Aniche et al. [2] curated a catalog of code smells for Model–View–Controller (MVC) architectures, showing that roles such as controllers and repositories follow distinct failure modes that generic smell catalogs may miss. While such domain knowledge remains valuable, the current approach diverges methodologically by using automated learning over minimaps to detect recurrent stereotypes from data itself, rather than relying on expert-defined catalogs. This data-driven route seeks scalability and adaptability to evolving practices, complementing architecture-specific insights reported in [2].

A parallel thread relevant to the methodology is the translation of non-visual phenomena into images that admit texture-based analysis and machine learning. Costa et al. [7] transformed audio signals into spectrograms for music genre classification, illustrating how time–frequency structure can be made amenable to visual pattern recognition. Subsequent work applied Local Binary Patterns

(LBP) with shallow classifiers such as SVMs to those spectrograms [6], underscoring the utility of texture descriptors in image-like encodings. The present contribution follows the same high-level premise—representing complex artifacts as images—but relies on end-to-end deep models (e.g., ResNet, EfficientNet, ViT) trained on fixed-size source code minimaps, coupled with Integrated Gradients to make class-relevant evidence explicit. This shift from hand-crafted features to learned representations enables scaling across heterogeneous codebases while retaining interpretability via XAI.

3 Dataset

The dataset used in this study was designed to provide large-scale and diverse resources for evaluating source code minimaps. It comprises a total of 685,906 minimap images extracted from 100 open-source projects, covering 63 distinct file types. In addition, contributions from 97 different authors were included, ensuring that the dataset reflects a variety of ecosystems, coding conventions, and development practices. This diversity enables the evaluation of machine learning models under realistic and heterogeneous conditions.

Projects and file types were selected to maximize representativeness across domains and programming paradigms. Figure 2 shows the Top-20 class frequencies for each objective. The most frequent types are `java` (88,249 samples), `ruby` (59,216), `c-header` (47,909), `c` (45,814), and `python` (42,581). Among projects, the Linux kernel dominates with 76,011 samples, followed by GitLabHQ (52,918), Elasticsearch (31,493), Moodle (24,484), and Kubernetes (23,435). In terms of authorship, Linus Torvalds accounts for 76,011 samples, with additional large contributions from automated accounts such as GitLab Bot (52,918) and Kubernetes Prow Robot (23,435). These numbers highlight the scale and breadth of the dataset, combining human and automated contributions across ecosystems.

The generation of minimaps followed the methodology established in our previous work [3], with extensions for scalability. Each project was scanned recursively; files were identified based on their extension and annotations; and anonymization was applied before transformation into minimaps. Two versions were produced: variable-size images (reflecting the original file layout) and fixed-size images of 128×128 pixels. The fixed-size setting ensures compatibility with convolutional neural networks and other deep learning architectures that require consistent input dimensions, enabling fair comparison between texture-based descriptors and modern deep learning methods.

To protect confidentiality, an anonymization process replaced identifiers (e.g., class names, variables, numeric constants) with canonical symbols while preserving structural tokens (e.g., brackets, operators, keywords). As shown in previous experiments [3], anonymization does not significantly affect classification accuracy while enabling datasets derived from industrial or proprietary projects to be safely shared, balancing privacy and fidelity.

In contrast to our previous study, this release *explicitly ships a file-level catalog* that accompanies the minimap images. The catalog CSV (Table 1) provides, for each sample, the image identifier, the inferred file *type* (e.g., `java`, `ruby`, `c-header`), the *project* label (e.g., `linux`, `gitlabhq`), the original *file_path* and *file_name*, and the *author*. This structured index enables reproducible joins across objectives (*type*, *project*, *author*) and supports simple aggregations (e.g., language distributions per project). In the provided catalog snapshot, we observe 685,906 entries, with 63 file types, 100 projects, and 97 authors.

Table 1. Catalog fields shipped with the dataset (CSV).

Field	Type	Description	Example
`image`	string	Minimap image filename (relative to the images root); unique per sample.	`00012345.png`
`type`	categorical	File type/stereotype inferred from extension and heuristics.	`java`
`project`	categorical	Repository/project identifier (normalized, lowercase).	`linux`
`file_path`	string	Relative path to the original source file inside the repository.	`kernel/sched/core.c`
`file_name`	string	Base filename of the source file.	`core.c`
`author`	categorical	VCS author name/handle associated with the sample.	`Linus Torvalds`

3.1 Dataset Description

As expected, the *author* task presents the largest number of classes and the highest imbalance, whereas *type* is comparatively more balanced and structurally distinct. The dataset is therefore suitable for studying imbalance-aware learning strategies, as well as for assessing model sensitivity across different classification objectives.

Cross-Objective Comparison. Table 2 consolidates cardinality and imbalance for direct comparison across objectives.

We characterize the corpus introduced in Sect. 3 using class-frequency summaries per objective and a compact cross-objective comparison. The release contains **685,906** fixed-size minimaps (128×128) annotated along three complementary objectives: *author* (97 classes), *type* (63), and *project* (99). Following best practice for imbalanced data , we report dispersion and two imbalance indicators: Global Imbalance Ratio (IR_{global}), defined in Eq. 1, and Average Imbalance Ratio (IR_{avg}), defined in Eq. 2. IR_{global} is about the extremes (worst case), and IR_{avg} is about the overall imbalance level [8,12].

$$IR_{global} = \frac{max_i\ n_i}{min_i\ n_i} \tag{1}$$

where: n_i is the number of samples in class i. If IR_{global} is equal to 1, the dataset is perfectly balanced. Larger values indicate stronger imbalance.

$$IR_{avg} = \frac{1}{K} \sum_{1}^{K} \frac{n_{\max}}{n_i} \tag{2}$$

where: K is the number of classes, $n_{max} = max_i \; n_i$.

More sensitive to how minority classes compare against the majority. To avoid redundancy while preserving intuition about the head of each distribution, Figures 2a, 2c and 2b show the Top-20 class frequencies for each objective. For a concise side-by-side view of cardinality and imbalance, Table 2 summarizes the per-objective statistics used throughout the paper.

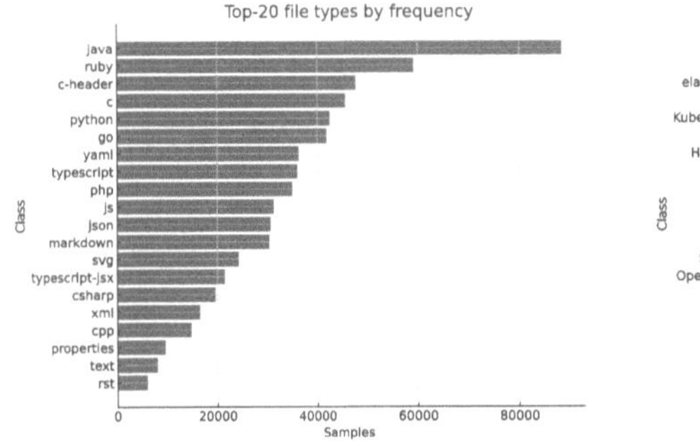

(a) **Type:** Frequency of file-role categories used in the type objective.

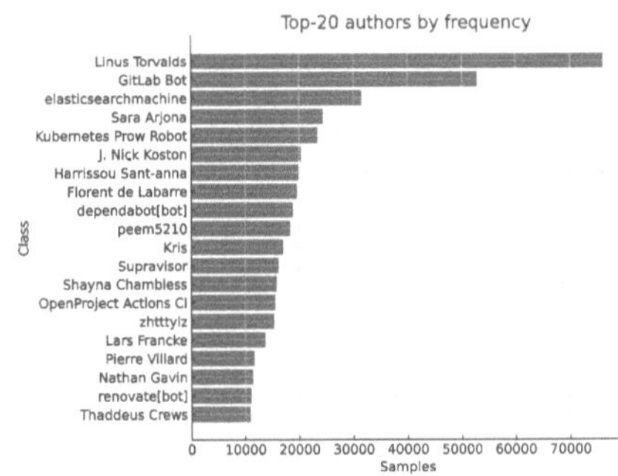

(b) **Author:** Frequency of last-committer identities used in the author objective.

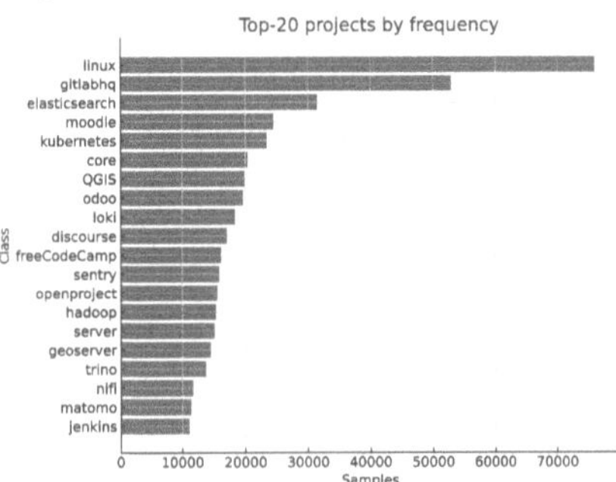

(c) **Project:** Frequency of repository identifiers used in the project objective.

Fig. 2. Top-20 classes by frequency for the three objectives (*Type*, *Author*, *Project*). Head classes are highlighted to contextualize class imbalance across tasks.

4 Training Protocol and Results

Table 2. Comparison across objectives (concise summary).

Objective	Classes	Samples	Median	Max	IR_{global}	IR_{avg}
Author	97	685,906	2,857	76,011	10,858.71	177.31
Type	63	685,906	1,120	88,249	88,249.00	4,402.12
Project	99	685,906	2,857	76,011	10,858.71	179.40

We adopt a unified protocol across objectives and backbones to ensure comparability and statistical stability. Inputs are fixed-size minimaps (128×128 grayscale) with input normalization to $[0, 1]$; models are fine-tuned end-to-end with task-specific linear heads; mixed-precision training (`torch.amp`) reduces memory footprint. Optimization uses Adam (10^{-3}) or SGD (momentum 0.9, 10^{-2}) with weight decay 10^{-4}. Training proceeds for at most 100 epochs with early stopping (patience = 10) based on validation accuracy; the best checkpoint is retained. We evaluate ResNet-18 (from-scratch and ImageNet-pretrained) and EfficientNet-B0 (ImageNet-pretrained) on three objectives: *type* (63 classes), *project* (99), and *author* (97); reporting Accuracy, ROC-AUC (macro/micro), and discussing macro-F1 and weighted-F1 as complementary measures under imbalance.

Metrics under Imbalance. While accuracy and ROC-AUC provide a broad view of separability, class imbalance motivates macro-averaged measures. Macro-F1 equally weighs all classes, penalizing errors on minority classes; weighted-F1 accounts for support and tracks practical performance when head classes dominate. In our setting, macro-F1 typically lags behind micro metrics when distributions are long-tailed, especially for *project* and *author*, whereas *type* benefits from clearer structural cues (Sect. 3).

Consolidated Training Artifacts. The training overview figures summarize per-task validation metrics, learning curves (loss/accuracy), confusion matrices, and ROC curves for representative runs.

Quantitative Summary. Table 3 consolidates validation metrics for two representative runs. In *type*, accuracies are ≥0.9236 with micro ROC-AUC ≥0.9988, indicating strong separability of file-level structures. In *project*, accuracies lie in [0.8513, 0.8557] with micro ROC-AUC ≈ 0.997, reflecting heterogeneous repository conventions. In *author*, accuracies lie in [0.8619, 0.8646], consistent with stylistic variability even under anonymization. At the adopted precision, optimizer differences are negligible; ImageNet pretraining yields consistent but modest gains in *type* and *project*.

Table 3. Validation metrics per objective for two representative runs.

Run	Task	Acc.	ROC-AUC (macro/micro)
run-20250817-053841	type	0.9236	0.9974 / 0.9989
run-20250817-053841	proj_orig	0.8513	0.9950 / 0.9964
run-20250817-053841	author	0.8646	0.9947 / 0.9963
run-20250817-012528	type	0.9263	0.9975 / 0.9988
run-20250817-012528	proj_orig	0.8557	0.9954 / 0.9968
run-20250817-012528	author	0.8619	0.9947 / 0.9963

Between-Run Differences. To make the comparison explicit, Table 4 reports the between-run deltas (run 20250817-012528 minus run 20250817-053841). We observe small but systematic improvements for *project*, and a slight trade-off in *author*; *type* remains consistently high with negligible AUC variation.

Table 4. Between-run deltas: run-20250817-012528 minus run-20250817-053841.

Task	Δ Acc.	Δ ROC-AUC (macro)	Δ ROC-AUC (micro)
type	+0.0027	+0.0001	−0.0001
proj_orig	+0.0044	+0.0004	+0.0004
author	−0.0027	0.0000	0.0000

Discussion. Convergence profiles indicate that fixed-size minimaps and end-to-end fine-tuning are sufficient to capture discriminative, multi-scale patterns. The gap between *type* and *project*/*author* aligns with class-cardinality and imbalance (Sect. 3), reinforcing the use of macro-averaged metrics. The minimal impact of the optimizer suggests that early stopping dominates optimization dynamics under the present data scale and regularization. Consistent, modest gains attributable to pretraining are observed for *type* and *project*, whereas *author* remains more variable, likely due to residual stylistic noise after anonymization; macro-F1 and weighted-F1 are therefore more informative for this objective and should be prioritized in future analyses.

We report per-task validation metrics for successful runs, aggregating by parameter configuration rather than run identifiers. Configurations are defined by the tuple *(backbone, initialization, optimizer, learning rate, weight decay).* Beyond Accuracy and ROC-AUC (macro/micro), we include macro-F1 and weighted-F1 to account for class imbalance.

Table 5 summarizes mean ± std across successful runs that share the same configuration. We observe that ImageNet pretraining consistently improves *type* and *project* in macro-F1 and AUC, while *author* remains more variable, consistent with residual stylistic noise even after anonymization. Optimizer choice (Adam vs. SGD with momentum) has a smaller effect than initialization and backbone depth; differences appear mostly as small variations in convergence stability (std across runs). Table 6 highlights, for each task, the best macro-F1 configuration; these settings are used for subsequent qualitative analyses (Sect. 5).

Table 5. Validation metrics over successful runs grouped by parameter configuration (mean ± std).

<table>
<tr><th>Task</th><th>Configuration</th><th>#Runs</th><th>Acc.
σ</th><th>F1
(macro)</th><th>F1
(weigh.)</th><th>AUC
(macro)</th><th>AUC
(micro)</th></tr>
<tr><td>author</td><td>– | pretrained | –@– | wd=–</td><td>4</td><td>0.8662
± 0.0050</td><td>0.8432
± 0.0055</td><td>0.8660
± 0.0049</td><td>0.9951</td><td>0.9965</td></tr>
<tr><td>author</td><td>– | scratch | –@– | wd=–</td><td>4</td><td>0.8646</td><td>0.8421</td><td>0.8655</td><td>0.9947</td><td>0.9963</td></tr>
<tr><td>proj_orig</td><td>– | pretrained | –@– | wd=–</td><td>4</td><td>0.8580
± 0.0027</td><td>0.8346
± 0.0025</td><td>0.8571
± 0.0017</td><td>0.9955</td><td>0.9969</td></tr>
<tr><td>proj_orig</td><td>– | scratch | –@– | wd=–</td><td>4</td><td>0.8513</td><td>0.8222</td><td>0.8521</td><td>0.9950</td><td>0.9964</td></tr>
<tr><td>type</td><td>– | pretrained | –@– | wd=–</td><td>4</td><td>0.9298
± 0.0040</td><td>0.8669
± 0.0031</td><td>0.9294
± 0.0033</td><td>0.9977</td><td>0.9989</td></tr>
<tr><td>type</td><td>– | scratch | –@– | wd=–</td><td>4</td><td>0.9236</td><td>0.8634</td><td>0.9238</td><td>0.9974</td><td>0.9989</td></tr>
</table>

Overall, macro-F1 tracks Accuracy for *type*, but diverges for *project* and *author*, reinforcing the need for macro-averaged metrics under long-tailed distributions.

Table 6. Best configurations per task (by macro-F1) among successful runs.

<table>
<tr><th>Task</th><th>Configuration</th><th>F1
(macro)</th><th>Acc.</th><th>AUC
(macro)</th><th>AUC
(micro)</th></tr>
<tr><td>author</td><td>– | pretrained | –@– | wd=–</td><td>0.8432 ± 0.0055</td><td>0.8662</td><td>0.9951</td><td>0.9965</td></tr>
<tr><td>proj_orig</td><td>– | pretrained | –@– | wd=–</td><td>0.8346 ± 0.0025</td><td>0.8580</td><td>0.9955</td><td>0.9969</td></tr>
<tr><td>type</td><td>– | pretrained | –@– | wd=–</td><td>0.8669 ± 0.0031</td><td>0.9298</td><td>0.9977</td><td>0.9989</td></tr>
</table>

5 Explainability (XAI)

We compute attribution maps with Integrated Gradients (IG) using Captum. For each task, we sample up to N validation images per class (configured via xai.per_class_samples) and attribute with respect to the model's predicted label. We overlay normalized IG attributions onto the input minimaps and include these panels in the run report. XAI is used solely for qualitative interpretation; it does not influence training or model selection.

Qualitative Findings. Figure 3 illustrates qualitative attribution maps obtained with Integrated Gradients for the three classification objectives: (A) *author*, (B) *project*, and (C) *type*. (1) In *type*, IG concentrates on structural bands (indentation fields, delimiters), mirroring the strong quantitative separability; (2) in *project*, salient regions align with repository-specific conventions (headers/boilerplate), consistent with moderate accuracies and high ROC-AUC; (3) in *author*, attributions are more diffuse, reflecting stylistic factors that persist despite anonymization. Cross-run consistency in the saliency layout supports that the models rely on stable visual cues rather than dataset artifacts. These insights inform the design of imbalance-aware objectives and regularizers that penalize over-reliance on boilerplate while preserving structural signals.

Implications. The XAI evidence explains the relative task difficulty and suggests concrete avenues: (i) leverage pretraining for *type*/*project* where structural cues dominate; (ii) incorporate mixup/cutmix or focal loss to mitigate head dominance; (iii) add saliency-based constraints to reduce dependence on non-semantic artifacts (e.g., template headers) that may impair cross-repository generalization.

Fig. 3. Integrated Gradients (IG) attribution maps for the three classification objectives: (a) *type*, (b) *project*, and (c) *author*. IG uses a black baseline and 50 steps with input normalization matched to training; attributions are rescaled to $[0, 1]$ and converted to a negative imagem for better visualization.

Integrated Gradients (IG) attributions overlaid on minimaps. IG uses a white baseline and 50 steps with input normalization matched to training. Attributions are rescaled to $[1, 0]$ for visualization. For the *type* objective, concentrated bands often indicate structural elements characteristic of a file role; for *project*, recurring boilerplate regions highlight repository-specific conventions; for *author*, more diffuse patterns reflect stylistic regularities rather than isolated tokens.

6 Concluding Remarks

This paper presented a large-scale study of *source code minimaps* spanning 685,906 images from 100 open-source projects and 10 programming languages, accompanied by a file-level catalog that enables reproducible joins across three objectives (*type*, *project*, *author*). A unified training protocol (i.e. fixed 128×128 inputs, end-to-end fine-tuning, early stopping, and macro/micro ROC-AUC reporting) supported systematic comparisons across backbones. Among the evaluated models, EfficientNet-B0 with ImageNet pretraining achieved the highest validation accuracies, with representative results of 0.9333 (*type*), 0.8604 (*project*), and 0.8705 (*author*), and ROC-AUC (macro/micro) consistently near or above 0.995. The relative difficulty across objectives aligns with corpus statistics: *type* benefits from strong structural cues, while *project* and *author* are moderated by cross-repository conventions and stylistic variability. Attribution analyses using Integrated Gradients yielded coherent, task-specific saliency patterns: structural bands and indentation fields for *type*; boilerplate and header regions for *project*; and more diffuse stylistic regions for *author*. Together, these findings indicate that minimaps preserve discriminative signals at multiple scales and that modern CNNs can leverage them effectively under severe class imbalance.

Several limitations remain. First, the *type* objective exhibits extreme skew, which, despite macro-averaged evaluation, can mask minority-class behavior. Second, author labels may conflate human and organizational/bot accounts, potentially introducing confounds at the style level. Third, fixed-resolution inputs standardize learning but may underutilize information present in especially long files. Finally, the present analysis focused on convolutional backbones; broader coverage (e.g., lightweight vision transformers) and cross-architecture ensembles were not explored here.

Beyond empirical results, the approach lends itself to everyday software-engineering workflows. Periodic, repository-wide snapshots of minimaps can help teams maintain architectural conformance by revealing small but systematic shifts in configuration templates, import patterns, or build scripts—signals of drift that are often missed when scanning raw diffs. The same visual regularities that characterize each repository can also guide continuous integration and code review: when a pull request introduces files whose minimaps diverge markedly from established repository signatures, reviewers obtain an immediate cue to prioritize scrutiny or to request justification for the deviation. Finally, aggregated author- and style-related regularities may assist incident triage by grouping suspicious changes during forensics; however, such signals are imperfect proxies and should be handled with safeguards and clear ethical guidance to avoid misattribution.

These scenarios align directly with the three supervised objectives. Repository-specific signatures captured by the *project* objective underpin both drift monitoring and CI triage; the *type* objective clarifies whether unusual combinations of file roles are being modified together; and the *author* objective—used cautiously—offers additional context about stylistic departure. In short, minimaps provide compact, reproducible cues that integrate naturally with existing pipelines while keeping human oversight central to decision making.

6.1 Future Work

Building on these results, we outline concrete directions:

- **Imbalance-aware learning.** Evaluate class-balanced reweighting, focal loss, logit-adjustment, mixup/cutmix, and deferred reweighting schedules specifically for the long-tail regimes observed in *type*. Report per-class and macro-averaged calibration alongside ROC-AUC and F1 to quantify minority-class reliability.
- **Architectures and pretraining.** Extend the benchmark with vision transformers and hybrid CNN-Transformer models, including self-supervised pretraining (e.g., contrastive objectives) directly on minimaps to reduce reliance on natural-image priors.
- **Resolution and multi-scale inputs.** Compare fixed 128×128 to pyramidal or patch-based inputs, and investigate adaptive cropping strategies guided by XAI to emphasize structurally salient regions without inflating training cost.
- **Data curation and splits.** Refine label hygiene for *author* by separating organizational bots from human contributors; design time-aware and repository-disjoint splits to control leakage across *project*/*author*. Leverage the catalog to produce task-specific balanced subsets and to probe domain shift across ecosystems.
- **XAI-guided modeling.** Incorporate attribution-driven regularizers (e.g., penalties for over-reliance on boilerplate regions) and concept-based analyses (e.g., TCAV) to connect saliency with interpretable code concepts. Use counterfactual editing on minimaps to test causal sensitivity to formatting and header templates.
- **Broader tasks and metadata.** Explore multi-task learning across *type*, *project*, and *author*; integrate repository metadata (contributors, activity, size) as auxiliary signals; and extend evaluation to quality-related tasks (e.g., defect-proneness proxies, architectural stereotype detection).
- **Reproducibility and release.** Package preprocessing, catalog generation, and reporting pipelines as a public toolkit, facilitating community extensions beyond the 100-project corpus and enabling standardized comparisons.

An additional direction for future work is the integration of a preprocessing step that removes comments from the source code before generating minimaps. By filtering out natural language text embedded in comments, the resulting visual signatures will focus solely on software artifacts (structural and syntactic elements). This refinement is expected to reduce noise, avoid spurious attribution to stylistic comment patterns, and provide a cleaner basis for both classification and interpretability analyses.

In summary, the dataset, protocol, and analyses establish a reproducible reference for learning with visual abstractions of code. Scaling imbalance-aware training, diversifying architectures and pretraining, and closing the loop between XAI and model design are promising steps toward robust, interpretable software-visualization pipelines.

Acknowledgments. We thank the Brazilian research support agencies: Coordination for the Improvement of Higher Education Personnel - Brazil (CAPES) - Finance Code 001, Fundação Araucária for funding (Grant No. PRD2023361000043) and National Council for Scientific and Technological Development (CNPq) for their financial support.

Disclosure of Interests. The authors declare no conflicts of interest.

References

1. Adadi, A., Berrada, M.: Peeking inside the black-box: a survey on explainable artificial intelligence (XAI). IEEE Access **6**, 52138–52160 (2018). https://doi.org/10.1109/ACCESS.2018.2870052
2. Aniche, M., Bavota, G., Treude, C., Gerosa, M.A., van Deursen, A.: Code smells for model-view-controller architectures. Empir. Softw. Eng. **23**(4), 2121–2157 (2017). https://doi.org/10.1007/s10664-017-9540-2
3. Anonymized Anonymized, A., Anonymized: Anonymized. In: Anonymized. Anonymized (Anonymized), anonymized
4. Costa, Y.M., Oliveira, L.S., Koerich, A.L., Gouyon, F.: Music genre recognition using spectrograms. In: International Conference on Systems, Signals and Image Processing (IWSSIP), pp. 1–4. IEEE (2011)
5. Costa, Y.M., Oliveira, L.S., Koerich, A.L., Gouyon, F., Martins, J.G.: Music genre classification using LBP textural features. Signal Process. **92**(11), 2723–2737 (2012)
6. Costa, Y.M., Oliveira, L., Koerich, A.L., Gouyon, F., Martins, J.G.: Music genre classification using LBP textural features. Signal Process. **92**(11), 2723–2737 (2012)
7. Costa, Y.M., Oliveira, L.S., Koerich, A.L., Gouyon, F.: Music genre recognition using spectrograms. In: 2011 18th International Conference on Systems, Signals and image processing, pp. 1–4. IEEE (2011)
8. He, H., Garcia, E.A.: Learning from imbalanced data. IEEE Trans. Knowl. Data Eng. **21**(9), 1263–1284 (2009). https://doi.org/10.1109/TKDE.2008.239
9. Kobayashi, K., Kamimura, M., Yano, K., Kato, K., Matsuo, A.: Sarf map: visualizing software architecture from feature and layer viewpoints. In: IEEE International Conference on Program Comprehension (ICPC), pp. 43–52. IEEE (2013)
10. Novais, R.L., Nunes, C., Garcia, A., Mendonça, M.: Sourceminer evolution: a tool for supporting feature evolution comprehension. In: IEEE International Conference on Software Maintenance (ICSM), pp. 508–517. IEEE (2013)
11. Novais, R.L., Nunes, C., Garcia, A., Mendonça, M.: Sourceminer evolution: a tool for supporting feature evolution comprehension. In: 2013 IEEE International Conference on Software Maintenance (ICSM), pp. 508–517. IEEE (2013). https://doi.org/10.1109/ICSM.2013.83
12. Ortigosa-Hernández, J., Inza, I., Lozano, J.A.: Measuring the class-imbalance extent of multi-class problems. Pattern Recogn. Lett. **98**, 32–38 (2017)
13. Sundararajan, M., Taly, A., Yan, Q.: Axiomatic attribution for deep networks. In: Proceedings of the 34th International Conference on Machine Learning (ICML). Proceedings of Machine Learning Research, vol. 70, pp. 3319–3328. PMLR (2017). http://proceedings.mlr.press/v70/sundararajan17a.html

Fast and Robust Lane Detection for 1:10 Scale Autonomous Cars with a Lightweight CNN

Jose M. Celaya-Padilla[1](✉), Rafael Reveles-Martínez[1], Javier Saldivar-Perez[1], Hamurabi Gamboa Rosales[1], Jorge I. Galván-Tejada[1], Huizilopoztli Luna-García[1], Selene Castañeda-Burciaga[2], Martha A. Ramírez-Salazar[2,3], Carlos E. Galván-Tejada[1], and Juan R. Delgado-Contreras[1]

[1] Unidad Académica de Ingeniería Eléctrica, Universidad Autónoma de Zacatecas, Jardín Juárez 147, Centro, Zacatecas 98000, Mexico
jose.celaya@uaz.edu.mx

[2] Department of Industrial Engineering, Universidad Politécnica de Zacatecas, Plan de Pardillo Sn, Parque Industrial, Fresnillo, 99059 Zacatecas, Mexico

[3] Postgraduate Department, Universidad Autónoma de Fresnillo, Carretera Federal Panamericana, Linda Vista, Ejidal 4, Fresnillo, 99039 Zacatecas, Mexico

Abstract. This paper presents a lightweight and real-time lane detection pipeline tailored for 1:10 scale autonomous vehicles operating under strict computational constraints. A compact convolutional neural network (CNN) is proposed for classifying regions of interest (ROIs) within infrared input images, achieving a classification accuracy of 98.67% and an F_1 score of 97.86% across 10,800 test ROIs. The complete pipeline achieves an average per-frame runtime of 19.8 ms (50 FPS) on a dual-core Intel i3-8109U CPU, demonstrating real-time suitability for embedded platforms. This work provides an effective and reproducible approach for scaled autonomous driving scenarios, as exemplified by the AutoModelCar and the FIRA competitions, and lays the groundwork for future integration into closed-loop control and multi-modal perception systems.

Keywords: Lane Detection · CNN · Autonomous Vehicles · Real-Time Perception · 1:10 Scale Robotics

1 Introduction

Autonomous vehicles rely on robust perception systems to navigate safely. Lane detection in particular is widely recognized as a pivotal element of autonomous driving and Advanced Driver-Assistance Systems (ADAS), as identifying lane markings is essential for maintaining safe vehicle positioning [19,20]. Many ADAS features, such as lane-keeping assistance and lane departure warning, fundamentally depend on reliable lane tracking [20]. By providing the vehicle

D. Chaves et al. (Eds.): CIARP 2025, LNCS 16528, pp. 364–379, 2026.
https://doi.org/10.1007/978-3-032-23161-1_26

with the geometry and location of lane boundaries, a lane detection subsystem enables the car's control system to understand its position relative to the road's lanes, which is critical for safe maneuvering and accident prevention.

Lane detection methods have evolved significantly, from classical vision-based algorithms to modern deep learning approaches. Traditional computer vision techniques typically involve hand-crafted pipelines, including edge detection and Hough transforms, but these often require careful tuning and perform sub-optimally in challenging conditions [20]. In recent years, Convolutional Neural Networks (CNNs) have increasingly dominated lane detection tasks, outperforming classical algorithms in robustness and accuracy [7,20].

Despite algorithmic advances, developing and testing lane detection systems on full-scale (1:1) autonomous vehicles poses practical challenges. Real-world deployment is hindered by high costs and significant safety risks. Experimental driving algorithms require costly instrumentation and extensive safety protocols, slowing down innovation cycles [15].

To mitigate these issues, researchers have increasingly turned to scaled-down autonomous vehicles. Using a 1:10 scale autonomous car offers a cost-effective and safe alternative for prototyping and testing. This approach reduces costs drastically while emulating the critical aspects of full-scale driving scenarios [15]. The Mexican Robotics Tournament 2025 (TMR2025) further exemplifies the value of scaled autonomous vehicles by dedicating a competitive category specifically to 1:10 scale autonomous driving.

This study employs a 1:10 scale vehicle to develop and evaluate a custom lightweight CNN model designed explicitly for real-time lane detection under low-power constraints. A dataset aligning with the TMR2025 AutoModel Car category specifications was created to train and test this model, ensuring realistic and relevant scenarios for lane detection tasks at this scale. The developed model achieves the necessary accuracy and inference speed suitable for deployment in embedded, resource-constrained environments, thereby demonstrating the practicality of sophisticated lane detection techniques even within strict computational limits [19].

2 Related Work

Lane detection is a fundamental research topic in the domain of autonomous vehicles, extensively explored due to its importance in ensuring the safety and reliability of Advanced Driver-Assistance Systems (ADAS) and self-driving capabilities [2,21]. The literature typically distinguishes lane detection approaches into classical vision-based and deep learning-based methodologies.

Classical vision-based techniques often exploit feature extraction algorithms, such as the Hough transform and Kalman filtering, along with gradient-based methods to enhance lane-marking contrasts and reduce sensitivity to illumination changes and shadows [2,18]. Fan et al. [2] proposed a method utilizing top-hat transformations and constrained Hough transforms, achieving a detection accuracy rate of 95.63% under varying lighting conditions. Similarly, Youjin et al. [18] leveraged vanishing point estimation with line segment detectors (LSD) and directional priority search methods, demonstrating robustness in complex structured road scenes.

Recent advances in deep learning significantly improved lane detection performance by leveraging CNN and end-to-end semantic segmentation frameworks [6,8,21]. Zhang et al. [21] introduced a spatio-temporal network employing double Convolutional Gated Recurrent Units (ConvGRUs), achieving approximately 5% accuracy improvement on the TuSimple dataset. Ko et al. [6] proposed PINet, achieving 96.75% accuracy and 99.5% specificity on the TuSimple dataset. Li et al. [8] leveraged 3D topology reasoning, achieving a 62.6% F-score on the OpenLane benchmark.

Several other notable deep learning-based methods have been developed. CondLaneNet [10] introduced a conditional convolution and Recurrent Instance Module (RIM) framework to handle complex lane topologies, achieving state-of-the-art performance with an F1 score of 79.48% and real-time inference at 220 FPS on the CULane dataset. Feng et al. [4] proposed an efficient curve modeling approach, significantly improving computational efficiency and achieving competitive accuracy rates. Neven et al. [11] proposed LaneNet, a real-time instance segmentation-based method achieving robust lane detection performance at 50 FPS on the TuSimple dataset. Ultra Fast Structure-aware Deep Lane Detection [13] provided a highly efficient method achieving an impressive speed of 300+ FPS with state-of-the-art accuracy on the CULane dataset.

However, a notable gap in existing research arises from the lack of consideration for small-scale implementations, specifically at the 1:10 scale. Such constraints introduce unique challenges, including limited computational resources, reduced sensor resolution, narrower fields-of-view, and lower power availability—constraints typically absent in full-scale autonomous vehicle experiments [2,4,8,10,11,13,21]. Consequently, established lane detection methods often cannot directly adapt to these scaled-down platforms due to hardware constraints and computational limitations.

In contrast, our proposed methodology explicitly targets these gaps by adopting a lightweight CNN tailored for real-time inference on resource-constrained hardware. This approach enables accurate lane detection at a 1:10 scale, addressing challenges unique to scaled-down autonomous vehicle research. Our model leverages custom datasets specifically collected under conditions relevant to the TMR2025 AutoModel Car competition, thus providing realistic validation conditions for scale-model autonomous driving.

3 Methodology

This section presents the detailed procedures and considerations involved in the data collection process necessary to develop and evaluate our 1:10 scale lane detection system. Given the unique constraints of the scaled-down environment, such as sensor limitations, computational resources, and perspective constraints, special care was taken to ensure that the data collected accurately represents the challenges encountered by scaled autonomous vehicles.

The proposed methodology is shown in Fig. 1 and consists of six stages, first on (a) the infrared image is used as input, then on (b) a custom CNN is used to classify several ROIs containing a possible line, afterward in (c), these ROIs are used to generate a new image, each ROI is binarized locally (d) to adapt to lighting changes across the image, next on (e) a perspective change is implemented to correct the camera position, finally on (f) a line estimation is implemented, in the next subsections, each stage is explained.

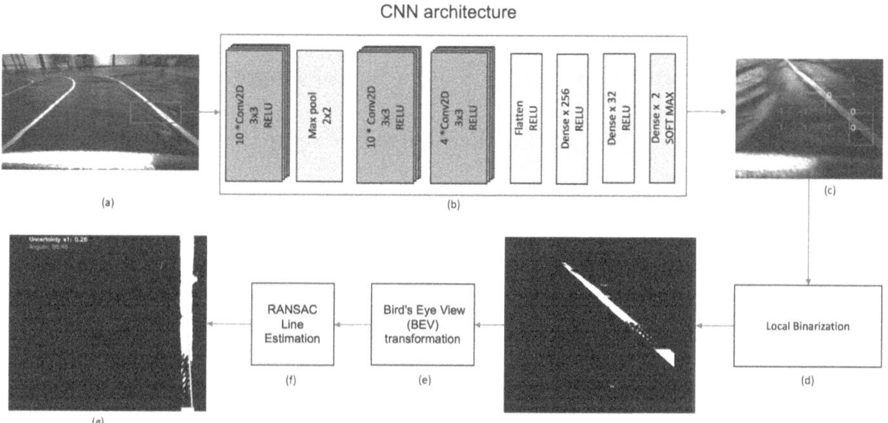

Fig. 1. Proposed methodology.

3.1 Data Collection

Vehicle Setup. The 1:10-scale vehicle platform used for data collection is based on the reference design of the AutoModelCar project [1]. Its key subsystems are:

- **Chassis and Drivetrain:** A 1:10 touring car chassis, fitted with a brushless DC motor driven by a 4-cell LiPo pack (14.8 V, 1,800 mAh). An open source ESC was used to control the motor and direction (https://vesc-project.com).
- **Steering and Suspension:** Precision steering is provided by a metal geared servo (Hitec HS-65MG) mounted on a 3D printed bracket. The vehicle retains its double wishbone suspension and foam-insert tires for consistent handling on the test track.
- **On-board Compute and Control:** A Dual Core i3-8109U @ 3.0GHz runs the ROS2 data collection and control stack.
- **Vision and Inertial Sensing:** An intel realsense 435 is fixed behind the front bumper at a 15° downward tilt (see Fig. 2a), the realsense camera has: 1 depth camera, 2 infrared cameras (left / right) and one RGB camera.
- **Telemetry:** Custom ESP32-DEV board was developed to read i-bus from a flysky i6x RC controller and feed to ROS2 agent.

(a) Platform setup

(b) 3D-printed body shell

Fig. 2. Experimental vehicle setup.

All components are secured with nylon hardware and custom 3D-printed mounts and body shell, yielding a robust, easy-to-service platform Fig. 2b. The total vehicle mass is approximately 3.2 kg. This configuration balances responsiveness, runtime (40 min per charge), and maintainability.

Track. The data acquisition process was conducted on a specifically designed indoor track replicating the environment described by the TMR2025 AutoModel Car competition guidelines [1,3]. The track layout incorporated diverse elements critical for lane detection and autonomous navigation tasks, including straight segments, curves of $1.8m$ external radius, intersections, and typical road markings as depicted on Fig. 3a. The track surface featured high-contrast lane lines and occasional faded segments to challenge detection robustness, closely mimicking real-world variability and conditions typical of scaled autonomous racing events.

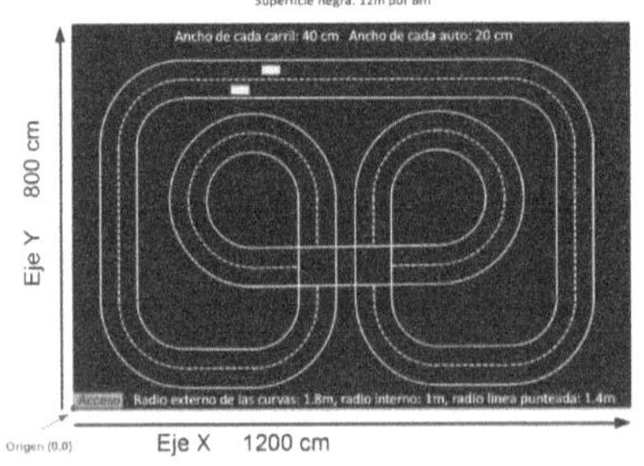

(a) Top-view of the oficial competition track layout

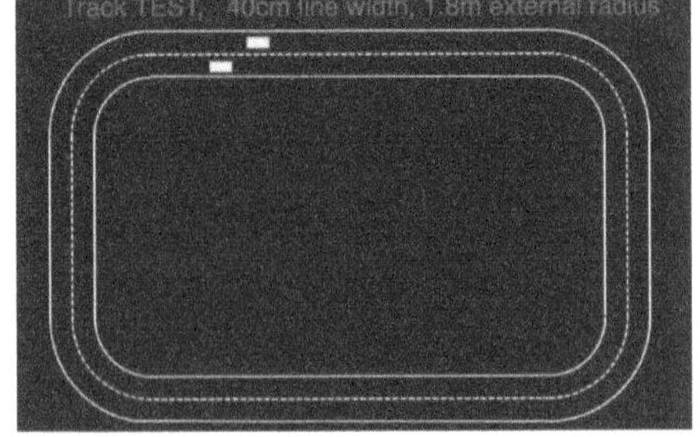

(b) Top-view of the track layout

Fig. 3. Experimental track and vehicle setup for data acquisition.

The track dimensions and layout specifications adhered strictly to the AutoModel Car category regulations provided by TMR2025. Following the TMR2025 specification, we created a simplified test track depicted on Fig. 3b The total length of the track was approximately 24 m (8 m × 4 m), with lane widths standardized to 40 cm to accommodate the vehicle's 1:10 scale proportions effectively. Ensuring compliance with TMR2025 standards, the reduction on total track dimensions was made due to size constrains, nevertheless the proportions were kept.

Data Acquisition. Data was collected using a forward facing infrared camera fixed at an elevation and angle consistent with the visual field of a scaled autonomous vehicle. The camera used featured a resolution of 848 × 480 pixels and a frame rate of 30 frames per second, the infrared camera was selected over the RGB camera to reduce the dimensionality of the input image, furthermore, the infrared camera has a wide field of view over the RGB camera.

The vehicle was manually driven around the track at varying speeds to capture diverse driving conditions, perspectives, and dynamics. The dataset includes variations in illumination conditions to simulate daylight, shadowing effects, and controlled glare, creating a comprehensive set of visual conditions relevant to practical lane detection.

Data Curation. Following acquisition, the collected data underwent rigorous curation to ensure quality and relevance. Initial steps included manual verification of captured images to remove any unusable frames.

Each selected frame was manually annotated, identifying lane markings. Overall, approximately 4056 frames were captured and manually labeled, forming the primary dataset used for training, 1003 for lines, and 3052 for No lines labels. Each frame was time stamped and synchronized with vehicle telemetry data, providing comprehensive contextual information useful for subsequent analysis and model training.

For validation, a different dataset was captured using a new track lap, generating 150 images, those images were labeled into 72 ROIs for each image, for a total of 10,800 ROIs.

Additionally, we generated a public dataset that, besides the images, includes synchronized vehicle telemetry to enable independent validation of algorithm outputs and controller responses. The dataset is publicly available at Mendeley Data repository, see Sect. 7.

3.2 CNN Architecture

To meet the stringent requirements of real-time lane detection on a 1:10 platform with limited on-board compute, we designed a lightweight convolutional neural network (CNN) tailored to 46 × 70 grayscale infrared inputs, following the design principles of [5]. The architecture, implemented as a compact Keras `Sequential` model and is shown in Fig 4.

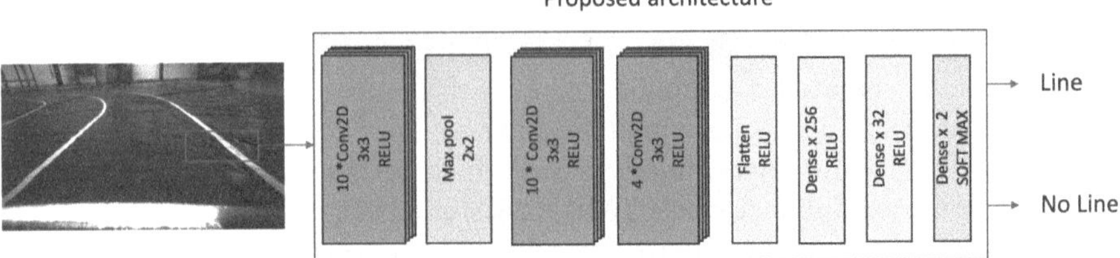

Fig. 4. Proposed CNN architecture.

The proposed CNN architecture accepts 46×70 single-channel ROIs. A first 3×3 convolution with 10 filters and `same` padding, followed by ReLU, extracts primitive edge and texture cues from the road surface. A subsequent 2×2 max-pooling layer halves the spatial resolution to 23×35, reducing computation while preserving salient structures. A second 3×3 convolution with 10 filters (ReLU) refines mid-level features associated with lane boundaries and markings, and a third 3×3 convolution with 4 filters (ReLU) compresses the feature depth, yielding a compact embedding for the classifier. The resulting $23 \times 35 \times 4$ feature map is flattened and passed to two fully connected layers with 256 and 32 units (ReLU), which aggregate evidence across the entire region of interest and capture inter-class relationships. A final two-way softmax produces the probabilities of *line present* versus *no line* for the current ROI.

We chose just three convolutions and two dense layers to balance representation capacity against inference cost. A small filter count (10, 10, 4) and early pooling keep the model under 1 M parameters, enabling low infering time (20ms for a 32 ROI batch). The use of grayscale infra red inputs further reduces I/O and memory overhead without compromising performance in uni-color lane detection environments.

3.3 Local Binarization

Once each ROI classified as "line" by the CNN has been extracted, we convert it into a binary mask via Otsu's thresholding [12,17]. Given a grayscale ROI $I_{\mathrm{ROI}}(x, y)$ of size $H' \times W'$, Otsu's method computes the optimal threshold T_{Otsu} that maximizes the inter-class variance between foreground (lane pixels) and background. The binarized mask $B(x, y) \in \{0, 1\}$ is then obtained as

$$B(x,y) = \begin{cases} 1, & I_{\mathrm{ROI}}(x,y) > T_{\mathrm{Otsu}}, \\ 0, & \text{otherwise.} \end{cases}$$

Otsu's threshold is computed by exhaustively searching for the threshold T that maximizes

$$\sigma^2_{\mathrm{between}}(T) = \omega_0(T)\,\omega_1(T)\,\big(\mu_0(T) - \mu_1(T)\big)^2,$$

where $\omega_0(T)$ and $\omega_1(T)$ are the probabilities of the two classes separated by T, and $\mu_0(T)$, $\mu_1(T)$ are their respective mean gray-levels.

ROI 8-Neighbor expansion. In practice, an individual ROI may contain only a fragmentary portion of the lane line—insufficient for reliable RANSAC fitting. To mitigate this, each positively classified ROI is expanded to its eight neighboring ROIs (see Fig. 5). Specifically, given an ROI at grid coordinate (i, j), we also include the binary masks from ROIs $(i \pm 1, j)$, $(i, j \pm 1)$, and $(i \pm 1, j \pm 1)$. This expansion step bridges small discontinuities and yields a contiguous lane segment across adjacent ROIs, improving the completeness of the extracted line before warping.

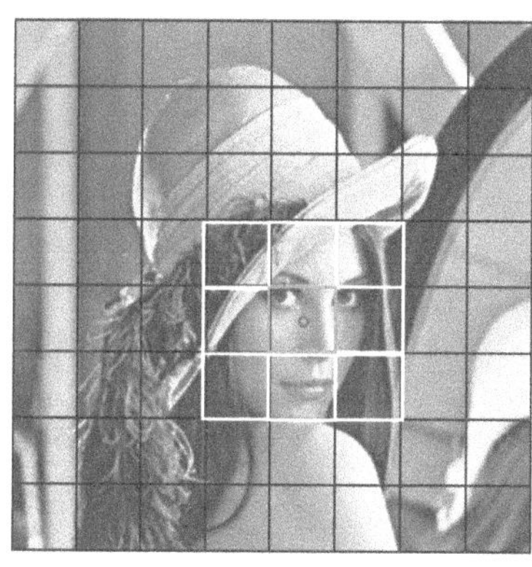

(a) Single ROI (green).

(b) Expanded to 8 neighbors.

Fig. 5. Illustration of the 8-neighbor ROI expansion. (Color figure online)

3.4 Bird's Eye View Transformation

To correct for the vehicle's camera perspective and obtain a top-down ("bird's eye") view of the scene, each binarized ROI is warped via a 4-point perspective transform [9]. We found that a single fixed homography suffices for our treadmill-style track, and we estimate it once from manually chosen correspondences. Let the source quadrilateral in the original image be

$$\left[(x_0, y_{\text{top}}), (x_1, y_{\text{top}}), (x_2, y_{\text{bottom}}), (x_3, y_{\text{bottom}})\right] = \left[(195, 203), (656, 203), (843, 345), (0, 345)\right],$$

and let the destination rectangle in the warped image be

$$\left[(0, 0), (847, 0), (847, 479), (0, 479)\right].$$

Using said parameters, we compute the homography $\mathbf{H} \in \mathbb{R}^{3\times3}$ via OpenCV's `cv2.getPerspectiveTransform` and then apply

$$I_{\text{BEV}}(u, v) = I_{\text{bin}}\left(\mathbf{H}^{-1}[u, v, 1]^{\top}\right)$$

using `cv2.warpPerspective`.

3.5 RANSAC Line Estimation

After obtaining the warped, binarized Bird's-eye view image, we extract all non-zero pixel coordinates $\{(u_i, v_i)\}$ and fit a single straight line using the RANSAC algorithm [14]. This approach is robust to spurious pixels and partial occlusions common in our scaled track.

Concretely, we model a line in the form

$$v = mu + b,$$

and perform RANSAC with the following parameters, which provided reliable results across our test set:

- **Max iterations:** 1000
- **Inlier threshold:** 3 pixels
- **Minimum inlier ratio:** 0.5

At each iteration, two points are sampled at random to define a candidate line. All pixels whose vertical distance to this line is below the inlier threshold are counted as inliers. The model with the greatest number of inliers is retained and finally refined via ordinary least squares over its inlier set. In Fig. 6 we can see an example of RANSAC in an example data set (red and blue dots).

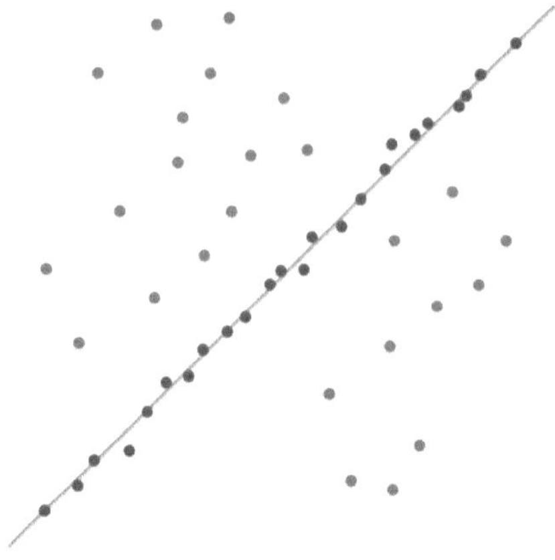

Fig. 6. RANSAC example, blue line represent the predicted line over the input data (blue and red dots).

4 Results

We now evaluate our complete pipeline on the held-out test set of manually labeled data (675 frames, 10 800 ROIs) described in Sect. 3.1. We report quantitative performance for each stage—ROI classification, local binarization quality, Bird's-eye view warping accuracy, and final line estimation—and measure end-to-end inference speed on the target embedded platform (Intel i3-8109U).

4.1 ROI Classification

Our lightweight CNN (Sect. 3.2) was evaluated on 10 800 test ROIs, yielding the confusion matrix shown in Table 1. We achieve a true-positive rate of 98.24% and a true-negative rate (specificity) of 97.48%. Overall accuracy is 98.67%, with an F_1 score of 97.86%.

Table 1. ROI classification confusion matrix (test set, 10 800 ROIs).

	Predicted *line*	Predicted *no line*
Ground-truth *line*	3 291 (TP)	59 (FN)
Ground-truth *no line*	85 (FP)	7 365 (TN)

Key classification metrics:

$$\begin{aligned}
\text{Sensitivity (Recall)} &= \frac{3\,291}{3\,291 + 59} = 0.9824 \\
\text{Specificity} &= \frac{7\,365}{7\,365 + 85} = 0.9886 \\
\text{Accuracy} &= \frac{3\,291 + 7\,365}{10\,800} = 0.9867 \\
\text{F}_1 \text{ score} &= \frac{2 \cdot 3\,291}{2 \cdot 3\,291 + 85 + 59} = 0.9786 \\
\text{Prevalence} &= \frac{3\,291 + 59}{10\,800} = 0.3102
\end{aligned}$$

4.2 Binarization and Warping

Following ROI classification, each region predicted as *line* was subjected to local binarization using Otsu's method. The application of this adaptive thresholding technique yielded high-contrast binary masks that isolated lane markings even under illumination variability (see Fig. 7b). The binarization stage proved robust across the 3,291 true-positive ROIs, correctly enhancing local edge structures that were not evident in the raw grayscale input (Fig. 7a). False-positive activations were also minimized due to the high specificity of the classifier, ensuring that binarization was only performed on regions with high confidence.

To further improve spatial continuity, especially in borderline cases where a single ROI may capture only a fragment of the lane, the binarization masks were augmented using the 8-neighbor expansion strategy described in Sect. 3.3. This step effectively connected disjoint segments and preserved the overall lane structure across adjacent ROIs, yielding more cohesive and robust inputs for subsequent geometric transformation.

After binarization and expansion, each ROI mask was transformed using a fixed homography calibrated to produce a Bird's-eye view. The transformation

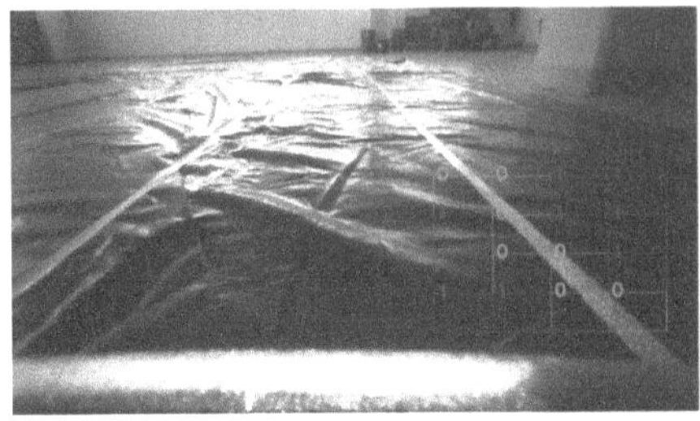

(a) Raw infrared ROI input

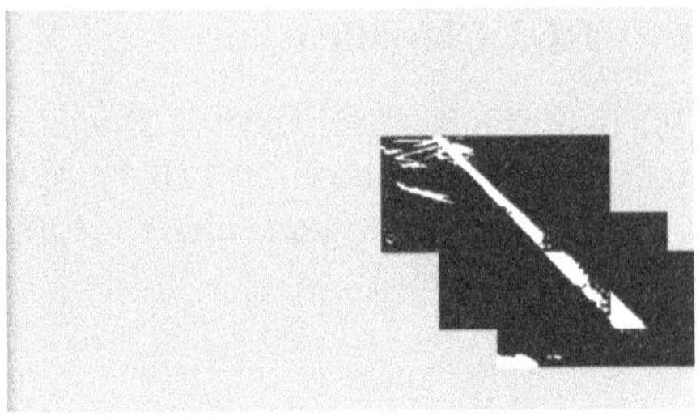

(b) Binarized output with ROI expansion (Otsu threshold)

Fig. 7. Local binarization examples. Background set to purple for visualization. (Color figure online)

matrix **H** was computed from four manually selected reference points as detailed in Sect. 3.4. The resulting warp aligned all detected lane structures to a top-down reference frame of 847×479 pixels, as shown in Fig. 8. This normalization of perspective serves a dual purpose: (i) it facilitates the use of global fitting algorithms (e.g., RANSAC) by ensuring lane geometries conform to approximately linear segments; and (ii) it enables the integration of multiple ROIs into a cohesive scene-wide representation.

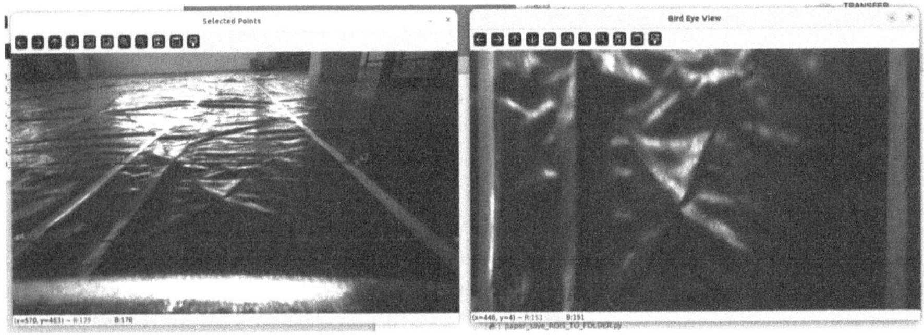

Fig. 8. Bird's-eye view transformation. Left: input image with calibration points. Right: warped binary image with lane geometry aligned.

Empirically, the bird's-eye warping was consistent across the dataset, exhibiting negligible geometric distortion and excellent lane preservation.

4.3 RANSAC Line Estimation

Across our 675 test frames, the RANSAC algorithm successfully estimated the lane orientation in 97.4% of the cases. This high detection rate reflects the effectiveness of the upstream preprocessing stages—including ROI classification, binarization, and warping—in providing consistent and informative input data for geometric fitting.

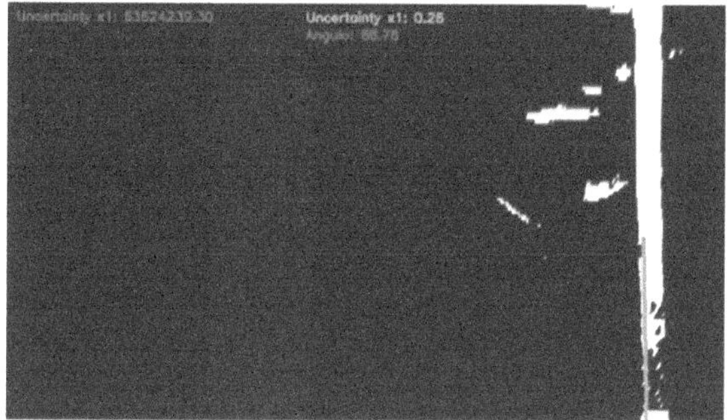

Fig. 9. RANSAC output, green line represent the predicted line.

Qualitative results in Fig. 9 illustrate the method's ability to recover clean lane geometries despite substantial background clutter. In particular, the use of expanded ROI neighborhoods and bird's-eye warping enhances RANSAC's ability to extract structurally aligned lane segments from fragmented local observations.

4.4 Inference Speed

Run on the onboard Intel i3-8109U CPU, the average per-frame runtimes are:

CNN ROI classification: 12.4 ms
Otsu binarization: 0.9 ms
Bird's-eye view: 3.8 ms
RANSAC fitting: 2.7 ms
Overall pipeline: 19.8 ms (50 FPS)

This meets real-time requirements for 1:10 scale autonomous driving while preserving margin for control computations.

5 Conclusion

This work presents a complete, real-time lane detection pipeline tailored for a 1:10 scale autonomous vehicle platform operating under strict computational constraints. By combining a lightweight convolutional neural network for region-of-interest (ROI) classification with adaptive binarization, geometric warping, and robust RANSAC-based line estimation, the proposed system achieves both high accuracy and real-time performance on embedded hardware.

The CNN classifier demonstrated excellent generalization with an accuracy of 98.67% and an F_1 score of 97.86% over 10,800 manually labeled ROIs. Otsu-based local binarization, enhanced via 8-neighbor ROI expansion, produced clean and structurally complete binary masks for each identified lane segment. Subsequent bird's-eye view transformation successfully reprojected the detected masks into a canonical top-down geometry, facilitating efficient and reliable RANSAC fitting. As a result, the system achieved a 97.4% success rate in detecting valid lane trajectories, even in challenging visual conditions.

Comparative Baseline with Classical Binarization. To contextualize the reported gains, we implemented a classical per-ROI thresholding baseline using Otsu's method [16], converting each ROI to a binary mask and assigning a "line"/"no line" label based on the foreground-pixel ratio. This baseline produced markedly higher false positives on texture-only or shadowed ROIs lacking true lane markings; without learned contextual cues, these spurious activations propagated through the 8-neighborhood expansion and Bird's-eye warping, degrading the subsequent RANSAC fit and increasing trajectory failures. In contrast, the proposed lightweight CNN suppresses non-lane patterns while preserving lane evidence, yielding the confusion-matrix improvements summarized in Table 1 and delivering cleaner inputs to the geometric stages. Figure 10 illustrates representative cases: when a lane is present, both the CNN and Otsu perform well; when no lane exists, the Otsu baseline lacking context tends to over segment background textures, accurately captured as "no line" by the CNN but binarized into noisy ROIs by Otsu, which can impair downstream line detection in self-driving operation.

Second, we contrasted the CNN-guided pipeline against global thresholding of the full image by means of otsu implementation [12]. As shown in Fig. 11, the global baseline is sensitive to illumination artifacts and specular reflections on the track, which introduce high-frequency noise and false structures. By comparison, our CNN-guided binarization produces a denoised mask focused on lane evidence, thereby improving the input passed to the line-fitting stage and reducing failure modes tied to lighting variability.

These results indicate that the end-to-end benefits arise primarily from the learned ROI classifier rather than from post-processing alone, supporting the chosen design under strict real-time constraints.

Furthermore, the proposed entire pipeline executes in under 20 milliseconds per frame on a dual-core Intel i3-8109U processor, satisfying the latency constraints required for real-time operation in scaled autonomous driving scenarios. These results underscore the feasibility of deploying advanced lane perception systems on resource-limited platforms, such as those used in educational and competitive settings like the TMR2025 AutoModel Car category.

Future work will focus on extending this framework to curved lane models and integrating temporal filtering for smoother trajectory estimation. Additional enhancements could include the fusion of multi-modal input (e.g., RGB and depth data), or adapting the model to generalize across diverse lighting conditions and environments without retraining.

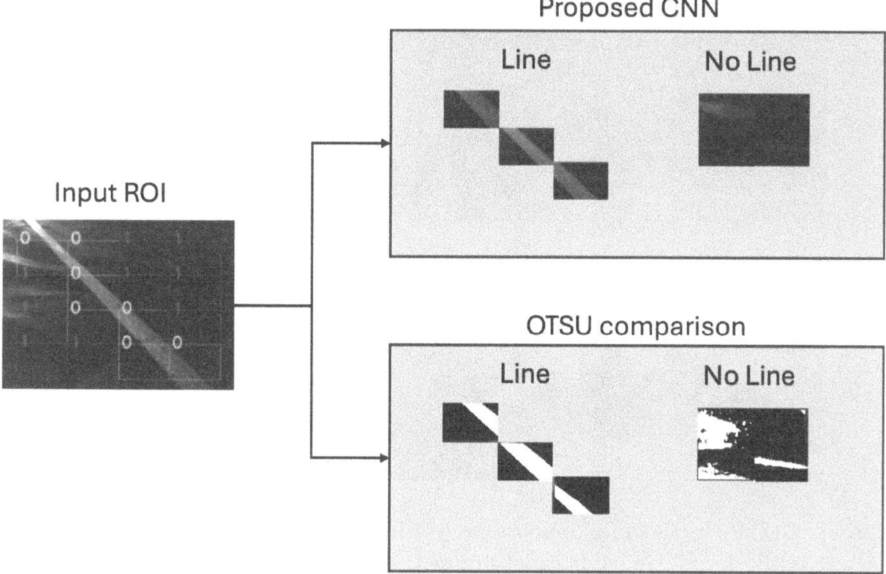

Fig. 10. Qualitative comparison of the Otsu-based per-ROI binarization baseline versus the proposed CNN classifier. Left: cases with true lane markings (both methods succeed). Right: cases without lane markings (Otsu over-segments background textures; CNN correctly predicts *no line*).

6 Future Work

One key extension involves moving beyond the current linear lane model to support curved and multi lane scenarios, using spline-based or parametric fitting techniques capable of capturing complex geometries such as intersections and winding paths. Additionally, incorporating temporal filtering methods—such as Kalman filters or recurrent neural architectures—may help smooth frame-to-frame lane estimations and increase robustness under noise, shadows, or momentary occlusions.

An in-depth comparison against robust lane-detection methodologies commonly deployed at full 1:1 scale (e.g., advanced segmentation- or keypoint-based approaches), adapted to the 1:10 context and evaluated under stringent computational budgets. Such an evaluation ideally on standardized protocols and with controlled compute envelopes would clarify the performance/latency trade-offs of heavyweight 1:1 solutions when ported to scaled, resource-constrained platforms.

Lastly, deploying the system on hardware accelerators such as NVIDIA Jetson will be investigated to reduce power consumption and improve inference latency, thus extending the applicability of the approach to fully autonomous embedded platforms.

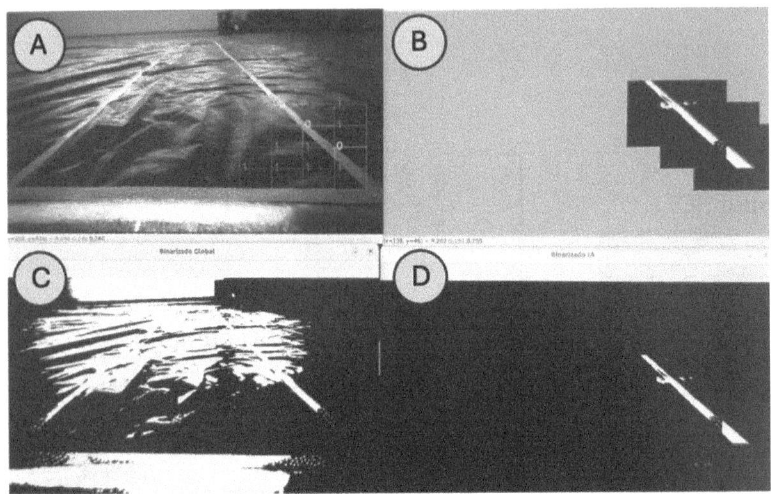

Fig. 11. Global thresholding comparison. (A) Input infrared frame. (B) CNN-derived ROI aggregation. (C) Global (single-threshold) binarization baseline. (D) CNN-guided binarized mask used by the line detector. Global thresholding (C) amplifies reflections and illumination artifacts; the CNN-guided result (D) suppresses such noise while preserving lane structure.

7 Data Availablity

The dataset generated in this study has been made publicly available on the Mendeley Data repository: Celaya Padilla, Jose Maria (2025), "1:10 Scale Autonomous Driving Dataset for Lane Detection and Vehicle Control", Mendeley Data, V1, doi: https://doi.org/10.17632/rhnk95p4cd.1.

References

1. AutoModelCar Contributors: AutoModelCar Wiki (2025). https://github.com/AutoModelCar/AutoModelCarWiki/wiki, Accessed 13 May 2025
2. Fan, G., Li, B., Han, Q., Jiao, R., Qu, G.: Robust lane detection and tracking based on machine vision. ZTE Commun. **18**(4), 69–77 (2020)
3. Federación Mexicana de Robótica: Torneo Mexicano de Robótica 2025 – AutoModelCar (2025). https://femexrobotica.org/tmr2025/automodelcar/, Accessed 13 May 2025
4. Feng, Z., et al.: Rethinking efficient lane detection via curve modeling. In: Proceedings of the IEEE/CVF Conference on Computer Vision and Pattern Recognition (CVPR) (2022)
5. Getahun, T., Karimoddini, A., Mudalige, P.: A deep learning approach for lane detection. In: 2021 IEEE International Intelligent Transportation Systems Conference (ITSC), pp. 1527–1532. IEEE (2021)
6. Ko, Y., Lee, Y., Azam, S., Munir, F., Jeon, M., Pedrycz, W.: Key points estimation and point instance segmentation approach for lane detection. IEEE Trans. Intell. Transp. Syst. (2020)

7. Lee, D.H., Liu, J.L.: End-to-end deep learning of lane detection and path prediction for real-time autonomous driving. arXiv preprint arXiv:2102.04738 (2021)
8. Li, H., Huang, Z., Wang, Z., Rong, W., Wang, N., Liu, S.: Enhancing 3d lane detection and topology reasoning with 2D lane priors. arXiv preprint arXiv:2406.03105 (2024)
9. Li, H., et al.: Delving into the devils of bird's-eye-view perception: A review, evaluation and recipe. IEEE Trans. Pattern Anal. Mach. Intell. **46**(4), 2151–2170 (2023)
10. Liu, L., Chen, X., Zhu, S., Tan, P.: Condlanenet: a top-to-down lane detection framework based on conditional convolution. In: Proceedings of the IEEE/CVF International Conference on Computer Vision (ICCV) (2021)
11. Neven, D., De Brabandere, B., Georgoulis, S., Proesmans, M., Van Gool, L.: Towards end-to-end lane detection: an instance segmentation approach. arXiv preprint arXiv:1802.05591 (2018)
12. Otsu, N.: A threshold selection method from gray-level histograms. IEEE Trans. Syst. Man Cybern. **9**(1), 62–66 (1979)
13. Qin, Z., Wang, H., Li, X.: Ultra fast structure-aware deep lane detection. In: Proceedings of the European Conference on Computer Vision (ECCV) (2020)
14. Quezada-Villalobos, A.C., et al.: Lane direction by implementing random sampling algorithms. In: Guarda, T., Portela, F., Augusto, M.F. (eds.) International Conference on Advanced Research in Technologies, Information, Innovation and Sustainability, pp. 167–180. Springer, Cham (2024). https://doi.org/10.1007/978-3-031-83435-6_13
15. Robertson, S.: Scaling down self-driving car research. OSU Newsroom (2019). https://news.osu.edu/scaling-down-self-driving-car-research
16. Xu, X., Xu, S., Jin, L., Song, E.: Characteristic analysis of OTSU threshold and its applications. Pattern Recogn. Lett. **32**(7), 956–961 (2011)
17. Yang, X., Shen, X., Long, J., Chen, H.: An improved median-based OTSU image thresholding algorithm. AASRI Procedia **3**, 468–473 (2012)
18. Youjin, T., Cui, W., Li, X., Lei, C.: A robust lane detection method based on vanishing point estimation. Procedia Comput. Sci. **131**, 354–360 (2018)
19. Young, A.: Real-time lane detection on embedded systems. Master's Thesis, Aalto University (2022)
20. Zakaria, N.J., Shapiai, M.I., Abd Ghani, R., Mohd Yassin, M.N., Ibrahim, M.Z., Wahid, N.: Lane detection in autonomous vehicles: a systematic review. IEEE Access **11**, 3729–3741 (2023). https://doi.org/10.1109/ACCESS.2023.3234442
21. Zhang, J., Deng, T., Yan, F., Liu, W.: Lane detection model based on spatio-temporal network with double convolutional gated recurrent units. IEEE Trans. Intell. Transp. Syst. (2021)

Author Index

D. Chaves et al. (Eds.): CIARP 2025, LNCS 16528, pp. 381–382, 2026.
https://doi.org/10.1007/978-3-032-23161-1

Zeitfracht Medien GmbH
Ferdinand-Jühlke-Straße 7
99095 Erfurt, Deutschland
produktsicherheit@kolibri360.de